Systems Analysis and Modeling

SYSTEMS ANALYSIS AND MODELING

A Macro-to-Micro Approach with Multidisciplinary Applications

Donald W. Boyd
Montana State University

ACADEMIC PRESS
A Harcourt Science and Technology Company

San Diego San Francisco New York Boston
London Sydney Tokyo

This book is printed on acid-free paper. ∞

ACADEMIC PRESS
A Harcourt Science and Technology Company
525 B Street, Suite 1900, San Diego, CA 92101-4495, USA
http://www.academicpress.com

Academic Press
Harcourt Place, 32 Jamestown Road, London, NW1 7BY, UK

Library of Congress Catalog Number: 00-104661

ISBN: 0-12-121851-1

Printed in the United States of America
00 01 02 03 04 05 QW 9 8 7 6 5 4 3 2 1

Contents

Preface ix

I Introduction to Analysis and Modeling 1

1 Systems Analysis and Model Synthesis **3**
1.1 Introduction . . . 3
1.2 Threefold Nature of the Universe . . . 5
1.3 Entropy: Agent of Change . . . 10
1.4 Hierarchical Analysis . . . 13
1.5 Macro-to-Micro Analysis . . . 17
1.6 Analytical Approaches . . . 19
1.7 Comparison of Analytical Approaches . . . 28
Exercises . . . 32

2 Systems Modeling Principles **35**
2.1 Knowledge-Based Modeling . . . 35
2.2 Circumscription Principle . . . 36
2.3 Conservation Principle . . . 38
2.4 Correspondence Principle . . . 39
2.5 Classification of Model Variables . . . 39
2.6 Linear Relationships . . . 42
2.7 Time Series . . . 45
2.8 Calibration Principles . . . 46
2.9 Building-Block Principle . . . 49
2.10 Exogenous Dynamic Forms . . . 51
2.11 Endogenous Dynamic Forms . . . 52
2.12 Information Feedback . . . 66
2.13 Distortion Reduction . . . 66
Exercises . . . 69

3 Population Model: Calibration and Validation **75**

3.1 Introduction . . . 75
3.2 Population Knowledge Base . . . 76
3.3 Calibration of Population Model . . . 78
3.4 Validation . . . 81
3.5 Validation of Population Model . . . 84
3.6 Error Control . . . 88
Exercises . . . 90
Appendix A . . . 94

II Deterministic Models **101**

4 Modeling with Dynamic Forms **103**

4.1 Harmonic Oscillation Model . . . 103
4.2 Polynomial Models . . . 113
Exercises . . . 125
Appendix B . . . 127

5 Small Arms Exterior Ballistics Model **129**

5.1 Projectile Knowledge Base . . . 130
5.2 Calibration of Ballistics Model . . . 134
5.3 Verification and Validation of Model . . . 139
5.4 Conclusion and Discussion . . . 143
Exercises . . . 145
Appendix C . . . 147

6 Inventory Systems Models: Shaping Dynamic Response **149**

6.1 Inventory Accumulation Model . . . 149
6.2 Inventory Reduction Model . . . 152
6.3 Model of an Ideal Inventory System . . . 154
6.4 Order Delay Model . . . 159
6.5 Random Demand Model . . . 162
6.6 Order Policy Model . . . 168
6.7 Three-Policy Model . . . 171
Exercises . . . 175
Appendix D . . . 179

7 Modeling Corporate Assets **185**

7.1 Introduction . . . 185
7.2 Asset Growth Model . . . 186
7.3 Corporate Assets Planning Model . . . 192
Exercises . . . 204
Appendix E . . . 206

III Stochastic Models **209**

8 Stochastic Analysis **211**
8.1 Introduction 211
8.2 Stochastic Representation 212
8.3 Modeling Stochastic Processes 215
8.4 Trend Analysis 219
8.5 Knowledge Acquisition via Inference Matrix 222
8.6 Regression Analysis 224
Exercises 225

9 Work Physiology Model **229**
9.1 Introduction 229
9.2 Systems Analysis 231
9.3 Knowledge Base 231
9.4 Generic Model 233
9.5 Calibration and Validation 235
9.6 Conclusions and Comments 245
Exercises 246
Appendix F 248

10 Macro Model of Blue Glacier **255**
10.1 Introduction 255
10.2 Systems Analysis of Blue Glacier 256
10.3 Knowledge Base 256
10.4 Calibration 258
10.5 Model Validation 260
10.6 Static Relationships 262
10.7 Further Modeling 265
Exercises 265
Appendix G 267

11 State Water Planning Model **269**
11.1 Introduction 269
11.2 Systems Analysis of Yellowstone River Basin 272
11.3 Knowledge Base 273
11.4 Calibration 277
11.5 Model Validation 283
Exercises 290
Appendix H 292

12 Prototyping A Forest Systems Model **301**
12.1 Introduction 301
12.2 Knowledge Base 303
12.3 Calibration 308
12.4 Model Validation 313
12.5 Concluding Remarks 320
Exercises 321
Appendix I 323

13 Applications Unlimited **325**
13.1 Introduction 325
13.2 Virtual Manufacturing 325
13.3 Time Series Synthesis 331
13.4 Optimization 338
13.5 Large-Scale Systems 340
Exercises 342
Appendix J 344

References **347**

Glossary **351**

Index **361**

Preface

This book, in its first edition, is not garnished by bells and whistles that are deferred to later editions. Its no-nonsense style conveys sincerity, a welcome feature in an introductory edition, and it has received effusive compliments from peer reviewers. Furthermore, the author believes that overembellishment detracts from the message of the book—its content should be the foremost attraction. The book is directed toward seniors, graduate students, and professionals whose career goals or practice involves fields that are systems oriented. These include the following areas:

Systems Analysis	Resource Modeling
Systems Engineering	Planning Models
Systems Modeling	Population Modeling and Demographics
Systems Simulation	Global Systems Models
Virtual Manufacturing	Times Series Modeling and Forecasting
Ergonomics	Management Information Systems
Econometrics	Industrial Management
Environmental Engineering	Knowledge-Based Systems
Biological Systems	Numerical Methods

The list is not exhaustive, and each area can be expanded into subareas (macro to micro), as, for example, *Population Modeling*, ranging from *global, ethnic populations* to *microbial populations* encountered in biofilm engineering.

A new and unique application of systems analysis and modeling was pioneered by the author as an outgrowth of a four-year research project (1968–1972) that produced for Montana an operative, state water resources planning model, first among the 50 states. Analytically, systems detail can be represented by a spectrum ranging from micro to macro. His macro-to-micro (Mtm) approach stands in contrast to the conventional approach that starts with micro forms, infinitesimal entities in differential equations that necessitate integration from micro to macro (mtM) to achieve application. Differential equations induce infinitesimal displacements as *explicit* functions of time. In general, solution of these equations by integration yields *nonlinear dynamic forms*, ill suited to systems modeling. Except for the simplest of examples, integration is possible only if one is willing to make simplifying assumptions—for example, linear approximations, time invariant coefficients, reduced number of variables, and so on. Thus, systems analysis and modeling that follow the mtM approach are fraught with approximation errors.

In contrast, the Mtm approach reverses the procedure by synthesizing an initial model at the macro-detail end of the spectrum. Mental grasp is easy and leads to successively more micro levels of detail to the extent required. In further contrast, Mtm models consist of linear equations that yield finite displacements as *implicit* functions of Δt. This book introduces a unique *modular* system of *linear balance equations* and *linear dynamic forms* that precisely models the dynamic characteristics of the physical system. Trajectories through space and time exhibited by mass and/or their phenomena, whether linear/nonlinear or deterministic/stochastic, are accurately reproduced by Mtm models.

Ultimate accuracy is limited by the precision capability inherent to the combination of hardware and software used for computation. Model synthesis follows the same pattern at each level within the hierarchical, Mtm sequence. Namely, the physical system is broken down into subsystems, each subsystem is broken down into components of supply and demand, and each component is assigned one-to-one correspondence with variables of the model. Variables are broken down into levels and rates. Primary data values are available from nature as levels, obtained by measuring or counting. Rates are calculated by differencing pairs of levels separated by Δt increments of time.

Surprisingly, Mtm modeling does not depend on *primary data* to produce a model, but rather, it depends on *systems analysis and knowledge acquisition* to produce a knowledge base. Secondary data are synthesized by the model to produce a complete data base. Reverse regression is used in the absence of primary data: Given the coefficients (derived from the systems knowledge base), find the data. Data synthesis by the model stands in obvious contrast to forward regression: Given the data (derived from historical observations), find the coefficients. The model must be validated for each intended application. Validation is achieved by employing Turing Type tests and various simulation tests such as perturbation

Because Mtm systems models are *linear* (not linearized or piecewise linear), fast and efficient solutions are obtained by using standard programming software that features matrix multiplication, inversion, and data handling. Having common matrix structures, interacting systems are readily linked so that accurate, multi-system solutions are possible. Furthermore, models can be optimized using standard linear or quadratic programming algorithms.

The Mtm approach has undergone thorough testing over the many years since its inception. Mtm methods were integrated into the engineering courses taught by the author and in 1981, formed the basis for introducing a graduate course in systems modeling.

Although providing much descriptive information about systems and systems analysis, the book's main purpose is quantitative—synthesis of systems models.

This book should be on the desk of anyone engaged in building systems models and especially those in engineering, business, economics, biology, agriculture, and federal agencies that oversee the nation's resources. Furthermore, the methodology is highly suitable to development and installation of regional, executive planning models that are operated on-line in real time via satellite telemetry between earth data collection stations and a supercomputer site. Such installations have the potential for monitoring traffic, floodwaters, and forest fires through the use of systems models and interacting these models with resources models such as water, forest, game populations, energy, and labor pool.

The book is suitable for seniors and graduates of any engineering major and also for other majors that provide a background of at least two years of quantitative work. Desirable prerequisites include (1) an introductory course in statistics that covers regression analysis, (2) a working knowledge of matrix algebra, and (3) computer familiarity. Calculus is used in some places, but it is not essential in implementing the Mtm approach.

Part I provides an introduction to systems analysis and modeling. Chapter 1 introduces the systems philosophy on which the Mtm approach is based. Chapter 2 introduces the principles and dynamic forms that uniquely shape the Mtm approach. Chapter 3 introduces a simple example for which a systems model is defined, calibrated, and tested.

Part II, spanning Chapters 4 through 7, provides numerous examples of deterministic models, drawing from physics, numerical methods, ballistics, inventory theory, and corporate strategy.

Part III details applications to stochastic modeling via Chapters 8 through 13, with examples from ergonomics, glaciology, water resources planning, forestry, manufacturing processes, time series, operations research, and large-scale systems.

Chapters 1, 2, and 3, with selections from Parts II and III, are adequate for a 3-credit semester course. If modeling projects are included, the book will serve two semesters. Practical exercises are provided to emphasize key concepts, verify details slighted in the narrative, suggest library assignments, and to encourage mastery of Mtm concepts via a "hands-on" approach.

Appendixes are provided with each quantitative chapter. They contain data and graphs so that model results can be reproduced by the student before attempting his/her own applications. An instructor's solutions manual is available for use with the text; however, some exercises have open-ended solutions that encourage student resourcefulness and innovation.

Finally, the author wishes to acknowledge the encouragement received from colleagues and students throughout the years that this work was being compiled. Portions of the materials submitted by graduate students Larry Cawlfield, Greg Goltz, Terry Ross, and Robert Wehrman were included in the book, and special appreciation is extended to them.

Part I

Introduction to Analysis and Modeling

CHAPTER 1

Systems Analysis and Model Synthesis

1.1 INTRODUCTION

Two or more components comprise a system if they interact within a common domain to achieve a common goal or function. In general, a system is not isolated but rather is linked to adjacent systems through its inputs and outputs. Consequently, systems are dynamic, adjusting internally in response to external disturbances. A systems model consists of two or more interacting components. An effective systems model is capable of linking with models of adjacent systems. To possess such specific properties, a model must be designed, following a procedure that in many ways resembles engineering design. Engineering design depends on a growing body of knowledge. Knowledge begins with engineering education and continues to grow through practical experience. Usually, engineering design involves the following elements:

- *Analysis* Design object is *differentiated* into subcomponents and their interrelationships. Specifications are formulated.
- *Synthesis* Alternative systems designs are considered. Subcomponents are *integrated* into a functional whole to produce tentative designs.
- *Testing* Design specifications are *verified* for the chosen alternative, performance is monitored for *validity*, and functionality is checked for *credibility* with the user.
- *Implementation* Design object is *installed* and becomes *operational* for the user.

Recycling to previous steps generates new alternatives and leads to improvements in the design.

Similarly, model design depends on a growing body of knowledge. Each system domain is characterized by knowledge items. Knowledge items are acquired through study (analysis) of the system. Relevant items are synthesized into a systems model. Even a cursory study reveals that systems are characterized by *triads*. For example, this paragraph just introduced a systems triad of *knowledge acquisition, systems analysis*, and *model synthesis*. Furthermore, knowledge acquisition is a process that involves three entities, a general *information triad* composed of *source, medium*, and *message*. Here are some examples of information triads:

- *Textbook* author, paper, "theme"
- *Tracks* animal, snow, "a bear passed this way yesterday"
- *Fossil record* prehistoric being, bone, "a tyrannosaur lived here 100 million years ago"
- *Physical universe* designer, design, "axioms of existence"
- *Knowledge base* Table 1.1 displays a triad of information triads.

Table 1.1 Information Triads of the Knowledge Base

Source	Medium	Message
scientist	scientific inquiry	theory
statistician	statistical observation	statistical data
domain expert	interrogation	heuristics

Knowledge items are expressed in the *message* column, and are classified as hard or soft. For *hard* knowledge to be relevant, the message must consist of *true* facts that are applicable. Whether from scientific inquiry or statistical observation, a fact must be treated as a binary variable, a hypothesis to be tested. Facts can be established as true or false because they have the property of being observable by objective application of the physical senses. On the other hand, *soft* knowledge is subjective, not open to observation, and consists of heuristics: *experience* and *rules of thumb*. Knowledge items may be sought from each source to form the knowledge base utilized in systems modeling. Because a systems model synthesizes information out of a blend of hard and soft knowledge, its initial status is *tentative*. Therefore, each application of the model becomes a hypothesis to be tested. Just as in statistical hypotheses testing, two types of error are possible: (1) rejecting a true application (hypothesis) and (2) accepting a false application (hypothesis) [27].

Although Chapter 1 echoes the title of this book, its purpose is merely to provide an introduction to systems analysis and modeling. Beginning with a *macro* overview in Chapter 1 and principles in Chapter 2, ensuing chapters introduce finer detail, thus paralleling the method by which a newspaper journalist presents the news: *headline, sketch*, and *back page* (another system triad!). The book shows how to employ a *macro-to-micro* (Mtm) approach to systems analysis and modeling. The next section further explores the concept of system triads.

1.2 THREEFOLD NATURE OF THE UNIVERSE

Figure 1.1 displays an orthogonal axes system representing the physical universe of *space, mass*, and *time*. Macro analysis is extended in the micro direction by increasing detail along one or more of the axes of space, mass, and time. Subjectively defined, unit quantifiers provide for scaling along each axis.

Entities **L**, **M**, and **T** are used by scientists in an *axiomatic* sense to represent the existence of space, mass, and time, respectively, as components of the physical universe. Their axiomatic nature may be symbolized by the *universal quantifier*, ∃, meaning "there exists":

$$\exists \text{ space} \qquad \exists \text{ mass} \qquad \exists \text{ time}$$

Axiomatic entities are accepted as the starting point in any formal system of reasoning. Such entities retain their abstract nature unless they are coupled to observation and experience. Observed as a system, the universe is a triad of *space, mass,* and *time*, each entity also being a triad. Figures 1.2 and 1.3 portray this threefold nature of the universe.

1.2.1 Space

Geometrically, *space* is described in terms of *length*, *width*, and *height*, as depicted in Figure 1.3. Demarcation of space, to the exclusion of mass and time,

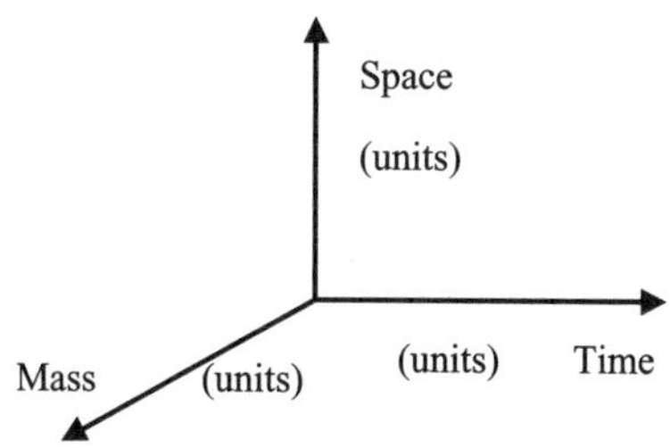

Figure 1.1 Fundamental Units of Extension

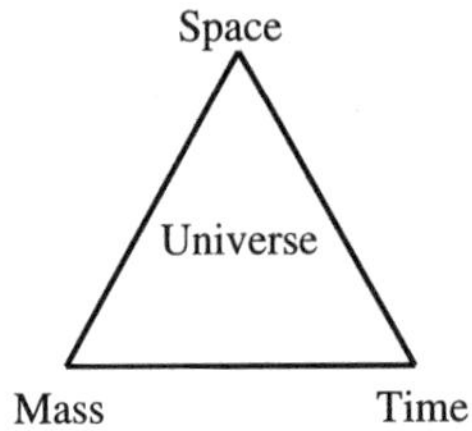

Figure 1.2 The Universe as a Triad

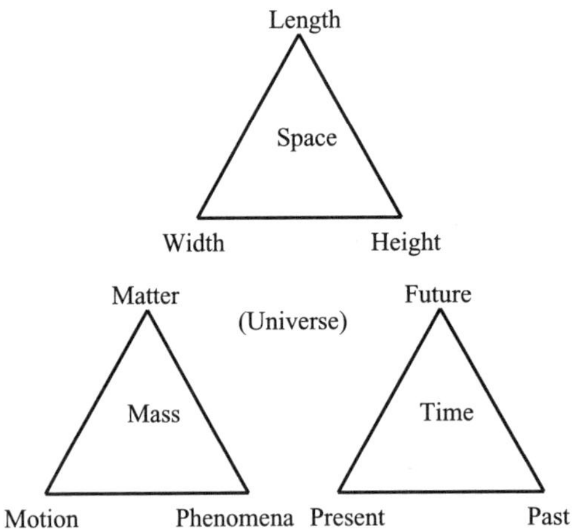

Figure 1.3 Space, Mass, and Time Triads

is senseless in systems analysis. Typical unit quantifiers of space are *universe, galaxy, solar system, Earth, hemisphere, continent, country, state, county, township, section, subsection, acre, plot, ..., increment* (Δx, Δy, Δz), *infinitesimal* (*dx, dy, dz*). Consequently, any system selected for Mtm analysis exists within some containing unit of geometric space. Furthermore, the size or *scale* of a system is most frequently classified with respect to the magnitude of its domain—that is, the amount of space *circumscribed* by the system.

Mtm concepts are applicable to systems of any scale, ranging from the *microscopic*, cellular level (e.g., microbial populations) to the *macroscopic* global level (e.g., human populations) and beyond to stellar systems. A *macroscope* [14] is the inverse of a microscope, obscuring all but general features of a system. For instance, a *large-scale system* impacts one or more human institutions and is extensive relative to the systems scale. Examples are the economy, government, health care, and so forth. Figure 1.4 presents the systems scale.

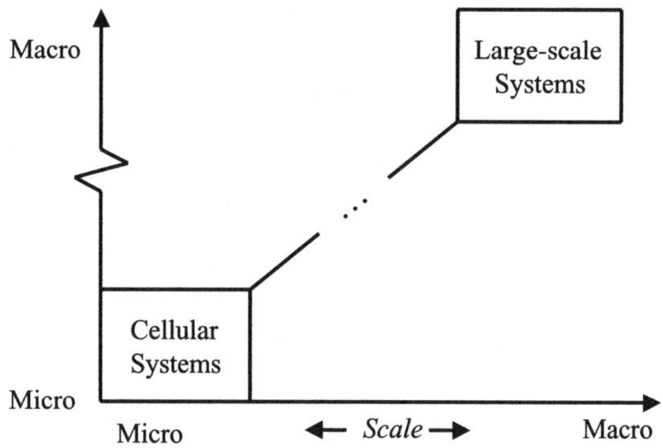

Figure 1.4 The Systems Scale

1.2.2 Mass

Physical objects are observable because of the properties of *mass: matter (in) motion (producing) phenomena.* Refer to Figure 1.3. Mass is the measure of difficulty by which an object can be accelerated [23]. Less precisely, mass is the quantity of matter in an object expressed in units of *weight*, for example: *megaton, kiloton, ton, kilogram, gram, centigram, milligram, ...,* Δm, dm. *Matter* conforms to a triad of natural states: gas, liquid, and solid. *Motion* resulting from constant acceleration is a triad expressed in terms of distance, velocity, and acceleration. From cosmological physics, *phenomena* are the result of a triad of primary fields: electric, magnetic, and gravitational.

Because of its physical properties, mass does not exist independently of space and time. Mass is related to space (volume) by its *density*: $\rho = m/V$. Mass (atomic) is also time (velocity) dependent, appreciable only at velocities comparable to the velocity of light [18].

Despite these dependencies, mass, as a quantifier for matter, satisfies the continuity or conservation principle expressed as a *mass balance equation* [30]. The same is true for related physical phenomena such as momentum, energy, and electricity. "Flow" of mass through time is modeled as a *conservative* system, portrayed in Figure 1.5, and defined as follows:

X_1 = initial level of mass
X_2 = terminal level of mass
X_3 = mass input rate
X_4 = mass output rate

At the end of each Δt period of time, the mass balance equation requires that mass be "conserved," that is, neither created nor destroyed:

$$X_1 + X_3\Delta t - (X_2 + X_4\Delta t) = 0.$$

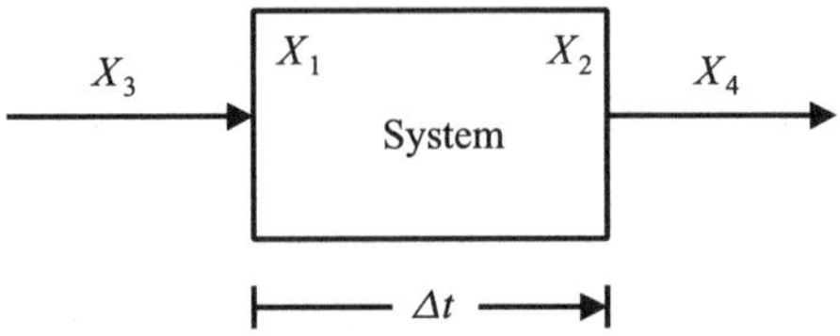

Figure 1.5 Conservation of Mass

1.2.3 Time

Time is a *physical* property and varies with respect to mass, acceleration, and gravity. Two objects exhibiting *mass* and separated by *space* are also separated by time. Time is related to space (distance) by the *velocity* of matter in motion: $v = S/t$. Thus, time completes the most basic physical triad, that of the *universe*. Time, also a triad, *future*, *present*, and *past*, is symbolized by the third triangle in Figure 1.3. Quantifiers of time are *aeon, ..., millennium, century, decade, year, season, month, week, day, hour, minute, second, ...,* Δt, dt.

The future consists of differentiated events and can be regarded in two different ways:

- *Deterministic* The future is a repetition of the past, and events come as no surprise.
- *Nondeterministic* The future is not a repetition of the past and events are differentiated only under risk or uncertainty.

Projecting from the present into the future is called *forecasting*. Projecting from the present into the past is called *backcasting*. Both projections are attended with exponentially increasing error whose magnitude becomes astronomical when projecting beyond the near present. Errors associated with time are as follows:

Error	Caused by
past	entropic degradation
present	imperfect observations
future	unpredictable perturbations

The future contains *differentiated* "bumps" and is divided from the past by the present. On the other hand, the past contains *integrated* bumps, past occurrences that have been blended with the passage of time through the *entropic* process of change. To illustrate, future precipitation will materialize in the present (be differentiated) as "storms." With the integration of precipitation over time into past occurrences, the storms as bumps are lost. Storms become integrated into macro bodies of water such as snowpack, streams, rivers, and eventually the ocean. Periodic measurement (monthly, daily, hourly, etc.) captures the result of this physical integration process. In contrast, analysts traditionally start with differential equations and the infinitesimal, dt, that describe flow at a point, as, for example, Darcy's law [21], and laboriously perform mathematical processes of integration to obtain approximate flow models and solutions. Why not allow the physical system to perform its natural process of integration to obtain the desired level of detail? Great ease and accuracy of solution is possible through circumscribing and modeling the physical system at a macro level. Linear equations, corresponding to the circumscription, model the natural process by integrating rates into levels over unit time, Δt.

Perception of *time* relates to *duration* and is found to vary widely among individuals. Thus, for objectivity, *time* is a process that requires calibration. Customary references to time involve triads: 12/7/41, month day hour, day hour minute, hour minute second. Does motion of a pendulum mark *time*? Yes, but its calibration is meaningless unless there is an observer. Likewise, the expanse of space and the properties of mass are unrecognized without an observer. Although space, mass, and time exist independently of any observer, their existence invites observation, thus leading to scientific inquiry.

Time, as similar to space, can be represented as an expanse. Mass is distributed by motion over both space and time. This suggests that a system is most basically characterized as the flow of mass over space and time. Time, as also similar to mass, can be represented as a flow. Which way does time flow? From the past through the present into the future? Or from the future through the present into the past? Thought provokingly, if the latter is true, then the future is older than the past, for it will take longer for the more distant future to arrive at the present! However, Einstein theorized that *time* is a vector field over *space* and furthermore that "the distinction between past, present, and future is only an illusion, however persistent" [16]. He further explained that time is not at all what it seems. It does not flow in only one direction, and the future exists simultaneously with the past.

Time measurements require a triad: an observer, a reference point, and a moving (usually rotating) object. Examples are sundials, hourglasses, clocks, and celestial bodies. Although time is axiomatic, "viewing" *time* from the basis of systems analysis reveals some interesting "working" definitions that may be expressed as follows:

Relational Definition Present time *is a process that converts one unit of* future time *into one unit of* past time *at an absolute rate (frequency) of one unit (cycle) per* 2π *radians.*

A radian is the angle subtended by an arc whose length is equal to the radius. Figure 1.6 expresses this relationship.

The process of converting future into past is schematically portrayed by a finite "time" wheel with unity radius, as seen in Figure 1.7. The wheel is rolling at an absolute frequency of $\omega = 1/2\pi$ cycles per radian along the expanse of time.

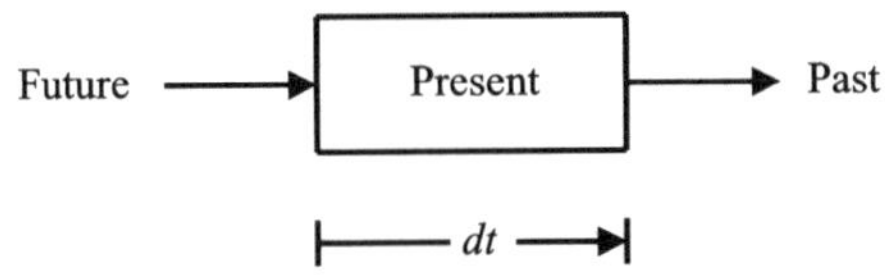

Figure 1.6 Relational Definition of Time

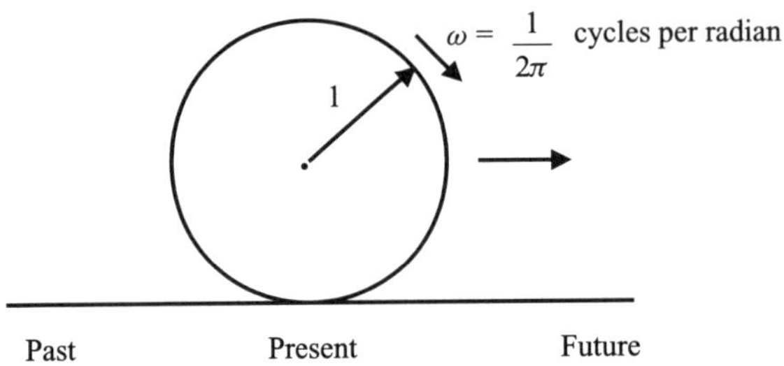

Figure 1.7 The Time Wheel

Absolute Definition *The absolute unit of time is defined by the reciprocal of absolute frequency: one time unit* = $1/\omega = 2\pi$ *radians per cycle.*

Consequently, the absolute unit of time is represented by the circumference of a time wheel with unit radius. Time measurements are relative to the absolute unit of time.

Example Let 52 weeks (one year) constitute one cycle. Then (2π radians) $\times$ $k = 52$ weeks per cycle, where k is a proportionality constant. $k = 52/2\pi$ *weeks/(radian cycle)* is defined as the radius, R. Thus, one year is represented by the circumference of a time wheel having radius $52/2\pi$: $C = 2\pi R = 2\pi \cdot (52)/(2\pi) = 52$ *weeks per cycle.*

In general, the circumference of the time wheel represents n units of time: $n\Delta t = 2\pi R(\textit{time units})/(\textit{cycle})$, where Δt is unit time. Then since $\Delta t = 2\pi/n \cdot R$, it follows that $\lim_{R\to 0} 2\pi/n \cdot R = dt$, an infinitesimal point in time; and $\lim_{R\to\infty} 2\pi/n \cdot R$ yields *eternity*, the unbounded expanse of time.

1.3 ENTROPY: AGENT OF CHANGE

Flow of mass produces a spatial-temporal growth in the level of entropy. *Entropy* is the "exhaust" of an energy-consuming universe. Entropy involves the triad of *age*, *decay*, and *chaos*. Thus, as systems evolve over time, they exhibit a *dynamic* triad: *force*, *motion*, and *change*.

Entropy has been credibly established as a *law* by the observation-based process of scientific inquiry. Observation is possible because entropy is present *now* and is not a speculation of the *past*. In its simplest form, the *Second Law of Thermodynamics* acknowledges that physical processes are accompanied by an increase in entropy of the universe, or equivalently, by a decrease in available energy [5]. Entropy per se is a process that carries out the "irreversible" law of degradation of energy and matter. With respect to time, entropy exhibits both *levels* and *rates*. Over time, any given *level* of *orderliness* is subject to degradation, resulting in a decreased *level* of orderliness. More generally, entropy is illustrated throughout the whole of human history as *levels* of civilization rise and fall. Civilizations are pockets of resistance against entropy, that is, negative entropy processes. A civilization's attack slows the *rate* of degradation. Thus civilizations rise (negative entropy) but, as under a curse, are destined to fall (positive entropy).

All problems are in some way related to the process of entropy. Systems analysis and modeling provide "weapons" in the battle against entropy. However, even the best "solutions" are temporary. Mtm methodology innovatively exploits the concepts of *negative* and *positive* entropy. Several more concepts relating to entropy are examined with the help of Figure 1.8.

The upper, horizontal-axis continuum represents the unbounded expanse of time, a cosmic scale covering aeons of time. A trend line depicting *evolution* is portrayed as the mirror image of a trend line depicting *entropy*, that is, evolution as a primeval form of entropy characterized by a *negative* rate. The lower, horizontal-

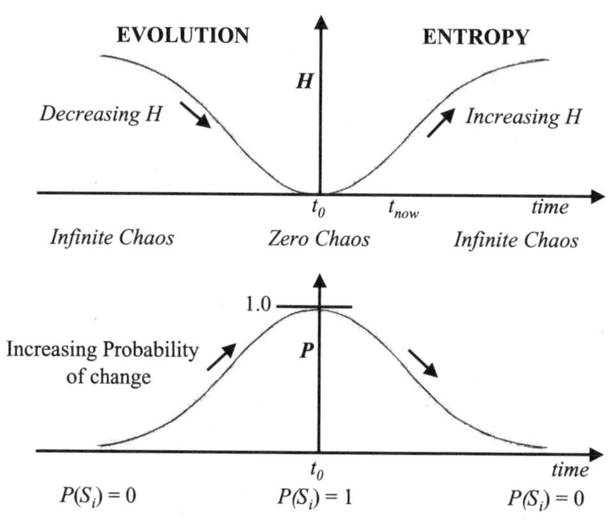

Figure 1.8 Evolution Relative to Entropy

axis continuum is keyed to the upper, where $(S_i)_t \in \{\textit{Set of all possible states}\}_i^n$ represents state space, and with $P(S_i)$ defined as the probability of a change in the ith state.

H is the state of disorderliness of the universe. At time t_0, perfect order is postulated for the universe. According to the hypothesis of evolution, aeons prior to t_0, the universe began to evolve from infinite chaos to its pristine state at t_0, from the *null* event $(P(S_i) = 0)$ to the *sure* event $(P(S_i) = 1.0)$. Beyond t_0 in time, *positive* rate entropy prevailed and the universe began to experience degradation, ultimately to achieve a state of infinite chaos some aeons in the future. Thus, during the time span $t \leq t_0$, the *hypothesis of decreasing evolution* is equivalent to the *hypothesis of increasing probability of change*; and during the time span $t > t_0$, the *law of increasing entropy* is equivalent to the *law of decreasing probability of change*.

Figure 1.9 illustrates the law of decreasing probability. A climber places a flat rock at position A on the very top of a mountain peak. Its balance is precarious, and the probability (P_0) of a change in state is very high. Next year, the climber returns and observes that the rock has toppled and now occupies position B. Probability of subsequent change is still relatively high but such that $P_1 < P_0$. Returning a year later, the rock has shifted to position C. As time continues to bring change, the rock shifts to lower elevations, is broken into pieces, pieces reach a stream, and eventually end up in the ocean as grains of sand so that the probability of change becomes minuscule: $P_0 > P_1 > \cdots > P_n$, approaching zero as n continues to increase. Placing the rock at position A illustrates negative entropy, but immediately thereafter the law of entropy prevails. Similarly, lifting a stone against the law of gravity illustrates reverse gravity. As soon as the stone is released, the law of gravity prevails, and the rock immediately begins to fall.

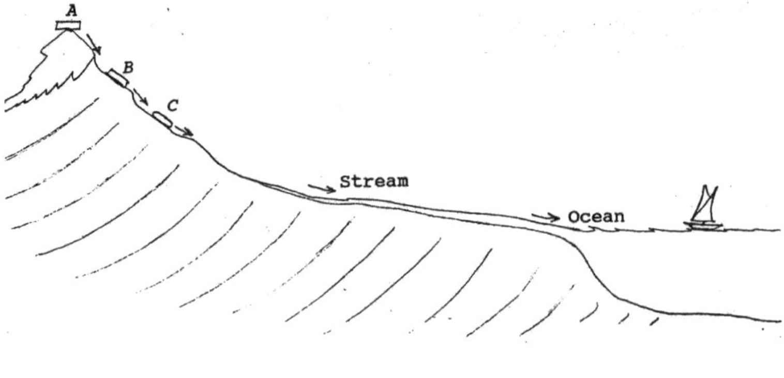

Figure 1.9 Entropy and Probability of Change

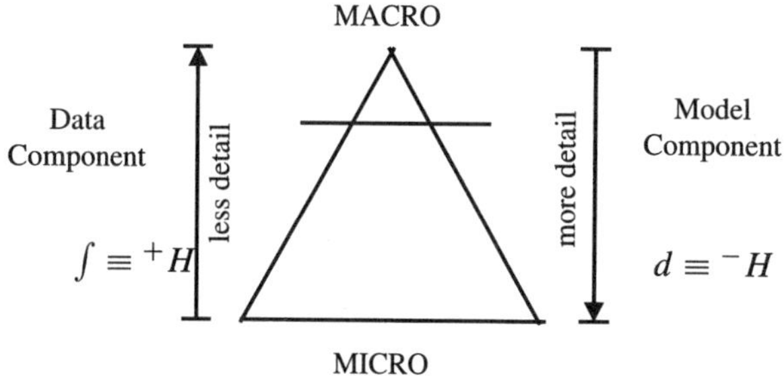

Figure 1.10 Entropy and the MTM Method

1.3.1 Implication for Data and Modeling

From a conceptual perspective, the law of entropy is applicable to both systems analysis and modeling (see Figure 1.10). *Macro* implies the greatest absence of individual identity (or detail) and coincides with $P(S_i) = 0$ (of Figure 1.8); no further integration is possible, only differentiation. *Micro* implies the greatest presence of individual identity and matches $P(S_i) = 1.0$; no further differentiation is possible, only integration.

- Data components of systems analysis are analogous to *positive* entropy. Levels become quantified as data as rates are accumulated to levels with passage of time. Thus, physical integration, from micro to select macro levels, is accomplished by the system as a natural consequence of entropy. Compilation of data at each level provides a historical record of each integration.
- Levels of articulation in model synthesis are analogous to *negative* entropy. The systems triad of knowledge acquisition, systems analysis, and model synthesis begins at a macro level of detail. Models involve imaginary differentiations from macro to lesser macro levels of detail. Thus, a *hierarchical* series of models is a mathematical record of each differentiation.

Consequently, systems analysis and modeling can be viewed from a hierarchical perspective.

1.4 HIERARCHICAL ANALYSIS

Calculus is a mathematical discipline that employs hierarchical orders of *differentiation* and *integration*. In addition to evolution and entropy, mathematical differentiation and integration also provide useful analogs in conceptualizing *systems analysis* and *model synthesis* [10]. For perspective, *systems analysis* and

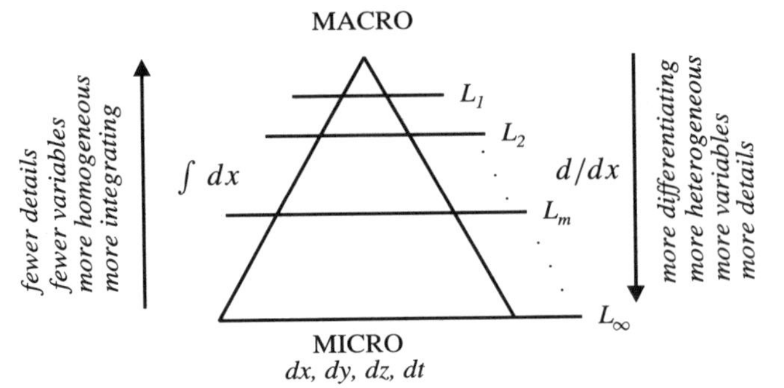

Figure 1.11 Modeling Tree

model synthesis can follow either of two hierarchical approaches: (1) proceeding *micro-to-macro* (mtM), that is, from the bottom up as an analog to integration or (2) proceeding *macro-to-micro* (Mtm), that is, from the top down as an analog to differentiation. Model synthesis utilizes hierarchical analysis to *articulate* system *details* as a series of mathematical models. Hierarchical analysis is formalized as follows:

Principle *Hierarchical analysis divides a system into homogeneous levels of detail,* L_i, *for* $i = 1, 2, \ldots$.

Hierarchical analysis is symbolized by the *triangle* in Figure 1.11, appropriately captioned a *modeling tree*, since each level represents a systems model. *Macro* is represented by the vertex of the triangle, a single point inferring the ultimate level of system integration. In contrast, *micro* is represented by the base of the triangle, an infinity of points typifying the ultimate level of system differentiation. Elements of analysis are represented by differentials, for example, dx, dy, dz for *space* and dt for *time*. $L_1, L_2, \ldots, L_m$ represent analytic levels of analysis and provide a scheme of reference. Both systems analysis and model synthesis are hierarchical activities. Hence, four combinations are possible as shown in Table 1.2. Combinations 1 and 4 are consistent because of their micro-macro matches. Combinations 2 and 3 are not consistent because of their micro-macro mismatches.

1.4.1 Theoretical Approach

Combination 1 represents the *theoretical approach* traditionally utilized by scientists and engineers to generate models. The *finite element method* (FEM) is very typical of this approach [26]. A system is first analyzed from below at the micro level with theory that expresses point phenomenon (growth, flow, motion, etc.) in terms of *micro forms*, that is, differentials and differential equations.

Table 1.2 Hierarchical Combinations for Systems Analysis and Modeling

Combination	Systems Analysis	Model Synthesis
1	mtM	mtM
2	mtM	Mtm
3	Mtm	mtM
4	Mtm	Mtm

Consequently, systems analysis and model synthesis must be performed as *serial* activities. Synthesis of a systems model depends on being able to obtain *macro forms* by integrating the selected micro forms to an acceptable aggregate or macro level of detail. Expertise from the *mathematical domain* is required. Amalgamation of *physical system* components from micro to macro is analogous to the mathematical process of *integration*, indicated by the upward arrow in Figure 1.11. Relationships that apply to points at the micro level are extrapolated to an upper, macro-level solution space consisting of finite segments. Each segment must be defined mathematically, and most often these relationships are nonlinear. Unless the scope of analysis is adequately restricted, the resulting mathematical articulation is characterized by a proliferation of arbitrary variables. To the extent that simplifying assumptions and estimation techniques are required, solutions bear only approximate resemblance to physical system response. Thus, theoretical analysis may be more useful as a *source* of systems knowledge than as a *method* for systems analysis and model synthesis.

1.4.2 Regression Approach

Because of micro-macro mismatches, a *parallel* progression of analysis and synthesis is not possible for combinations 2 and 3. Although either combination provides a basis for systems analysis, lumped together, they form a single, *multihierarchical* class characterized by heterogeneous levels of detail. Statistical modeling techniques that utilize the *regression approach* [15] are an example of this class that requires an existing data base. Modeling expertise relies mainly on the *statistical domain*. Regression variables are not restricted in number nor to a *homogeneous level of detail*, and multi-hierarchical grouping of variables is permitted. This stands in contrast to the *system mass balance equation* that spans an exact number of variables and that applies to a single level of the hierarchy. However, regression analysis, if restricted to a homogeneous level of detail, forms a major subclass of combination 4.

1.4.3 Mtm Approach

Combination 4, *macro-to-micro (Mtm)*, stands in contrast to combination 1 *(mtM)* as a clear departure from tradition. For example, *Mtm* methodology reverses the order of the *mtM* FEM in producing water planning models [11]. Furthermore, the U.S. Air Force Program for Integrated Computer Aided Manufacturing utilizes a Mtm hierarchy of diagram names and node numbers called a *node tree* [46].

Mtm analysis starts from above with *macro forms* that incorporate system components. Macro-level data are a record of natural integration to this level by the physical system as it functions over time. Breakdown of the *physical system* from macro to micro is thus analogous to the mathematical process of *differentiation*, symbolized by the downward arrow of Figure 1.11. Systems domain knowledge includes expertise from both the mathematical and statistical domains.

Systems analysis conducted at any homogeneous level of detail enables synthesis of a linear systems model for that level. Thus, systems analysis and model synthesis are *concurrent* activities that iterate toward the micro until differentiation has produced adequate detail. Concurrent development is portrayed by the triangles of Figure 1.12. Level of detail is a matter of choice and subject to definition. L_1 defines a first order of systems differentiation for analysis and synthesis. L_2 defines a second order of differentiation and, thus, distinguishes additional detail. Iteration continues until the desired level of detail is achieved at L_n. Consequently, combination 4 gives rise to the premise of this book, stated as follows:

> *The macro-to-micro approach provides a practical, effective, and natural framework for systems analysis and model synthesis.*

Model synthesis is based on systems domain knowledge that may only incidentally include a data base. Although the message of this book could be expressed with mathematical rigor, theorems and proofs are left to others. Knowledge-based systems modeling is presented through concepts, principles, and examples.

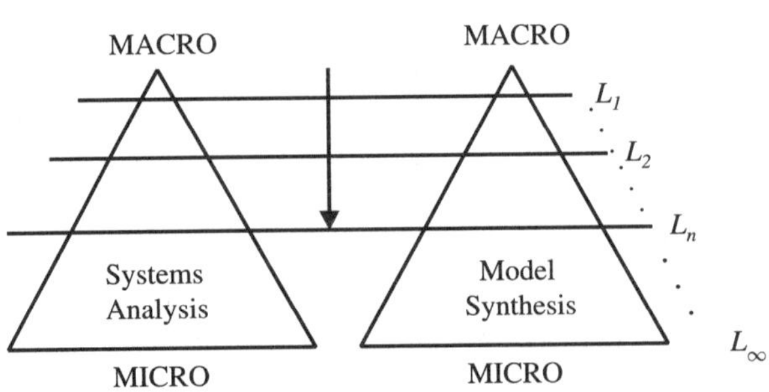

Figure 1.12 Concurrent Development

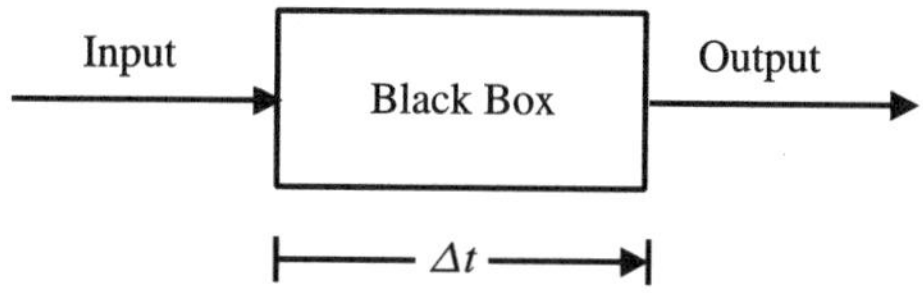

Figure 1.13 A System as a Black Box

1.5 MACRO-TO-MICRO ANALYSIS

In general, Mtm analysis can be likened to opening a series of *black boxes*. At any given level of detail, the *state* (or level) of a system can be determined by considering only its inputs and outputs. Thus, the system itself is obscure. The first black box is opened only to discover it contains several black boxes. As each of these black boxes is opened, it is discovered that they also contain additional black boxes. The process is that of differentiating (or articulating) detail. At each level of detail, input and output rates are integrated (or accumulated) over time by a corresponding model to levels that match the system's response over time. Black-box representation of a system is illustrated in Figure 1.13. Associated with each black box is one *state* equation and equations for each input and output. Calibration schemes provide values for nonunity coefficients of the equations.

Exploration of Mtm systems analysis and modeling concepts began with the four-year development of an operational water planning model for the state of Montana, the first of its kind in all 50 states [7]. A first-level model was synthesized six months before data became available for its calibration and validation. Eight additional Mtm models were synthesized, each displaying an increased level of detail. Although data were acquired as a knowledge item, model synthesis did not depend upon a pre-existing data base [12]. But rather, Mtm systems analysis and model synthesis proceeded in parallel.

Formalization of Mtm analysis followed the discovery that analytical techniques used to synthesize planning models had general applicability to systems modeling. Systems analysis was found to provide knowledge items that could be incorporated into a model whose (knowledge-derived) output became an acceptable substitute for data. Thus, a systems model is a type of *knowledge-based system* that captures the current level of Mtm systems analysis. Finer mathematical articulation requires acquisition of finer systems details. An overview of the analytic structures for systems analysis and model synthesis, common to all levels of detail, is presented in Figures 1.14 and 1.15, where *input* and *output* (I/O) *rates* and *initial* and *terminal* (I/T) *levels* are relative to Δt of Figure1.13.

Comparison of system and model structures reveal their similarities. As examples, *Subsystems* correspond to *Subsystem Models, Components/Relationships*

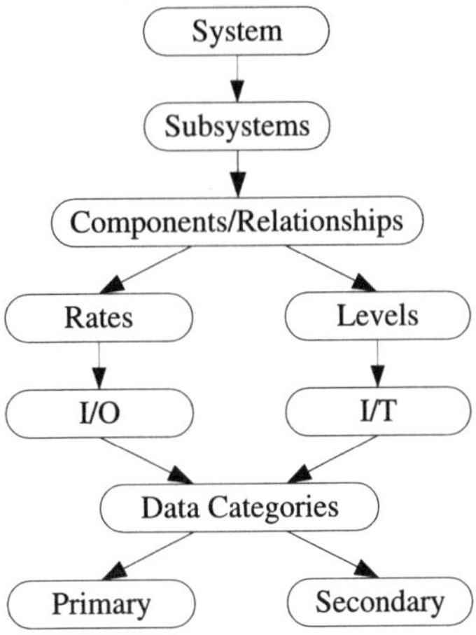

Figure 1.14 Structure for Systems Analysis

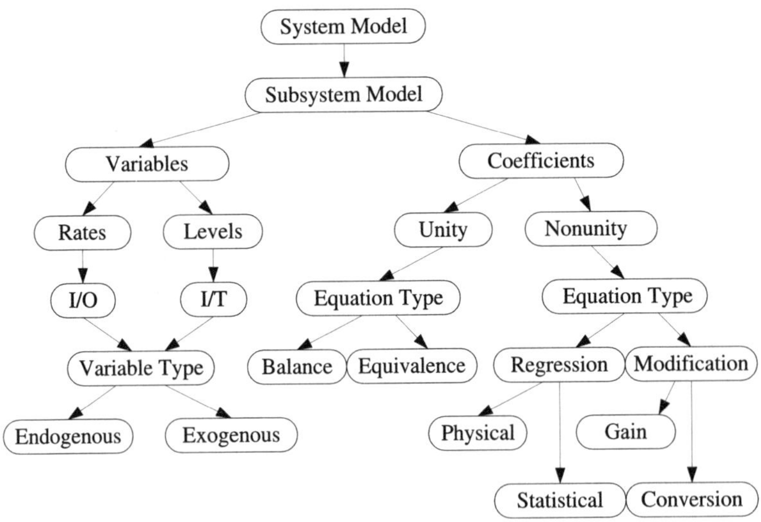

Figure 1.15 Structure for Model Synthesis

correspond to *Variables* and *Coefficients*, and *Data Categories* correspond to *Equation Types* of the system model. The system data base identifies two data categories: *primary* and *secondary*. Data obtained from historical records are categorized as *primary data*. Primary data supply values for the *exogenous* variables of the model. Data initially generated by the model to supplement primary data are categorized as *secondary data*. *Endogenous* (solution) variables of the model supply values for these secondary data. The higher the number of primary variables, the lower

Table 1.3 Population Data Base

Year	X_1	X_2	X_3	X_4
1	10	—	—	—
2	—	—	—	—
⋮				
10	—	—	—	—

the initial number of secondary variables to be generated by the model, and vice versa. If there are no primary data, the model is structured to generate a complete data base. Systems modeling is usually attended by a sparsity of primary data. A *reverse regression* technique is utilized to convert system knowledge items into regression coefficients. Then, *given the knowledge-derived coefficients, reverse regression yields the corresponding data.*

Model coefficients are categorized by value: having *unity* or *nonunity* values. Unity coefficients occur in two different types of equations, *balance* and *equivalence*. Nonunity coefficients occur in *regression* or *modification* equations. These four equation types are defined in Chapter 2.

1.6 ANALYTICAL APPROACHES

Models can be synthesized through three distinctly different approaches: *theoretical analysis, regression analysis,* or *systems analysis*. To better enable understanding and to display their differences, a model will be synthesized for one level of a simple nonlinear system by each of the three analytic approaches. In preview, theoretical analysis and regression analysis serve to provide knowledge items for expanding the knowledge base required by the systems analysis approach.

1.6.1 Population Example

An exponential growth pattern results from an unconstrained population system. Bacteria, mice, and predators are typical of such systems. Mathematically, population level at any time t is expressed by

$$x_t = x_o e^{(GR-DR)t}. \tag{1.1}$$

Net growth results when $GR > DR$. Figure 1.16 illustrates a rapid-growth population system, showing a 10-year exponential increase from an initial population *level* of 10 million.

As typical of most systems, the *data base* is incomplete; its format is presented in Table 1.3. In Equation 1.1, $x_0 = 10$ million. Other knowledge items are

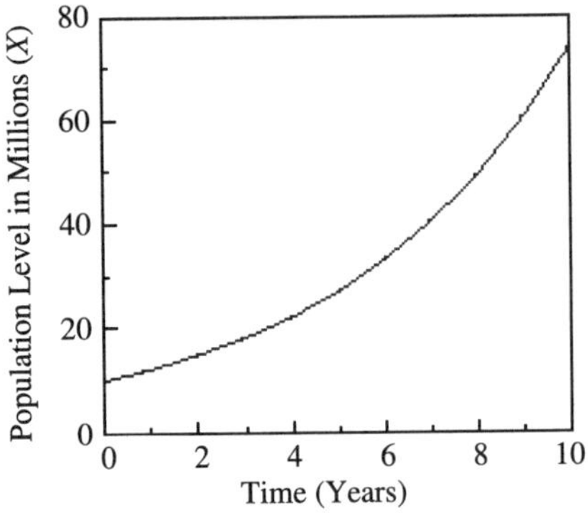

Figure 1.16 Exponential Growth

$GR = 0.30$ birth per unit of the population
$DR = 0.10$ death per unit of the population

Population system levels and rates are parameterized by four time-dependent variables defined as follows:

X_1 = population level initial to Δt
X_2 = population level terminal to Δt
X_3 = annual number of births during Δt
X_4 = annual number of deaths during Δt
$\Delta t = T = 1$ year

A system block diagram is presented in Figure 1.17. Based on these parameters, simple illustrations of each of the three analytical approaches follow.

1.6.2 Theoretical Analysis

Synthesis of a population model draws on mathematical expertise to derive a *micro form* for point rate of change in population level with respect to time t. From Equation 1.1,

$$\frac{dx_t}{dt} = (GR - DR)x_o e^{(GR-DR)t}$$

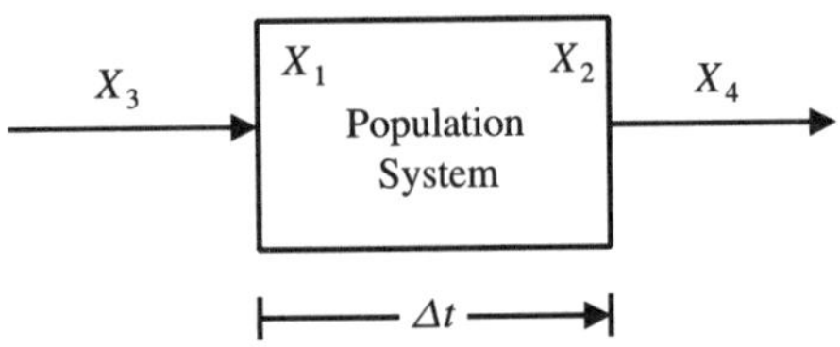

Figure 1.17 Population System

$$\frac{dx_t}{dt} = (GR - DR)\,x_t. \tag{1.2}$$

However, the differential equation of Equation 1.2 does not adequately model the population. It is a micro form that defines only point rate-of-change for the population with respect to time. Because this system is simple, Equation 1.2 can easily be solved by separation of variables and integration over a finite interval of time, as in Figure 1.18, to obtain an equation for exponential growth during time T:

$$\int_{X_1}^{X_2} \frac{dx}{x} = \int_0^T (GR - DR)\,dt$$

$$\ln X_2 - \ln X_1 = (GR - DR)T$$

$$\ln \frac{X_2}{X_1} = (GR - DR)T$$

$$e^{\ln(X_2/X_1} = e^{(GR-DR)T}$$

$$X_2 = X_1 e^{(GR-DR)T}. \tag{1.3}$$

As x_t increases from X_1 to X_2 during time $0 \le t \le T$, the point rate of change in number of births, dx_3/dt, increases in direct proportion to x_t:

$$\frac{dx_3}{dt} = GRx_t.$$

Then, by Equation 1.1 with $x_o = X_1$,

$$\frac{dx_3}{dt} = GR \cdot X_1 e^{(GR-DR)t}.$$

Solve for number of births *during* time T, as number of births increases from 0 to X_3, by integrating over the interval from 0 to T:

$$\int_0^{X_3} dx_3 = GR \cdot X_1 \int_0^T e^{(GR-DR)t}\,dt$$

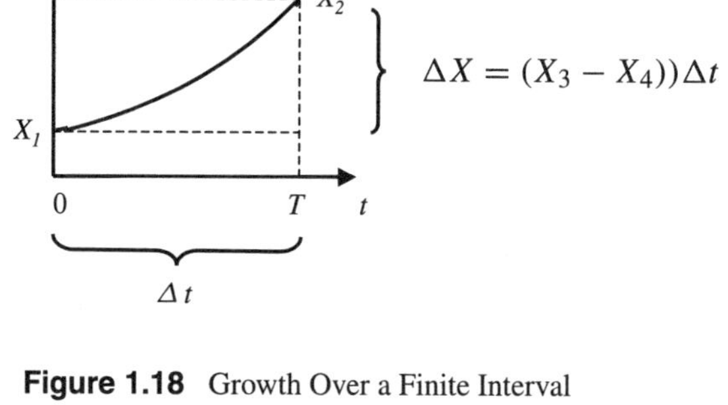

Figure 1.18 Growth Over a Finite Interval

$$= \frac{GR}{GR - DR} X_1 e^{(GR-DR)t} |_0^T$$

$$X_3 = \frac{GR}{GR - DR} X_1 (e^{(GR-DR)T} - 1). \tag{1.4}$$

Similarly, the point rate of change in number of deaths is directly proportional to x_t:

$$\frac{dx_4}{dt} = DR\, x_t.$$

Then by analogy to X_3,

$$X_4 = \frac{GR}{GR - DR} X_1\, (e^{(GR-DR)T} - 1). \tag{1.5}$$

Equations 1.3, 1.4, and 1.5 constitute a nonlinear system model of a population system experiencing net exponential growth. Calibration requires supplying relevant values for exponential parameters GR and DR, often obtained by scaling slopes from plots on exponential (semilog) paper [53]. Values for X_1, GR, and DR from the sample population were substituted into the model for $T = \Delta t = 1$. Computer calculations, carrying eight decimal place accuracy, were performed *serially* to yield solutions for X_2, X_3, and X_4 for the first Δt. For the next Δt computation, X_2 provided the initial X_1 value. In like manner, a 10-year data base was obtained for the population. These data are presented in Table 1.4.

REMARKS

1. In general, population systems are linked with other systems. But when equations of a systems model are *nonlinear*, as is the case with Equations 1.3, 1.4, and 1.5, linking to the equations of other systems models for *simultaneous* solution is not possible. Instead, systems containing nonlinear equations that must be solved together require *serial* solution involving iterative approximation methods.

2. Examination of Table 1.4 reveals that its data satisfy two types of *linear* relationships:

Initialization
$$X_1 = \begin{cases} X_1^0 & \text{initially,} \\ X_2 & \text{of previous year thereafter} \end{cases} \tag{1.6}$$

Balance
$$X_1 - X_2 + X_3 - X_4 = 0 \tag{1.7}$$

3. Replacing nonlinear Equation 1.3 by unity-coefficient Equations 1.6 and 1.7 produces a hybrid system model, containing both linear and nonlinear relationships. This suggests the possibility of discovering linear equations to replace model Equations 1.4 and 1.5 and thereby to achieve a linear system model for nonlinear population growth.

Table 1.4 Population Data from Theoretical Model

Year	X_1	X_2	X_3	X_4
1	10.000000	12.214028	3.321041	1.107014
2	12.214028	14.918247	4.056329	1.352110
3	14.918247	18.221188	4.954412	1.651471
4	18.221188	22.255410	6.051332	2.017111
5	22.255410	27.182819	7.391114	2.463705
6	27.182819	33.201170	9.027527	3.009176
7	33.201170	40.552001	11.026246	3.675415
8	40.552001	49.530326	13.467488	4.489162
9	49.530326	60.496477	16.449227	5.483075
10	60.496477	73.890565	20.091131	6.697043

1.6.3 Regression Analysis

In contrast to theoretical analysis, regression analysis can only be applied to an existing data base. Consequently, data were "borrowed" from the X_1 column of Table 1.4 for analysis by regression. Examining X_1 revealed a nonlinear growth pattern, as in Figure 1.16.

1.6.3.1 Nonlinear Regression

As suggested by the type of nonlinearity in the growth pattern, an *exponential predicting equation* that expressed population level as a function of time was utilized:

$$x_t = be^{at}.$$

Log transformation produced a linear predicting equation:

$$\ln x_t = \ln b + at.$$

Standard least-squares regression [27] yielded a best-fitting regression equation:

$$\ln x_t = 2.302585 + 0.20t.$$

The net growth rate, $GR - DR$, is identified by $a = 0.20$. Exponential transformation of $\ln x_t$ reintroduced the exponential form:

$$x_t = e^{(2.302585+0.20t)},$$

or

$$x_t = 10.00e^{0.20t}. \tag{1.8}$$

Note that Equation 1.8 is identical to Equation 1.3 if $X_1 = 10.0$, $GR = 0.30$, and $DR = 0.10$ are substituted.

REMARKS

1. Dynamic problems can be solved by other means, rather than by integrating differential equations. Fitting an exponential predicting equation was equivalent to solving $(dx_t)/(dt) = (GR - DR)x_t$ for x_t.

2. Because Equation 1.8 is *nonlinear*, it is not suitable for inclusion into a system of equations for simultaneous solution.

3. In this instance, nonlinear regression was able to match the information provided by theoretical analysis to systems modeling.

1.6.3.2 Linear Regression

Nonlinearity is the consequence of expressing population growth as an *explicit* function of time. Alternatively, if future population *level* is regressed on current population *level* as an *implicit* function of time, the predicting equation is *linear*:

$$X_2 = c_1 X_1.$$

X_1 and X_2 are separated in time by Δt. Data for X_1 and X_2 were "borrowed" from Table 1.4 to obtain a linear regression equation of comparable accuracy:

$$X_2 = 1.2214027618 X_1 \tag{1.9}$$

With Equations 1.6 and 1.9 comprising a two-equation model, population levels were computed for each of the 10 years. These match those displayed in the X_1, X_2 columns of Table 1.4.

REMARKS

1. To obtain comparable accuracy for X_1 and X_2 values, 10 decimal places were used. Moreover, if the data had come from actual observations, regression error would have greatly diminished the accuracy.

2. Comparison of Equation 1.8 with $t = \Delta t$ and Equation 1.9 yielded a useful item of knowledge: regression coefficient c_1 can be obtained with similar accuracy by a *calibration equation*:

$$c_1 = e^{0.20} = 1.2214027618,$$

or to generalize,

$$c_1 = e^{(GR-DR)}.$$

3. From time series analysis [40], X_1 and X_2 of Table 1.4 data are a realization of a nonstationary time series, and Equation 1.9 is identical to a deterministic ARIMA(1,1,0) process: $z_t = 1.2214027618 z_{t-1}$, where $\phi = 1.2214027618 > 1$ implies nonstationarity.

Population *rates* are also *correlated* with population *levels* as *implicit* functions of time for each Δt:

$$X_3 = c_2 X_1, \text{ and } X_4 = c_3 X_1$$

Again, "borrowing" data from Table 1.4, values for regression coefficients c_2 and c_3 providing comparable accuracy were obtained:

$$X_3 = 0.3321041402 X_1 \tag{1.10}$$

$$X_4 = 0.1107013784\, X_1 \tag{1.11}$$

Equations 1.6, 1.9, 1.10, and 1.11 were grouped to form a linear regression model of the exponential population system. Computer calculations provided a match for the 10-year data base of Table 1.4.

REMARKS

1. To obtain comparable accuracy from the regression model for X_3 and X_4 values, 10 decimal places were used.

2. Note that c_2 appears to be an adjusted representation of $GR = 0.30$, and c_3, of $DR = 0.10$.

3. Linear Equations 1.10 and 1.11 provide an alternative to nonlinear Equations 1.4 and 1.5, and suggest the possibility that better linear regression equations might be found.

4. Substituting Equations 1.10 and 1.11 into Equation 1.7 yields Equation 1.9. Equation 1.7 is totally compatible to system dynamics and should be included in a systems model.

5. A regression equation restricted to a single hierarchical level is equivalent to an incomplete balance equation. Incompleteness is observed by comparing the coefficients of the population variables in balance Equation 1.7 and regression Equation 1.9:

Equation 1.7: $X_2 = 1.0000000000 X_1 + 1.0(X_3 - X_4)$
Equation 1.9: $X_2 = 1.2214027618\ X_1 + 0.0(X_3 - X_4)$

In Equation 1.7, incompleteness occurs if X_3 and X_4 are omitted. In Equation 1.9, X_2 is regressed upon X_1, the *unity* coefficient of X_1 has been adjusted to

$c_1 = 1.2214027618$, a *nonunity* coefficient that compensates for the net effect of omission. Furthermore, c_1 bears resemblance to $1.0 + GR - DR = 1.20$, and correctly so, because 0.2214027618 is the exponential growth trend associated with the per capita net growth rate of 0.20. Therefore, in Equation 1.9 exponential adjustment has taken place, and c_1 contains the evidence $e^{(GR-DR)} = 1.2214027618$.

1.6.4 Systems Analysis

Systems analysis, rather than data, provides the basis for model synthesis. Data are highly desirable as a knowledge base item, useful for validation but only incidental to synthesis. Knowledge acquisition comprises a major activity of systems analysis and yields a *knowledge base* for each level of articulation.

1.6.4.1 Knowledge Base for L_1 Population Model

- System: Population exhibiting net exponential growth
- Fundamental units of analysis:

  Space: Population domain

  Mass: One unit of the population

  Time: One year (Δt)
- Number of subsystems: one
- System variables:

  X_1 = initial population level

  X_2 = terminal population level

  X_3 = number of births per year, a rate

  X_4 = number of deaths per year, a rate
- Data items:

  $X_1^0 = 10$ million (population level at $t = 0$),

  $GR = 0.30$, $DR = 0.10$
- Relationships for system balance and initialization
- Interacting dynamic forms that correlate population rates to population levels
- Schematic: See Figure 1.17.

1.6.4.2 Model Structure

Change in population level with respect to time is defined by the *macro form*, $(\Delta x)/(\Delta t) = (X_2 - X_1)/(\Delta t) = X_3 - X_4$, that for $\Delta t = 1$ yields the *system balance equation*:

$$X_1 + X_3 = X_2 + X_4 \tag{1.12}$$

Dynamic forms for exponential growth and decline, introduced in Chapter 2, provide *two linear regression equations* as replacements for nonlinear Equations 1.4 and 1.5:

$$X_3 = GR(w_1X_1 + w_2X_2) = c_1X_1 + c_2X_2 \tag{1.13}$$

$$X_4 = DR(w_1X_1 + w_2X_2) = c_3X_1 + c_4X_2 \tag{1.14}$$

Terminal population levels are transferred to initial population levels by an *initialization relationship*:

$$X_1 = \begin{cases} X_1^0 & \text{initially,} \\ X_2 & \text{of previous year thereafter.} \end{cases} \tag{1.15}$$

1.6.4.3 Model Calibration

Calibration principles are presented in Chapter 2, and calibration schemes for this L_1 population model are fully explained in Chapter 3. Dynamic form components of the model, Equations 1.13 and 1.14, were calibrated by substituting $GR = 0.30$ and $DR = 0.10$ into the following auxiliary calibration equations:

$$\begin{aligned} c_1 + c_2 &= GR \\ c_3 + c_4 &= DR \\ c_1 &= 3c_3 \\ \frac{1 + c_1 - c_3}{1 - c_2 + c_4} &= e^{(GR-DR)} \end{aligned}$$

Accuracy to eight decimal places was retained in solving these equations:

$$\begin{aligned} c_1 &= 0.15499664 \\ c_2 &= 0.14500336 \\ c_3 &= 0.05166555 \\ c_4 &= 0.04833445 \end{aligned}$$

Solution Starting with $X_1^0 = 10.0$, yearly, simultaneous solutions of linear Equations 1.12, 1.13, 1.14, and 1.15 were obtained for 10 years. Computer calculations were carried out using standard matrix algebra and sufficient accuracy to match values in Table 1.4 to six decimal places. Thus, the linear equation model projects

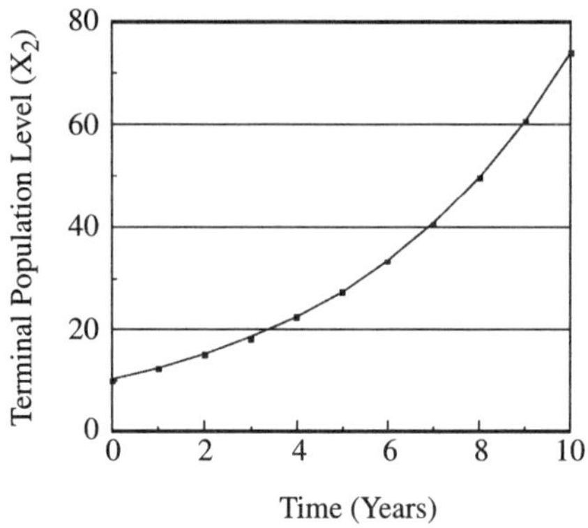

Figure 1.19 Population Level versus Year

nonlinear, exponential growth with great accuracy. X_2 versus *Time* is plotted in Figure 1.19.

REMARKS

1. Although linear models derived from systems analysis may match theoretical analysis for accuracy, they must retain tentative status until validation in one or more applications.

2. Matrix algebra is applicable because the equation system is linear. Simultaneous solutions for interacting systems are easy to obtain, simply by solving their linked matrix models.

1.7 COMPARISON OF ANALYTICAL APPROACHES

Analytical approaches produce quantitative results, but are they truly objective? Unfortunately, no! Selection of an approach is based on subjective evaluation, deciding which works *best* for a specific application. What if the application is systems modeling? Information presented in the form of *advantages vs disadvantages* for each of the three approaches provides that subjective basis for comparison.

1.7.1 Theoretical Analysis

Analysis begins with the infinitesimal components of a physical system, *micro forms*, that describe the rate of change of each component in terms of differential equations. To obtain a systems model, the equations are transformed to macro forms by integration. Single or independent micro forms are integrated separately.

If several micro forms exist and interact with each other, simultaneous solution of differential equations by integration is required.

1.7.1.1 Advantages

1. Grounded in commonly accepted theory.
2. Provides precise, continuous definition of dynamic systems response at the micro-form level.

1.7.1.2 Disadvantages

1. For other than trivial problems, an enormous number of variables and differential equations are required to describe a system; and, although integration to a macro level is usually possible, the model is unwieldy because of its extensiveness.
2. In general, differential equation systems are difficult to integrate unless they are linear and have constant coefficients.
3. Reduction of difficulty requires approximations that remove nonlinearity and limit amount of detail, that is, that simplify reality.
4. Errors, due to missing variables and/or ignored interactions, restrict the usefulness of systems models derived from a micro basis.
5. Differential equations and integration schemes are incomprehensible to a nonmathematician, seriously impairing credibility of the systems model.

1.7.2 Regression Analysis

One or more regression equations are fitted to historical data related to the system domain. Contiguous data are required, and detail need not be homogeneous.

1.7.2.1 Advantages

1. Data are required for as few as two variables, and one of these can be time.
2. Regression equations are easy to solve simultaneously if they are linear.
3. Linear equations are more comprehensible to a nonmathematician than are nonlinear equations.

1.7.2.2 Disadvantages

1. Number of variables is limited by data availability.
2. Resulting models span one or more hierarchical levels (lack homogeneity in level of detail), are incomplete, and are often unrealistic, yielding statistical rather than physical correlation.
3. Regression models are accompanied by regression error.
4. Models may become totally useless outside the range established by the data.
5. Regression equations are difficult to solve simultaneously if they are non-linear.
6. Nonlinear equations are less comprehensible to a nonmathematician.

1.7.3 Systems Analysis

Analysis produces a systems knowledge base of homogeneous level. Variables of the systems model are set in one-to-one correspondence with components of the systems domain. Hierarchical, Mtm, systems models are obtained by "differentiation" of macro models, eventually incorporating deep levels of systems knowledge.

1.7.3.1 Advantages

1. Model synthesis is possible even in the absence of historical data.
2. Utilizes a knowledge base consisting of such items as units of analysis, system components, theoretical and regression relationships, schematics, and historical data (if available) to construct a linear model and to generate a complete data base.
3. Employs finite time increments and provides point-by-point tracking of non-linear trajectories.
4. Simultaneous solution of the model's system of linear equations is accomplished with ease.
5. Linear equations are more easily understood by nonmathematicians than other types of equations.
6. Linear-equation matrices of interacting systems are easily linked within a super matrix for simultaneous solution.
7. Level of detail is specified by the user.

8. Reproduces "exact" historical database values and permits perturbation of these values to simulate answers to *what if* questions.

9. Incorporates knowledge from human experts from mathematical, statistical, and systems domains.

1.7.3.2 Disadvantages

1. Systems approaches tend to overwhelm those modelers who are not systems oriented.

2. Systems analysis is too time consuming and is often bypassed in favor of "quick" solutions.

3. Domain expert(s) may not be available when their knowledge is required.

1.7.3.3 Conclusion

Modeling of a population system was illustrated by three different analytical approaches. Theoretical analysis was initiated at the micro level with a linear, first-order differential equation model defining point rates of change for the three basic components of a population system. The base line of the triangle in Figure 1.20 labelled L_∞ indicates the infinitesimal, articulation level of the differential equation model.For this simple system, these equations were easily integrated to obtain a population system model at the macro level, L_1:

$$\mathbf{X}_1 = (X_1, X_2, X_3, X_4).$$

To accommodate comparison, regression analysis and systems analysis approaches were also applied at the L_1 level.

The systems analysis approach included in its knowledge base several knowledge items obtained by preliminary theoretical and regression analysis. Although systems analysis and modeling were illustrated for only one level of detail, by

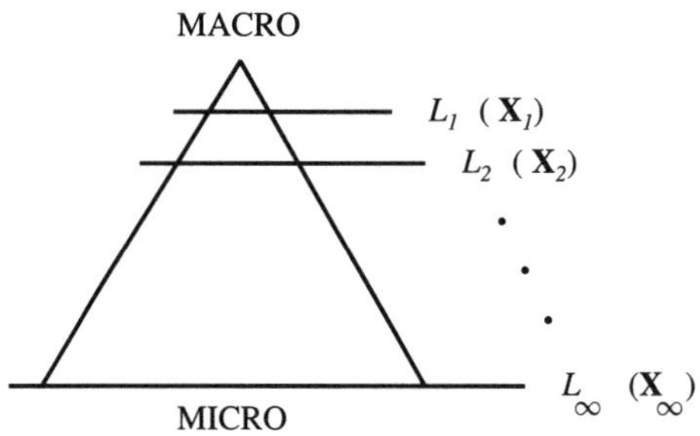

Figure 1.20 Mtm Location

following the Mtm approach, analysis is repeatable at the next lower level. To effect further analysis, L_1 ($\mathbf{X}_1$) is subdivided (differentiated) to $L_2(\mathbf{X}_2)$, where, for example, the population can be modeled in terms of age-segment subsystems and the flow of the population through these segments with passage of time.

Analyses based on the systems approach inherently overcomes major disadvantages of theoretical and regression analyses approaches and incorporates most of their advantages. Consequently, systems analysis is recommended as the "best" approach, and subsequent chapters present rationale in support of that claim.

EXERCISES

1. What elements are involved in engineering design?
2. What are the three elements of a general information triad?
3. Provide your own example of an information triad.
4. Distinguish between *hard* and *soft* knowledge.
5. What are the three axioms of existence and how are they related to the Mtm approach?
6. Using the de Rosnay reference [14], write a one- or two-page paper describing de Rosnay's "macroscope."
7. What is a
    a. cellular system?
    b. large-scale system?
8. What is the principle of conservation of mass and how is it expressed mathematically?
9. Calibrate the time wheel to a 24-hour day.
10. Using library resources, write a one- or two-page discussion of entropy.
11. How does entropy relate to
    a. evolution?
    b. probability?
    c. the Mtm method?
12. How do mathematical differentiation and integration provide useful analogs in conceptualizing systems analysis and model synthesis?

13. Using library resources, provide a one- or two-page discussion for each of the following:

    a. hierarchical analysis

    b. finite element analysis

    c. regression analysis

14. What is a modeling tree?

15. Describe the Mtm approach to concurrent systems analysis and modeling.

16. What is the black box approach to systems analysis and modeling?

17. Make a list showing the parallel between analytic structures for systems analysis and model synthesis.

18. Using equations from the theoretical model (Equations 1.3, 1.4, 1.5) and data from Table 1.4, calculate values for GR and DR for the population example.

19. Use data from Table 1.4, Equation 1.3 of the theoretical model, and semilog paper to estimate the value of $GR - DR$.

20. Referring to Equation 1.7, identify $X_3 - X_4$ as NG (net growth). Using data from Table 1.4, find the linear regression of NG on X_2. Replace $X_3 - X_4$ in Equation 1.7 by the regression equation for NG, and solve for X_2. Discuss the results.

21. Make a table of the *characteristics* that you consider to be *advantageous* to systems analysis and modeling. Include a second column in your table for comments in support of your choice.

CHAPTER 2

Systems Modeling Principles

2.1 KNOWLEDGE-BASED MODELING

Systems analysis results in the organization of a knowledge base that contains both *hard* and *soft* knowledge items. Hard knowledge consists of laws, theory, principles, system relationships, and for some systems, *primary data*. Soft knowledge is comprised of heuristics: the informal, experiential knowledge of one or more experts that constitutes *rules of good judgment* in the system domain. Each Mtm model is supported by a knowledge base that typically includes items such as these:

- Physical system identification and level of circumscription
- Fundamental units of analysis:

    Space

    Mass (or mass attribute)

    Time

- Number of subsystems
- System schematic
- System variables
- Primary data

- Table displaying structure of the model
- Dynamic forms
- Specifications
- Heuristics

2.2 CIRCUMSCRIPTION PRINCIPLE

Systemic thinking—systematic, step-by-step thinking—typifies the Mtm approach. Mtm systems models are synthesized in sequential order. Conceptually, the most macro-level model is a triad consisting of two variables and one equation, represented by the vertex of Figure 2.1. Because this model articulates zero detail and contains zero subsystems, it is designated by $L_{0,0}$. Its sole purpose is axiomatic: to define the state of equilibrium in the aggregate flow of *mass* unbounded by *space* and *time* via a *balance equation* that equates *supply* to *demand.*

Therefore, to initiate the Mtm approach, a *domain* must be defined that bounds or *circumscribes* the system at the most integrated level to be considered, at $L_{1,j}$. Figure 2.1 indicates progressively more differentiated models that articulate increasing levels of detail. Each model is articulated at a homogeneous level of detail. Model performance is referenced back to the *baseline* established by the previous level. Comparison of the output with that of the previous level is unique to Mtm modeling. Sequential expansion of detail could become an overwhelming hindrance to understanding system dynamics. Through circumscription, the Mtm approach limits expansion of detail by imposing a limit on the number of variables admitted at each level.

Knowledge acquisition (pursuit of knowledge), *systems analysis*, (source of knowledge), and *model synthesis* (deposition of knowledge) constitute a systems

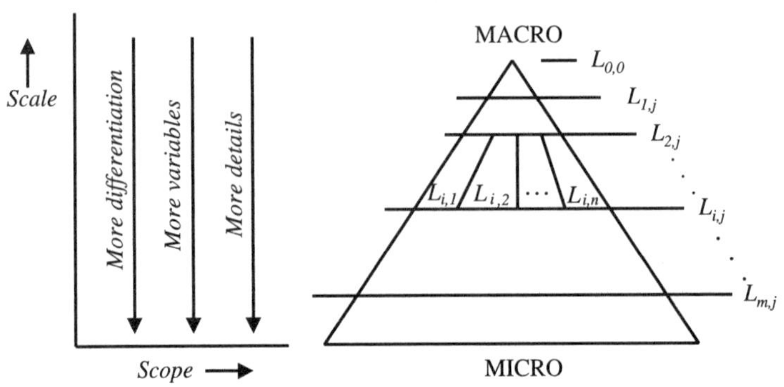

Figure 2.1 Levels of Circumscription

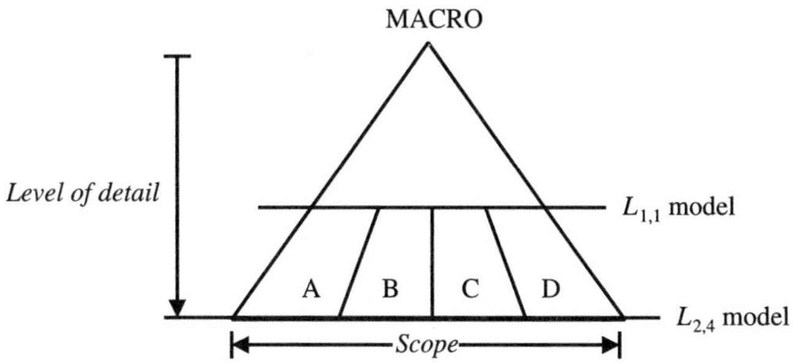

Figure 2.2 Circumscribed System

triad. At each level of articulation, the search for domain knowledge is limited in breadth and depth through the principle of circumscription.

> *System circumscription confines knowledge acquisition to a fixed scope of coverage and fixed level of detail.*

Thus, levels of circumscription may be viewed in terms of two orthogonal axes as depicted in Figure 2.1.

1. Horizontally, $L_{i,1}, L_{i,2}, \ldots, L_{i,n}$, as *scope* (or breadth) of coverage is extended.

2. Vertically, $L_{1,j}, L_{2,j}, \ldots, L_{m,j}$, as *level* (or depth) of detail is extended.

$L_{i,j}$ and its indices represent a succession of models extending from macro to micro. *Horizontal circumscription* involves limiting the number of subsystems, and hence number of variables, to be included in the model. *Vertical circumscription* involves constraining the variables to a homogeneous level of detail and fixation of each of three fundamental units of extension (or analysis)—*space, mass*, and *time*—as defined in Chapter 1. Thus, selecting the horizontal scope of coverage and vertical level of detail establishes the level of circumscription of the system. Relative levels of circumscription are indicated by values of the indices i, j. Analysis and modeling are extended until a target level of circumscription is achieved, a level of detail that affords definition and solution of system domain problems.

Consider for example, an $L_{2,4}$ circumscription of a grizzly bear population symbolized in Figure 2.2. The three units of analysis are fixed as follows:

Space: Yellowstone National Park
Mass: One grizzly bear
Time: One year

The scope of coverage includes four subsystems:

A = adult male

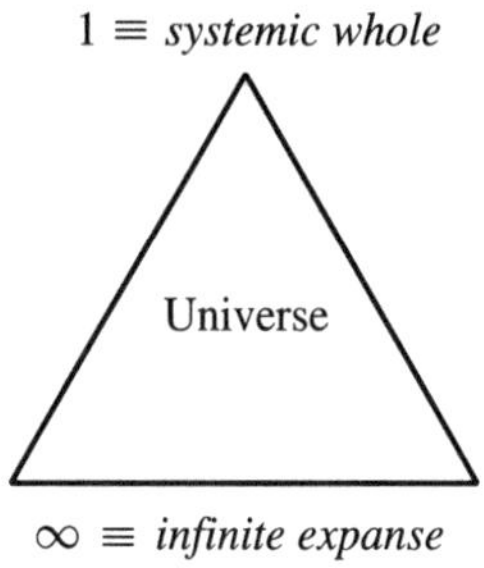

Figure 2.3 Symbolic Model of the Physical World

B = adult female
C = cub male
D = cub female

$L_{2,4}$ is a "model world" circumscription of the grizzly bear system, a subset of the universal model of the "physical world" symbolically represented by the triangle of Figure 2.3.

2.3 CONSERVATION PRINCIPLE

Physically, the universe is made up of interacting systems of mass that experience entropic changes of state. Laying nuances aside (see discussion of mass in Chapter 1), matter is indestructible, and units of mass can be used to define quantity of matter in a *conservative* system. Consequently, such systems of mass satisfy the conservation principle (first law of thermodynamics):

> *With the transition of time, mass (and its associated energy, momentum, etc.) cannot be created or destroyed; it can only be accumulated, transferred, or transformed.*

To apply the principle, each physical system circumscribed for analysis and modeling is characterized by incremental change in *level* of mass (or its attributes) during a finite transition in time, the increment Δt. Change involves a triad for each *cause* and *effect*: (1) one or more dynamic factors *cause* change from (2) an *initial level* to (3) a *terminal level*, the *effect*. Change can be *positive* or *negative*, and *continuous* or *discrete*. *Continuous* change is subject to *measurement; discrete* change to *counting*. Any system exhibiting multiple cause/effect relationships is circumscribed as two or more subsystems and leads to a systems model consisting of two or more corresponding submodels. Because the physical system is in mass balance, a system (subsystem) model will contain only the number of variables required to complete the *system (subsystem) mass balance equation*. These variables obtain their identity in one-to-one correspondence with components of the physical system (subsystem).

2.4 CORRESPONDENCE PRINCIPLE

Mtm systems analysis and model synthesis were presented as concurrent activities in Chapter 1. Refer back to Figures 1.14 and 1.15 to note structural similarities. Physically, systems of mass undergo entropic changes with respect to space and time. Thus, each physical system can be viewed as a triad consisting of (1) *transitional components of mass*, shaped by (2) *dynamic factors*, and, subject to (3) *mass balance*. Without question, close correspondence between the physical world and the model world is highly desirable. Such parallelism is possible by synthesizing a model world that is congruent to this triad of the physical world.

Mathematically, a systems model must consist of interacting systems of equations that simulate entropic changes in physical-world entities. Therefore, to establish congruence, each systems model must also be viewed as a triad consisting of (1) *time variant levels*, shaped by (2) *dynamic forms*, and, subject to (3) *mass balance equations*.

Triad congruency is summarized in the principle of correspondence.

> *For each circumscribed level $L_{i,j}$ of analysis, each variable of the systems model is assigned one-to-one correspondence with each transitional component of mass (or mass attributes) of the physical system.*

Effective systems modeling depends on establishing correspondence between the most basic components of a system and model variables. As a result, systems modeling enhances systemic understanding of goals, objectives, boundaries, resource flows, and the general management process. Modeling principles stated and developed in this chapter apply to all levels of modeling within the Mtm spectrum. Applications follow in later chapters.

2.5 CLASSIFICATION OF MODEL VARIABLES

Model variables are designated by the column vector, $\mathbf{X} = (X_1, X_2, \ldots, X_n)^T$. Each variable carries the same single subscript for each of $N\Delta t$'s. When a particular value of t is to be referenced in the text, a superscript is used: X^t_j. Three functions performed by variables during the modeling process form a basis for classification. Variables may be observed to do the following:

1. Comprise the most basic model *component*
2. Establish *data categories*
3. Fulfill *roles*

2.5.1 By Component

A system is composed of subsystems and variables. Therefore, a systems model must take into account three component levels: the system itself, *subsystems,* and *variables*. How many subsystems to include is determined by circumscription. The following rules of thumb are helpful in identifying system components:

1. Look for logical or naturally occurring division of function by subsystem.
2. Define the component that performs a functional role in the subsystem.
3. Determine if the component is connected in any way to other parts of the system.
4. Define any relationship between the system or any of its other components that affects the function of the component.

Variables comprise the most basic model component and may be further classified into naturally occurring *levels* and *rates* [19]. A systems model must provide a simple mathematical structure for applying one or more rates, over successive Δt increments of time, that accurately replicates changes in the level of a system (subsystem).

2.5.1.1 Levels

In Mtm modeling, rates become integrated into levels within the *balance equation*. An initial level relative to Δt, plus the change in level due to the net of input/output rates during Δt, results in a new terminal level relative to Δt. Thus, a systems model emulates continuity, with selection of Δt providing the degree of resolution. System *states* are defined by levels at discrete points in time. Changes in state occur as the system evolves over a series of Δt's. Given a current level X_I, the system will evolve to the level X_T during a finite increment of time, Δt. X_I and X_T form an *initial-terminal* or *IT* pair and are illustrated in Figure 2.4.

2.5.1.2 Rates

By defining levels as paired variables, rate of change is simply the difference between two levels separated by Δt. System level evolves from X_I to X_T in response to *single* or *net* rates of change applied as system input and/or output. For example, a single input rate, X_j, is represented by the macro form $X_j =$

$(X_T - X_I)/\Delta t$. In general, the value of X_j is unique to each Δt and varies with time as illustrated in Figure 2.5. Specific rates are defined by *dynamic forms*.

2.5.1.3 Interaction of Levels and Rates

Rates can produce changes in levels: $X_T = X_I + X_j \Delta t$. Levels can produce changes in rates: $X_j = c_1 X_I + c_2 X_T$. Thus, an initial level, X_I; a rate, X_j; and a terminal level, X_T comprise a *triad of variables*.

2.5.2 By Data Category

Each variable of an Mtm systems model defines a data category, either *primary* or *secondary*. Any variable whose values are established by historical data is a *primary variable*, and its data are called *primary data*. Very seldom do all variables qualify as primary. Yet despite deficiency in primary data, an operational systems model must be supported by a complete data base, complete in the sense that each data category contains data for the periods of interest. Consequently, the first application of a model is to complete the data base by solving for *secondary data*. These solution variables are referred to as *secondary variables*. Their tentative values form the basis for the first of a series of validation tests of the model.

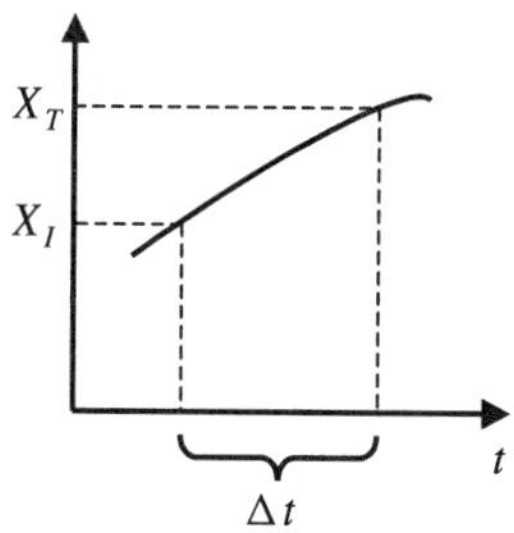

Figure 2.4 System Levels

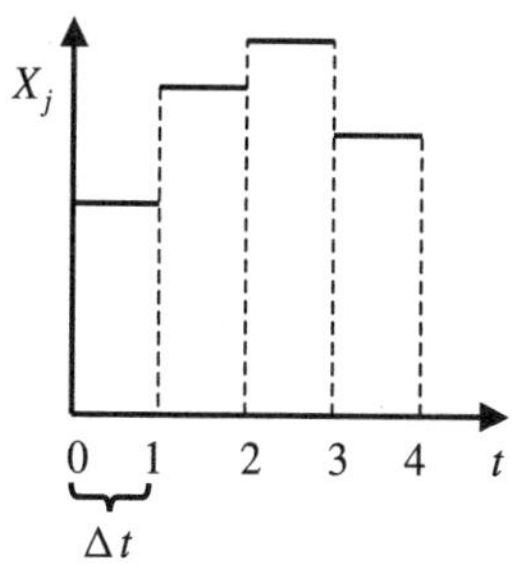

Figure 2.5 Variable Rates

2.5.3 By Role

Each level and rate variable fulfills a role in the model. External data series are processed by the model via *exogenous* variables. Primary data are always exogenous, but primary variables can assume endogenous roles. Dynamic response of the model is generated via *endogenous*—that is, its *solution*—variables. Various tests and applications of the model utilize *role inversion*, accomplished by interchanging pairs of variables in terms of their endogenous and exogenous roles.

2.6 LINEAR RELATIONSHIPS

Level and rate variables interact within a system of linear equations. Two different types of equations are used: *unity coefficient* and *nonunity coefficient* equations. Both types are expressible in homogeneous form

$$c_1X_1 + c_2X_2 + \cdots + c_nX_n = 0.$$

Nonunity coefficient equations require calibration. *Initialization relationships, balance equations*, and *equivalence equations* have unity coefficients and do not need calibration.

2.6.1 Initialization

Boundary conditions relative to a Δt period are introduced through the *initialization* relationship. Each system level is *initialized* at the outstart of modeling and is *reinitialized* at the beginning of each subsequent Δt iteration.

Illustration The IT pair (X_I, X_T) represents two levels or states of a system:

$$X_I = \text{ initial level relative to } \Delta t.$$

$$X_T = \text{ terminal level relative to } \Delta t.$$

$$X_I = \begin{cases} X_I^0 & \text{initially,} \\ X_T & \text{of previous } \Delta t \text{ period thereafter.} \end{cases}$$

2.6.2 Modification

Sometimes system variables require *modification* to achieve a specific system response pattern, accomplished by a simple *gain* relationship: *amplification* for a greater effect or *attenuation* for a lesser effect. Also, variables from diverse-mass subsystems may interact within the same equation, so that *modification* is required for *scaling* and to maintain *unit* consistency.

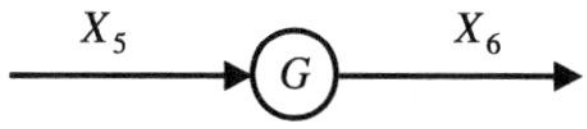

Figure 2.6 Simple Gain Relationship

Illustration In Figure 2.6, arbitrary variables X_5 and X_6 are related by a simple *gain*:

X_5 = variable subjected to *gain*
X_6 = modified version of X_5
G = coefficient modifier
$X_6 = GX_5$
$G > 1$ produces a gain. $G < 1$ produces an attenuation.
Values for G are assigned during the calibration process.

2.6.3 Balance Equations

Systems are either *conservative* or *nonconservative*. The term *systems model* generally relates to a conservative system. A conservative system must satisfy a *balance* equation for such entities as *mass, energy, momentum, and so forth.* A conservative system neither creates nor destroys entities. Entities are either accumulated, transferred, or transformed. On the other hand, *source* or *sink* systems are nonconservative. However, by providing appropriate software support, systems models can process exogenous data from or supply exogenous data to nonconservative models or submodels during runtime.

A *balance equation* or *difference equation* is the macro form of the solution of the differential equation of a conservative system (or subsystem). It is also known as a *continuity equation.* For example in a system (subsystem) experiencing mass flow over time, the balance equation expresses the time integrated *level* of mass accumulated during Δt and has this form:

$$\textit{Supply} - \textit{Demand} = 0.$$

In general, balance equations have unity coefficients: $c_j = +1$ for each supply variable, and $c_j = -1$ for each demand variable. An exception occurs when a subsystem interacts with another subsystem whose variables do not have the same units. Then $c_j = G$ is used for scaling and units conversion.

Illustration Both subsystems of the $L_{1,2}$ model of Figure 2.7 have balance equations:

$$\mathrm{SS}_1\text{: } X_1 - X_2 + X_5 - X_6 - X_9 + X_{10} = 0$$
$$\mathrm{SS}_2\text{: } X_3 - X_4 + X_7 - X_8 + X_9 - X_{10} = 0$$

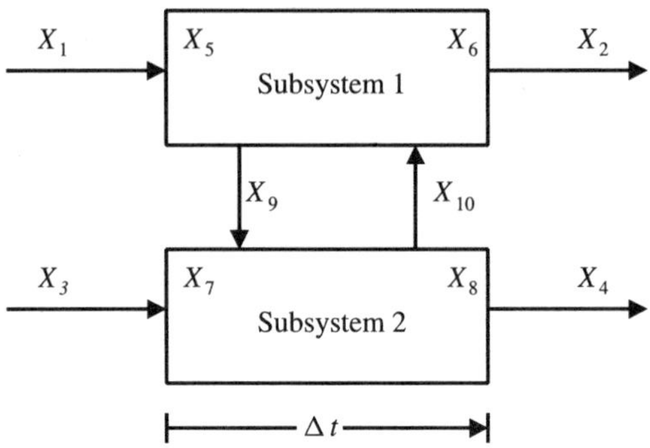

Figure 2.7 Two Subsystem Model

Figure 2.8 Summation Equivalence

Adding (or superimposing) these two subsystem balance equations produces this more macro $L_{1,1}$ system balance equation:

$$X_1 - X_2 + X_3 - X_4 + X_5 - X_6 + X_7 - X_8 = 0$$

Note that linking variables X_9 and X_{10} no longer exist; the two subsystems have been integrated into a single system.

2.6.4 Equivalence Equations

An *equivalence equation* also has unity coefficients and provides simplification through redefining.

Illustration Consider the relationship among the three model variables, X_1, X_2, and X_3, presented in Figure 2.8. This relationship is represented by an equivalence equation:

$$X_3 = X_1 + X_2, \text{ or } X_1 + X_2 - X_3 = 0.$$

Consequently, the sum $X_1 + X_2$ can be replaced by X_3 anywhere X_3 occurs in the model.

2.6.5 Regression Equations

Physical correlations that exist among and between variable components of a physical system embody the principle of *cause and effect.* In general, these correlations consist of causal elements that effect entropic changes, changes that are nonlinear with respect to time. However, in some instances variable components are related only by *statistical correlations.* Both types of correlation are represented quite simply at Δt increments of time by *linear regression equations* that express *physical* and/or *statistical correlations* between variables within the circumscribed system (model world). Multiple linear regression has the following form:

$$X_k = c_0 + \sum_{j=1}^{n} c_j X_j,$$

where $k \neq j$, and c_j need not be unity valued and can be zero. In standard statistical applications, regression relationships are selected from various *predicting equation forms*, for example, linear, nonlinear, exponential, and so on. In systems modeling, regression equations are selected from various *linear dynamic forms.*

System variables may also interact with one or more external variables (outside the circumscribed system) in establishing system levels. Although circumscription omits (by excluding) these variables, it may be advantageous to *integrate* their average (lumped) effect into the model, thereby to impact the system balance equation. Regression equations provide structure for *implicit* inclusion of external variables. The average, nonzero effect of omissions is modeled by the constant term c_0. As shown in Chapter 1, the regression equation can be viewed as an incomplete balance equation whose c_j values have been adjusted above or below unity to compensate for these omissions. An expanded (more micro) model is obtained by *differentiating* (dividing) the lumped effect of external variables into system variables for *explicit* inclusion in the balance equation(s) of additional subsystem(s).

2.7 TIME SERIES

Mtm systems modeling is supported by systems analysis that produces an initial knowledge base that may or may not contain historical data. To be useful, historical data must be compiled into sequences of primary data that are ordered in uniformly spaced increments of time. Primary data provide one of the sources of knowledge used in calibrating systems models that are then used to generate sequences of secondary data. Primary and secondary data sequences are commonly called *time series* [40]. Mathematically, the sequence $\{X_j^t\}$; for $t = 0, 1, 2, \ldots, N$; represents a *sample realization*, R, of a time series.

2.7.1 Classification

One or more *dynamic forms* are calibrated to introduce dynamic behavior into a model. Therefore, effective model calibration depends on classification of time series sequences in terms of dynamic behavior:

- **Randomness** Any sequence $\{X_j^t\}$ characterized by randomness comprises a *stochastic time series*, and exhibits a probability distribution. Otherwise, the sequence constitutes a *deterministic time series*.
- **Homogeneity** Behavior of any full sequence or segment of R is said to be *homogeneous* if the following properties hold:

  The *same* trend line (linear or nonlinear) prevails throughout; that is, behavior is shaped by the *same* dynamic force(s).

  If R is stochastic, deviation above or below the trend line does not exceed fixed limits, usually two or three standard deviations (σ).

- **Stationarity** Behavior of any stochastic sequence $\{X_j^t\} \in R$, or *segment* of R is said to be *stationary* if the mean (μ) and variance (σ^2) of the distribution are time invariant. A series is *nonstationary* if either varies with time.

REMARK Deterministic sequences, even though nonstationary, may also exhibit homogeneity. For example, exponential growth (decline) or oscillatory divergence (convergence) resulting from time invariant rates exhibit homogeneous nonstationarity. On the other hand, relatively large *discontinuities* will disrupt homogeneity.

2.8 CALIBRATION PRINCIPLES

The systems triad of activities—systems analysis, knowledge acquisition, and model synthesis— is performed at each Mtm level of circumscription. Model synthesis involves specifying two types of equations, those having unity coefficients and those having nonunity coefficients. Balance equations (representing levels) and equivalence equations (providing simplification) are characterized by unity coefficients. On the other hand, modification equations (providing scaling and consistency) and regression equations (representing rates) require nonunity coefficients. To enable a model to *track* nonlinear response of the physical system, it must be calibrated. Calibration is the process of specifying values for the model's coefficients.

Based on existence of primary data, calibration of systems models falls into one of three categories: (1) Total primary data; (2) some primary data; (3) no primary data. For category (1), *forward regression* is applicable. Categories (2) and (3) require *reverse regression*.

2.8.1 Calibration via Forward Regression

Model calibration by standard techniques requires a *primary* (historical) data base for key variables plus *statistical expertise*. *Knowledge* and *experience* are utilized to classify variables as dependent or independent. An equation fitting method is *subjectively* selected for regressing dependent variables upon independent variables. Conventional, *data-based regression* is emphasized in this context as "forward" regression. The principle is simply this:

> *Given adequate primary data, regression coefficients can be found that generate a best fit of equation to data.*

When primary data exist (thereby providing a standard), regression equation errors can be calculated. If primary data are available for all system (subsystem) variables, then a composite error term should be included in the system (subsystem) balance equation to account for inevitable algebraic imbalance.

2.8.2 Calibration via Reverse Regression

Knowledge is the sum and essence of systems analysis, the woof and fabric of model synthesis. But more often than not, hard knowledge in the form of primary data is sparse or missing. Consequently, one might ask, Can regression coefficients be synthesized from other than primary data? The answer is yes! This problem of data deficiency was solved upon discovery and application of *knowledge-based regression*, referred to as "reverse" regression, and formalized by the following principle:

> *Given adequate system knowledge, regression coefficients can be found that generate secondary data consistent with that knowledge.*

Linear regression equations function as *dynamic forms* that are defined in the next section. Calibration equations, synthesized from or associated with system knowledge items, provide values for regression coefficients. Thus, dynamic forms become a repository for system knowledge, deposited in the form of regression coefficients. Hard knowledge is often scarce or nonexistent and can be substituted by soft knowledge. *Knowledge acquisition*, an expert systems synthesis technique, involves transfer of knowledge from a domain expert [31]. Soft knowledge items such as data estimates, calibration relationships, and regression coefficient values can be obtained in this manner.

Systems engineers are experts in the calculus domain as well as specific systems domains. Systems modeling is not restricted to engineering domains and, in those instances, can require the assistance of other domain experts, such as accountants, biologists, economists, geologists, and sociologists. Acquisition and application of domain knowledge make it possible to select appropriate dynamic forms and to calibrate these as regression equations.

Reverse regression, based on knowledge of calculus, system components, internal and external relationships, system response, and so on enables synthesis of *secondary data* as a first test of the model. Although secondary data are linked by model structure to system knowledge that may even include primary data, reverse regression errors are inherently distributed over secondary data. Nonetheless, error reduction is possible during the process of validating secondary data. During this process, domain experts may provide their greatest assistance, with revision of coefficient values or even model structure. These data, together with any existing primary data, constitute a complete data base for access by the systems model.

2.8.3 Calibration Period for Reverse Regression

Calibration via reverse regression requires choosing one of two calibration principles. Which principle to apply depends on the type of dynamic behavior of the exogenous times series that comprise inputs and outputs of the systems model:

- **Homogeneous stationarity** For a stochastic time series realization, R, or any of its contiguous segments that exhibit homogeneous stationarity, the following calibration principle is applicable:

  Average Δt Period *If the systems model is calibrated to the* average Δt period *of the time series, then it is calibrated to any other Δt period.*

Note that the concept of average period is purely hypothetical; no such period in the historical record exhibits average levels and rates. See the chapter on stochastic analysis for further discussion of stochastic time series.

- **Homogeneous nonstationarity** For a stochastic or deterministic time series realization, R, or any of its contiguous segments that exhibit homogeneous nonstationarity, apply the following calibration principle:

  Typical Δt Period *If the systems model is calibrated to a* typical Δt period *of the time series, then it is calibrated to any other Δt period.*

2.8.4 Calibration Equations

In performing forward regression, coefficients that minimize regression error (deviation from known data values) for a standard regression equation are obtained by solving a set of *normal equations*. Conversely, in performing reverse regression, coefficients are obtained by solving a set of *calibration equations* that contains systems knowledge relating to the single Δt period of calibration. Depending on the calibration scheme, calibration equations may consist solely of *auxiliary equations*, or they may also include model equations. One calibration equation is required for each unknown; unknowns may include some combination of data and coefficients. Auxiliary equations are used only for calibration to the single Δt period and are not part of the model (excluding prototype models). Often, calibration can be performed one subsystem at a time, involving only auxiliary equations.

Consider, for example, a model consisting of one balance equation and one regression equation. Prior to calibration, there are four unknowns: X_2, X_3, c_1, and c_2. Coefficient values may be found by solving two calibration equations that apply to the single Δt period of calibration. Alternatively, trial values for X_2 and X_3 may also be desired. Consequently, four calibration equations are required. The additional two must come from the model. Calibration schemes are demonstrated through numerous applications to systems models in later chapters.

2.8.5 Tentative Nature of Calibration

An initial calibration is necessarily tentative because the model must yet be verified and its performance validated through testing. The following are typical questions asked about the model:

1. Do secondary data values meet the expectation of the domain expert(s)?
2. Do any rates or levels reverse in algebraic sign unexpectedly over time?
3. Can the roles of endogenous and exogenous variables be exchanged without producing excessive calculation error?
4. When hypothetical, exogenous series are input, is model response suggestive of behavior in the physical world as judged by the domain expert(s)?

2.9 BUILDING-BLOCK PRINCIPLE

Systems models are synthesized from simple, modular building blocks as defined by the following principle:

> *A system module consists of one difference (balance) equation and one or more dynamic forms.*

It follows that the following are true:

- A *module* is the smallest model entity that functions as a system. Minimally, a triad of variables is involved: two level variables and one (net or single) rate variable.
- A simple system model is comprised of one module.

- A complex systems model contains one or more *groups* of modules. A group consists of two or more modules. Thus, each module is a *subsystem.*

Figure 2.9 presents a simple, modular building block with a net-rate variable expanded into input and output rates, where

X_1 = initial level

X_2 = terminal level

X_3 = input rate during Δt

X_4 = output rate during Δt

Equating *supply* and *demand* components, the balance equation is

$$X_1 + X_3 = X_2 + X_4. \tag{2.1}$$

Balance equations are *static* with respect to change: never inducing change, only integrating change. For example, X_1 is given a value initial to Δt. If the rates (X_3, X_4) are zero during Δt, then terminal to Δt, $X_1 = X_2$ and no change has taken place. However, if during Δt either of the rates or both are nonzero, then the resulting *dynamic* change is integrated into X_2. Therefore, the entities that supply nonzero values to X_3 and X_4 are called *dynamic forms.* Two or more rates evoke a net response: As system response evolves over time, changes in level are positive if X_3 is dominant, or negative if X_4 is dominant.

2.9.1 Dynamic Forms

Entropic forces (as linear or nonlinear causes) bring about changes (as effects) in physical systems. Changes that occur during a given increment of time are attributed to rates that are continually being integrated into new levels. Changes are introduced into a mathematical model by either micro or macro *dynamic forms.* Micro dynamic forms consist of differential equations that define point phenomena as *explicit* functions of time. In general, integration of micro forms yields *nonlinear* macro forms, ill suited to systems modeling. In contrast, systems modules consist of linear equations that yield finite displacements as *implicit* functions of Δt. Dynamic characteristics of the system are modeled by macro dynamic forms. The

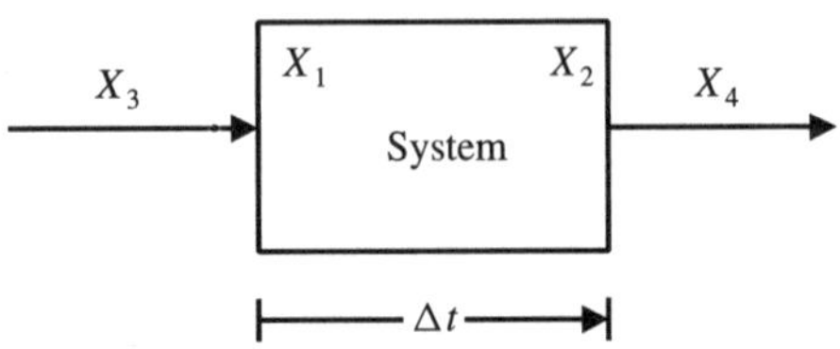

Figure 2.9 Modular Building Block

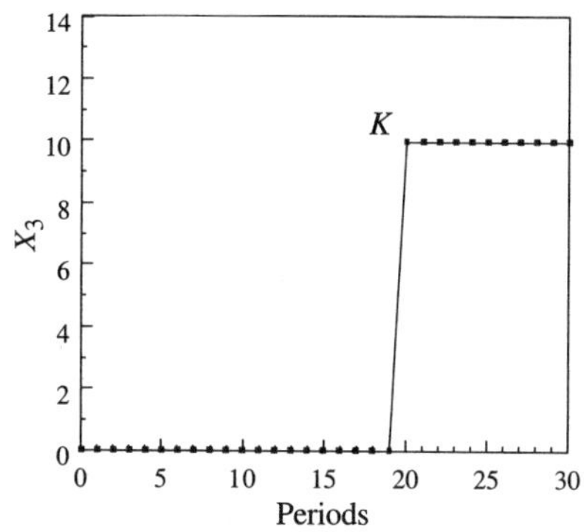

Figure 2.10 Deterministic Time Series

role of dynamic forms in Mtm systems modeling is summarized in the following principle:

> *Changes in levels and rates are induced through integration of dynamic forms by the system's balance equation(s).*

Dynamic forms are a repository for knowledge and become standardized through frequent use. Next, several, standard dynamic forms are defined or derived; context is provided by the generic four-variable module of Figure 2.9.

2.10 EXOGENOUS DYNAMIC FORMS

Times series supplied externally to the model to introduce input and output rates function as *exogenous dynamic forms*. For each X_j time series,

$$Xj = \begin{cases} \textit{datum} & \text{value at time } t = n\Delta t, \\ 0 & \text{otherwise.} \end{cases} \tag{2.2}$$

2.10.1 Deterministic Form

A *deterministic form* is characterized by a time series that does not exhibit randomness. Examples include ramp-, step-, oscillatory-, and polynomial functions. To illustrate the dynamic effect of input rate X_3 on level, let output rate $X_4 = 0$, so that the balance equation $X_2 = X_1 + X_3$ integrates the X_3 time series input into system level.

Illustration Figure 2.10 displays a deterministic time series, X_3, in the form of a step function. $X_3 = 0$ for $n = 1, 2, \ldots, 19$, but steps to K beginning with $n = 20$.

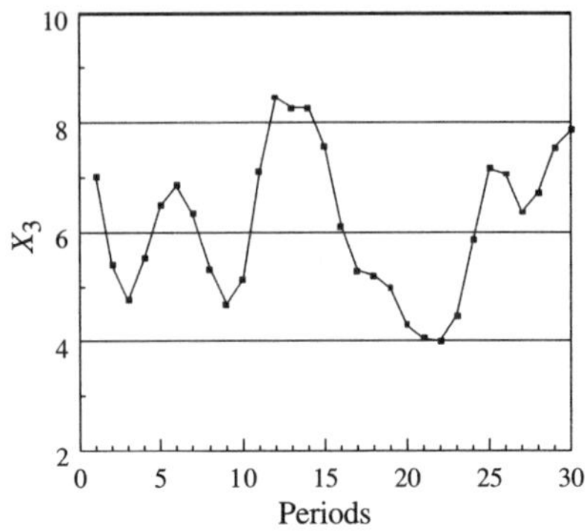

Figure 2.11 Stochastic Time Series

2.10.2 Stochastic Form

A *stochastic form* is characterized by a times series that does exhibit randomness. The series may come from historical data or may be generated using *Monte Carlo simulated sampling* (see Chapter 8).

Illustration Figure 2.11 displays a stochastic time series, X_3, as an input rate for $n = 1, 2, \ldots, 30$. X_3 is a *stochastic form* that causes the level of the system to change from X_1 to X_2 at each Δt. Again, let output rate $X_4 = 0$, so that the balance equation $X_2 = X_1 + X_3$ integrates stochastic growth into system level.

2.11 ENDOGENOUS DYNAMIC FORMS

Times series generated internally by the model as input or output rates function as *endogenous dynamic forms*. Each form is given a name descriptive of the function it provides.

2.11.1 Stochastic Form

Stochastic times series, as in Figure 2.11, may result from direct solution of the model's equations.

2.11.2 Constant Form

The *constant form* introduces a constant rate of change, K, that is to be integrated into system level during each Δt:

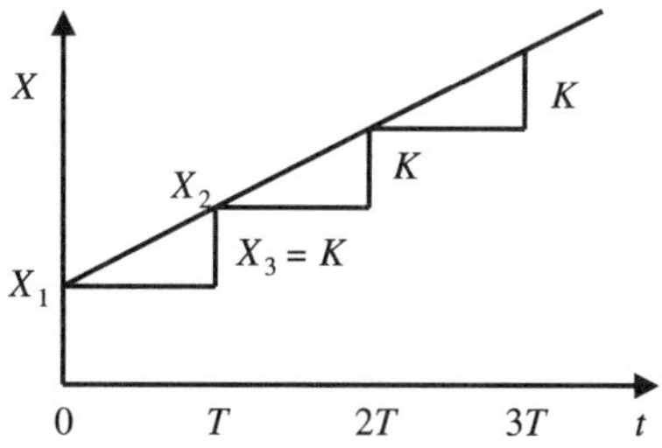

Figure 2.12 Linear Accumulator

$$X_j = \begin{cases} K > 0 & \text{produces linear growth,} \\ K < 0 & \text{produces linear decline.} \end{cases} \tag{2.3}$$

Illustration In Figure 2.12, $X_3 = K$ is a constant form. With $X_4 = 0$, the module serves as a *linear accumulator* so that $X_2 = X_1 + X_3$.

2.11.3 Logical Form

A *logical form* consists of these three components:

- Binary variable, X_j
- Threshold value, T
- Trigger variable, X_t

Mathematically, a logical form expresses binary logic

$$X_j = \begin{cases} K_1 & \text{if } X_t > T, \\ K_2 & \text{if } X_t \leq T. \end{cases}$$

To illustrate, let X_4 of Figure 2.9 be a binary variable whose value is determined by

$$X_4 = \begin{cases} 6 & \text{if } X_2 > 12, \\ 0 & \text{if } X_2 \leq 12. \end{cases}$$

2.11.4 Exponential Forms

Natural change is exponential and is expressed by the *exponential form*, related to the base of the *natural* logarithm, $e = 2.7182818284590452$. Change can be positive, resulting in *growth*, or negative, producing *decline*. In Chapter 1, the population model exhibited net growth, including dynamic forms for both types of change.

Exponential change is caused by *positive feedback*. Therefore, the dynamic form imposing the increment of exponential change is not constant for each Δt, but rather, directly proportional to a weighted average of initial and terminal levels, X_I and X_T. When exponential change is to be expressed by a regression equation, values for the weight coefficients are determined by solving *calibration equations*. In general, the values derived for the weights depend on individual applications. Although dynamic forms for exponential growth and decline are similar, their calibration equations are dissimilar. These dynamic forms and their calibration equations are presented next.

2.11.4.1 Exponential Growth

Let growth rate for $\Delta t = 1.0$ time unit be denoted by $GR > 0$, and exponential weights by $w_1, w_2 > 0$ that must satisfy $w_1 + w_2 = 1.0$. Exponential growth is expressed by the following dynamic form:

$$Xj = \frac{\Delta X}{\Delta t} = GR(w_1 X_I + w_2 X_T) = c_1 X_I + c_2 X_T \tag{2.4}$$

Calibration involves establishing values for w_1 and w_2, or equivalently, c_1 and c_2.

Calibration Equations for Exponential Growth

Consider a triad consisting of IT pair X_1, X_2, and single rate X_3. In Figure 2.13, exponential growth is experienced during time $\Delta t = T$ and for consecutive Δt increments. X_3 is defined by the dynamic form given as Equation 2.4.

The model for exponential growth is specified by the following vector table:

X_j	IT Pair	Exo	Endo	Equation Type	Equation No.
X_1	I	*		—	—
X_2	T		*	balance	1
X_3			*	regression	2

Although calibration can be performed with either the w_j's or c_j's, the latter are selected. Two unknowns (c_1 and c_2) require two calibration equations. Calibration equations are derived from the following knowledge items:

- Two module equations: Equation 2.1 with $X_4 = 0$, and Equation 2.4

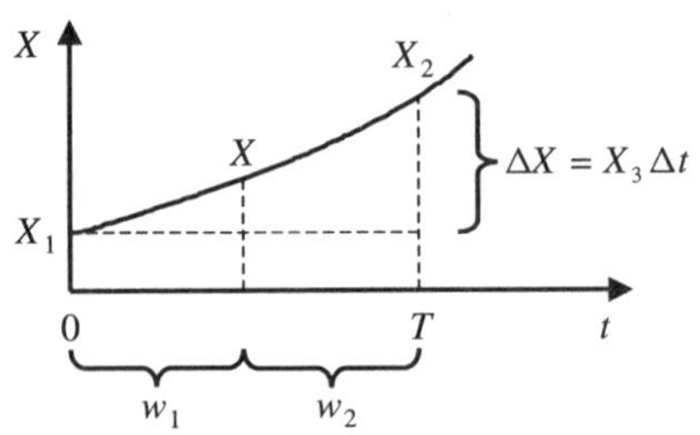

Figure 2.13 Exponentially Weighted Projections

- Knowledge of calculus from the *mathematical domain*
- *Regression of X_2 on X_1 is unique for exponential growth. Therefore, any two such expressions are equivalent.* Because the two expressions in Equation 2.4 are equivalent, their coefficients are equal. A derived knowledge item follows:

$$GR\, w_1 = c_1 \text{ and } GR\, w_2 = c_2$$

In Figure 2.13, X is an exponentially weighted projection of X_1

$$X = X_1 e^{GR \cdot w_1 T} = X_1 e^{c_1 T}.$$

Similarly, X_2 is an exponentially weighted projection of X

$$X_2 = X e^{GR \cdot w_2 T} = X e^{c_2 T}, \text{ or } X = X_2 e^{-c_2 T}.$$

Equate the two expressions for X and solve for X_2

$$X_2 = X_1 e^{(c_1 + c_2) T}. \tag{2.5}$$

In Equation 2.4, consider the $\lim_{\Delta t \to 0} (\Delta X)/(\Delta t)$ for the IT pair X_1, X_2 of Figure 2.13:

$$\begin{aligned} &X_1 \to X \\ &X_2 \to X \\ &\Delta X \to dX, \text{ and} \\ &\frac{\Delta X}{\Delta t} \to \frac{dX}{dt} = GR\,(w_1 X + w_2 X) = GR \cdot X, \end{aligned}$$

yielding the differential equation, $(dX)/X = GR \cdot dt$, whose solution is obtained by integration

$$\int_{X_1}^{X_2} \frac{dX}{X} = \int_0^T GR \cdot dt$$

$$\ln X_2 - \ln X_1 = GR \cdot T, \quad \ln \frac{X_2}{X_1} = GR \cdot T, \frac{X_2}{X_1} = e^{GR \cdot T}, \text{ or}$$

$$X_2 = X_1 e^{GR \cdot T}. \tag{2.6}$$

Equations 2.5 and 2.6 for growth period $T = \Delta t$ equivalently express the regression of X_2 on X_1. Therefore, set their exponents equal to obtain the first calibration equation

$$c_1 + c_2 = GR. \tag{2.7}$$

Another equivalent regression expression is obtained for X_2 by replacing X_3 in Equation 2.1 by the c_j version of Equation 2.4 ($X_4 = 0$):

$X_2 = X_1 + c_1 X_1 + c_2 X_2$. Solve for X_2:

$X_2(1 - c_2) = X_1(1 + c_1)$, yielding

$$X_2 = \frac{1 + c_1}{1 - c_2} \cdot X_1. \tag{2.8}$$

Compare Equations 2.8 and 2.6 for growth period $T = \Delta t$. Equate coefficients of X_1 to obtain the second calibration equation:

$$\frac{1 + c_1}{1 - c_2} = e^{GR}. \tag{2.9}$$

From Equations 2.4 and 2.7, the w_j' s are found:

$$w_1 = \frac{c_1}{c_1 + c_2}, \text{ and } w_2 = \frac{c_2}{c_1 + c_2}.$$

The two calibration equations for exponential growth are:

$$c_1 + c_2 = GR \text{ and } \frac{1 + c_1}{1 - c_2} = e^{GR}.$$

GR must be replaced by $GR \cdot \Delta t$ for $\Delta t \neq 1.0$.

2.11.4.2 Exponential Decline

Let decline rate for $\Delta t = 1.0$ time unit be denoted by $DR > 0$, and exponential weights $w_1, w_2 > 0$ that must satisfy $w_1 + w_2 = 1.0$. The dynamic form for exponential decline is given by:

$$X_j = \frac{\Delta X}{\Delta t} = DR(w_1 X_I + w_2 X_T) = c_1 X_I + c_2 X_T. \tag{2.10}$$

Calibration involves establishing values for w_1 and w_2, or equivalently, c_1 and c_2.

Calibration Equations for Exponential Decline

Consider a triad consisting of IT pair X_1, X_2, and single rate X_4. Exponential decline is experienced during time $T = \Delta t$ and for consecutive Δt increments. X_4 is defined by the *exponential decline form* presented as Equation 2.10.

The vector table is identical to that of the growth system but with X_4 replacing X_3. As before, two unknowns (c_1 and c_2) require two calibration equations. Calibration equations are derived from the following knowledge items:

- Two module equations: Equation 2.1 with $X_3 = 0$, and Equation 2.10.
- Knowledge of calculus from the *mathematical domain*
- *Regression of X_2 on X_1 is unique for exponential decline. Therefore, any two such expressions are equivalent.* Because the two expressions in Equation 2.10 are equivalent, their coefficients are equal. A derived knowledge item follows:

- $DR\ w_1 = c_1$ and $DR\ w_2 = c_2$.

Similar to that of exponential growth, but with X_4 replacing X_3 and DR replacing GR, the knowledge-based derivation leads to

$$X_2 = X_1 e^{-(c_1+c_2)T} \tag{2.11}$$

$$X_2 = X_1 e^{-DR \cdot T} \tag{2.12}$$

$$c_1 + c_2 = DR \tag{2.13}$$

$$X_2 = \frac{1-c_1}{1+c_2} X_1 \tag{2.14}$$

$$\frac{1-c_1}{1+c_2} = e^{-DR} \tag{2.15}$$

The w_j's for decline are obtained from Equations 2.10 and 2.13, and are found to be identical to those for growth

$$w_1 = \frac{c_1}{c_1+c_2}, \text{ and } w_2 = \frac{c_2}{c_1+c_2}.$$

Exponential decline calibration equations are

$$c_1 + c_2 = DR$$
$$\frac{1-c_1}{1+c_2} = e^{-DR}.$$

DR must be replaced by $DR \cdot \Delta t$ for $\Delta t \neq 1.0$.

2.11.5 Malthusian Form

Growth, of whatever sort, may in the "short" run be rapid and appear to be exponential. However, in the "long" run, entropic degradation imposes constraints so that growth eventually reaches a saturation level, or upper limit. T. R. Malthus, a British economist, propounded a theory in 1798 that population tends to outrun the means of subsistence [17]. More recently, D. H. Meadows et al., warned that global population could reach a saturation level if current trends prevail [34]. To model this saturation effect, a declining growth rate that approaches zero as the IT growth pair of the triad approaches its upper limit is defined by a *Malthusian form*

$$X_j = \frac{\Delta X}{\Delta t} = DR\,[X_{\max} - (w_1 X_I + w_2 X_T)] = c_0 - c_1 X_I - c_2 X_T \tag{2.16}$$

The Malthusian form is a composite of two dynamic forms

$$X_j = \underbrace{c_0}_{T_1} - \underbrace{(c_1 X_I + c_2 X_T)}_{T_2}$$

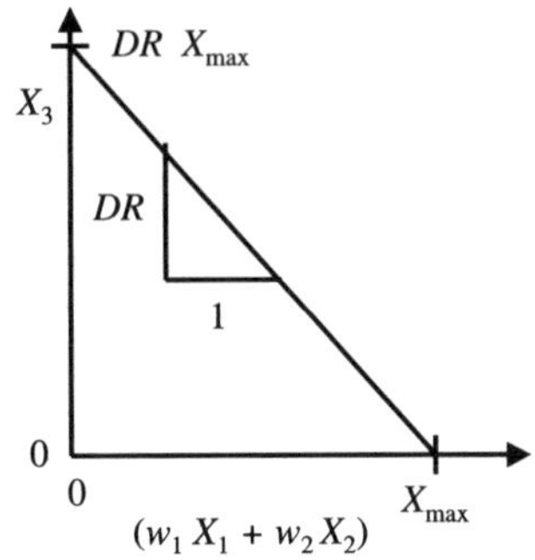

Figure 2.14 Linear Model of Diminishing Growth Rate

$T_1 = c_0 = DR\ X_{\max}$ is a *constant form* (see Equation 2.3). $T_2 = c_1 X_I + c_2 X_T = DR(w_1 X_I + w_2 X_T)$ is the *exponential decline form* (see Equation 2.10), where again, decline rate $DR > 0$. Exponential weights, w_1, $w_2 > 0$ are unknown but must satisfy $w_1 + w_2 = 1.0$. Thus, calibration involves establishing values for w_1 and w_2, or equivalently, c_1 and c_2, as for exponential decline.

2.11.5.1 Calibration Equations for the Malthusian Form

Consider a triad consisting of IT pair X_1, X_2, and single rate X_3 that varies inversely with weighted values of X_1 and X_2 for each Δt increment. This inverse relationship is portrayed by linear Equation 2.16 in Figure 2.14.

The vector table is identical to that of the exponential growth system. Two unknowns (c_1 and c_2) require two calibration equations. Calibration equations are derived from the following knowledge items

- Two module equations: Equation 2.1 with $X_4 = 0$, and Equation 2.16
- Knowledge of calculus from the *mathematical domain*
- *Regression relationships for the Malthusian form are unique. Therefore, any two such expressions are equivalent.* Because the two expressions in Equation 2.16 are equivalent, their coefficients are equal. Applying this knowledge produces the following derived knowledge item:
- $c_0 = DR\ X_{\max}$, $c_1 = DR\ w_1$, and $c_2 = DR\ w_2$.

Referring to Figure 2.15, Y experiences exponential decline and is analogous to X_1 of Equation 2.11 except that Y is referenced to $X_{\max}$ instead of the origin. Thus, as t makes its transition through $T = \Delta t$, Y goes from $X_{\max} - X_1$ to $X_{\max} - X_2$. Replacing X_1 by $Y = X_{\max} - X_1$ and replacing X_2 by $Y = X_{\max} - X_2$ in Equation 2.11 yields Equation 2.17

$$X_2 = X_{\max} - (X_{\max} - X_1)e^{-(c_1+c_2)T}. \tag{2.17}$$

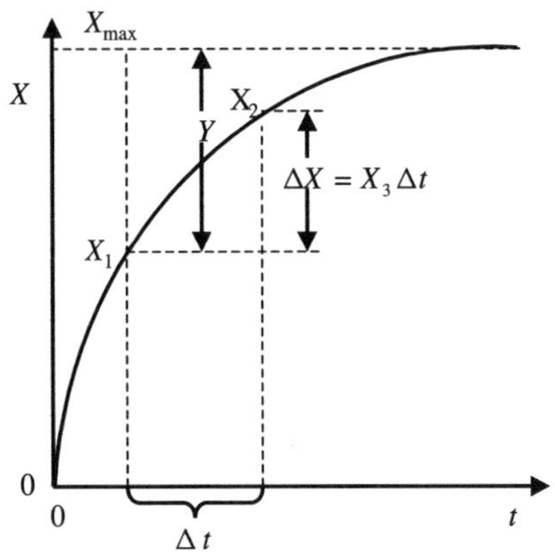

Figure 2.15 Diminishing Growth

In Equation 2.16, consider the $\lim_{\Delta t \to 0}(\Delta X)/(\Delta t)$ for IT pair X_1, X_2 of Figure 2.15

$$
\begin{aligned}
X_1 &\to X \\
X_2 &\to X \\
\Delta X &\to dX,
\end{aligned}
$$

and

$$
\begin{aligned}
\frac{\Delta X}{\Delta t} \to \frac{dX}{dt} &= DR \cdot X_{\max} - DR\,(w_1 X + w_2 X) \\
&= DR \cdot X_{\max} - DR \cdot X, \text{ or} \\
\frac{dX}{dt} + DR \cdot X &= DR \cdot X_{\max},
\end{aligned}
$$

a differential equation with *constant forcing function* of the form $(dy)/(dt)+ay = K$, whose solution $y = c^0 e^{-at} + K/a$ becomes constant at *steady state.*

Applying the solution to X, $X = c^0 e^{-DR\,t} + (DR\,X_{\max})/(DR)$. From boundary conditions, $X = X_1$ when $t = 0$, and $X = X_2$ when $t = T$.

$$
\begin{aligned}
X_1 &= c^0 + X_{\max} \text{ leads to } c^0 = X_1 - X_{\max}, \text{ and} \\
X_2 &= (X_1 - X_{\max})e^{-DR\,T} + X_{\max}, \text{ or} \\
X_2 &= X_{\max} - (X_{\max} - X_1)\,e^{-DR\,T}.
\end{aligned}
\tag{2.18}
$$

Equations 2.17 and 2.18 for growth period $T = \Delta t$ equivalently express the regression of X_2 on X_1. Therefore, equate their exponents to obtain the first calibration equation

$$c_1 + c_2 = DR. \tag{2.19}$$

Replacing X_3 in Equation 2.1 by the c_j version of Equation 2.16, c_0 by $DR \cdot X_{\max}$, and DR by $c_1 + c_2$ with $X_4 = 0$ produces another form of the regression equation for X_2:

$$X_2 = \frac{1 - c_1}{1 + c_2} X_1 + \frac{c_1 + c_2}{1 + c_2} X_{\max}. \tag{2.20}$$

Regrouping the terms of Equation 2.18 and equating coefficients of like terms in Equation 2.20 for one period of declining growth, $T = \Delta t$ yields a second calibration equation

$$\frac{1 - c_1}{1 + c_2} = e^{-DR}. \tag{2.21}$$

The w_j's are derived from Equations 2.16 and 2.19, and are identical to those for exponential growth and decline:

$$w_1 = \frac{c_1}{c_1 + c_2}, \text{ and } w_2 = \frac{c_2}{c_1 + c_2}.$$

Malthusian form calibration equations are

$$\begin{aligned} c_0 &= DR\ X_{\max} \\ c_1 + c_2 &= DR \\ \frac{1 - c_1}{1 + c_2} &= e^{-DR}. \end{aligned}$$

DR must be replaced by $DR \cdot \Delta t$ for $\Delta t \neq 1.0$.

2.11.6 Delay Form

A dynamic system is uniquely characterized by its *transfer function* that determines the type and amount of *delay*, the time displacement (or lag) which occurs between a change in system input and the system's response. Mass (or energy) is input to the system. The system transforms or shapes this input into a response over an interval of time. In general, there are two types of delay: *continuous delay* and *transportation delay* [1]. Transportation delay is a special case of continuous delay. Nonetheless, a single, dynamic form accommodates either type of delay in the systems model. A *delay form* is a regression equation that has been calibrated relative to parameters that account for type and amount of delay.

2.11.6.1 Continuous Delay

Continuous delay refers to a lagged period of continuous response, the length of time required to realize the full effect of an input (see Figure 2.16).

Derivation can proceed from either the mathematical (micro form, unit step function, and convolution integral) or systems domain (system module). In deference to practicality, consider the dynamic response of a simple system that consists of water flowing into the open end of a barrel and the simultaneous release of water from an orifice in the bottom of the barrel. Initially, the barrel is empty and outflow rate is zero but eventually catches up with the inflow rate. Identify the following system variables in units of gallons:

X_1 = initial volume $\qquad$ X_3 = inflow rate

X_2 = terminal volume $\qquad$ X_4 = outflow rate

A block diagram of the module is presented in Figure 2.17.

X_3 is an exogenous inflow that is integrated into the volume of water in the barrel. X_4 is an endogenous outflow that is regressed upon X_1 and X_2, initial and terminal levels of that volume

$$X_4 = c_1X_1 + c_2X_2.$$

However, outflow is delayed, and parameters of the dynamic form that shape delay must be coded into the coefficients of the regression equation for X_4. Two unknowns (c_1 and c_2) require two calibration equations. Calibration equations are derived from the following knowledge items:

- *Regression of X_4 on X_1 and X_2 is unique. Therefore, any two such expressions are equivalent.*
- Inflow becomes outflow following a time delay (or time lag), D, measurable in Δt time units.
- Outflow from the barrel in D time units is $X_4 \cdot D$ gallons.
- Outflow is regressed upon a weighted average of X_1 and X_2:

$$X_4 \cdot D = \alpha X_1 + (1 - \alpha)X_2,$$

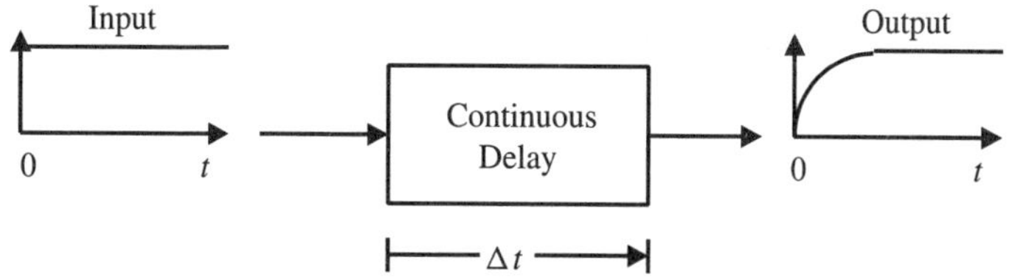

Figure 2.16 Continuous Time Delay

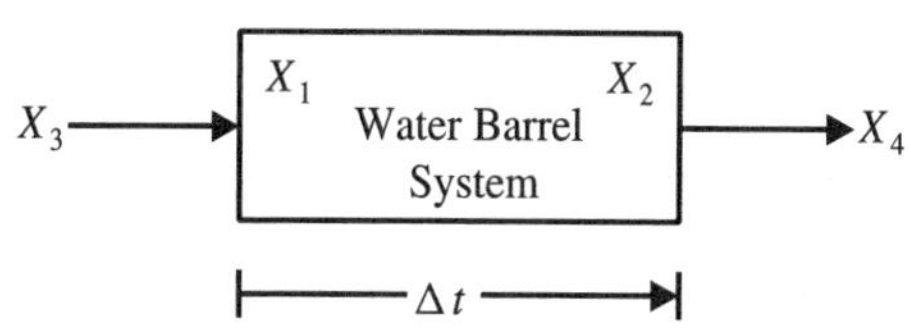

Figure 2.17 Water Barrel System

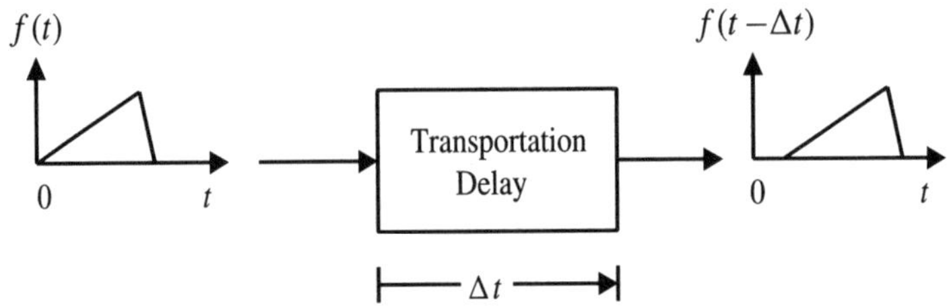

Figure 2.18 Transportation Time Delay

where $0 \le \alpha < 1$ is the *shaping parameter of delay.*

Hence, the *dynamic form* for *continuous delay* is

$$X_4 = \frac{1}{D}[\alpha X_1 + (1-\alpha)X_2] = c_1 X_1 + c_2 X_2 \tag{2.22}$$

Because the two expressions in Equation 2.22 are equivalent, their coefficients are equal. Thus, the two calibration equations for delay are:

$$c_1 = \frac{\alpha}{D} \qquad c_2 = \frac{1-\alpha}{D}$$

$D = 1/(c_1 + c_2)$ represents the *time-integrated steady-state delay* that input units experience as they flow through the system. And depending on context, D can refer to *longevity*, *transit time*, or *time constant.*

Although D interacts with α to shape the *transient response envelope*, primary control is attributable to α. For low inertia systems, response converges rapidly to steady state and α is close to unity. But for high inertia systems, convergence is relatively delayed and α is close to zero.

2.11.6.2 Transportation Delay

Transportation delay occurs when an interval of time elapses between an input and its appearance as an output response. Figure 2.18 illustrates *pure* delay: Output appears as undistorted, delayed input.

Transportation delay is a special case of continuous delay. Pure transportation delay is realized for D equal to *one* Δt and for α equal to *unity.* To achieve multiple Δt transportation delays, transportation delay subsystems are cascaded, that is, linked in series.

2.11.6.3 Applications to Modeling

Example 1 Illustrate the effect of delay parameters. Consider a typical, four-variable subsystem, the *work center* depicted in Figure 2.19. The center is characterized by *work in process (WIP).*

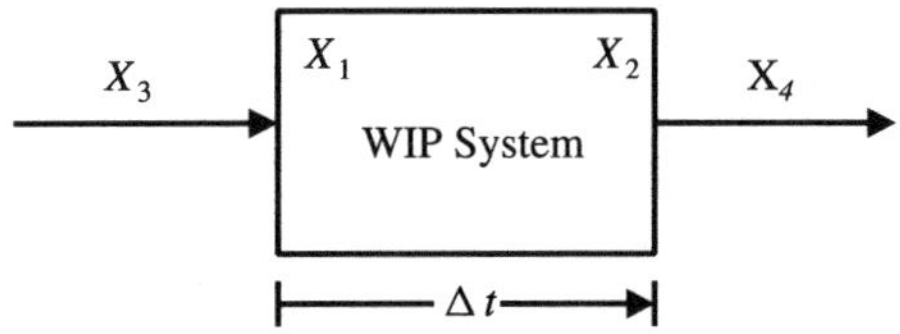

Figure 2.19 Work in Process System

X_1 = initial level of WIP, X_1^0 given

X_2 = terminal level of WIP

X_3 = input rate, given

X_4 = output rate

The vectors of the subsystem model are specified as follows:

X_j	IT Pair	Exo	Endo	Equation Type	Equation No.
X_1	I	*		—	—
X_2	T		*	balance	1
X_3		*		—	—
X_4			*	regression	2

Equating supply and demand variables, the balance equation is

$$X_1 + X_3 = X_2 + X_4. \tag{2.23}$$

The dynamic form for delay is

$$X_4 = \frac{1}{D}[\alpha X_1 + (1 - \alpha)X_2] = c_1 X_1 + c_2 X_2. \tag{2.24}$$

a. *Continuous time delay* will be illustrated for *multiple* Δt and constant dynamic form for X_3, a *step* function of input units

Period	1	2	3	4	5
X_3	6	6	6	6	6

Delay parameter $D = 2.0$ and shape parameter $\alpha = 1.0$. Thus, delay of the input units is embedded in the regression coefficients as $c_1 = \alpha/D = 1/2$, and $c_2 = (1 - \alpha)/D = 0$. Equations 2.23 and 2.24 revise to

$$X_2 = X_1 + X_3 - X_4 \text{ and } X_4 = \frac{1}{2}X_1,$$

and are used to generate data for five Δt periods.

Period	X_1	X_2	X_3	X_4	Units Delayed	Periods Delayed
1	0	6	6	0		
2	6	9	6	3	3	1
3	9	21/2	6	9/2	3	2
					3/2	1
4	21/2	45/4	6	21/4	9/2	2
					3/4	1
5	45/4	93/8	6	45/8	21/4	2
					3/8	1
Sum				147/8	147/8	

Average delay

$$= \frac{(3)(1) + (3)(2) + \left(\frac{3}{2}\right)(1) + \left(\frac{9}{2}\right)(2) + \left(\frac{3}{4}\right)(1) + \left(\frac{21}{4}\right)(2) + (3/8)(1)}{\frac{147}{8}}$$

$$= \frac{\frac{249}{8}}{\frac{147}{8}} = 1.694 \text{ periods.}$$

REMARKS

1. The delayed output, X_4, will approach the step input, $X_3 = 6$, as t increases.
2. The average delay of individual units, calculated to be 1.694 periods, will approach $D = 2.0$ as t increases.

b. *Transportation time delay* will be demonstrated by letting X_3 represent a *ramp* function.

Period	1	2	3	4	5
X_3	0	3	6	0	0

Transportation delay is a special case of continuous delay and results when $D = 1.0$ and $\alpha = 1.0$. Substituting these values into the regression coefficients yields $c_1 = 1.0$ and $c_2 = 0$. Equations 2.23 and 2.24 for this case become

$$X_2 = X_1 + X_3 - X_4 \text{ and } X_4 = X_1,$$

for which data are again generated for four Δt periods.

Period	X_1	X_2	X_3	X_4
1	0	0	0	0
2	0	3	3	0
3	3	6	6	3
4	6	0	0	6
5	0	0	0	0

Notice that the "0 3 6 0" input is output with a delay of one time unit.

c. *Fractional time delay*—The delay form will be calibrated for a *fractional* Δt. Let Δt = one working month (say, 20 days). The delay for one unit of WIP is one working day. This means that $D = (1\text{ day})/(20\text{ days/mo}) = 1/(20)\text{mo}$. Therefore, all initial WIP $= X_1$ plus a large proportion of work entering during a month, kX_3, will become output during the month. Hence, the output rate can be rewritten as

$$X_4 = X_1 + k\,X_3. \tag{2.25}$$

Substitute into Equation 2.23:

$$X_1 + X_3 = X_2 + X_1 + kX_3.$$

This leads to

$$(1-k)X_3 = X_2, \text{ or } X_3 = \frac{X_2}{1-k}.$$

Substitute into Equation 2.25

$$X_4 = X_1 + \frac{k}{1-k}X_2 = c_1X_1 + c_2X_2.$$

Referring to Equation 2.24 and equating coefficients of X_1,

$$c_1 = 1 = \frac{\alpha}{D}, \text{ leading to } \alpha = D = \frac{1}{20} = 0.05.$$

Equating coefficients of X_2,

$$c_2 = \frac{1-\alpha}{D} = \frac{k}{1-k} = \frac{1-\frac{1}{20}}{\frac{1}{20}} = 19,$$

from which

$$k = \frac{19}{20} = 0.95.$$

Thus, the calibrated version of Equation 2.24 becomes

$$X_4 = \frac{1}{\frac{1}{20}}\left[\left(\frac{1}{20}\right)X_1 + \left(\frac{19}{20}\right)X_2\right] = X_1 + 19X_2.$$

Example 2 Find D and α for the population model of Chapter 1 from the regression equation for number of deaths:

$$X_4 = DR(w_1X_1 + w_2X_2),$$

where $DR = 0.10$, $w_1 = 0.516656$ and $w_2 = 0.483344$. Number of deaths represented by X_4 is a delayed output for which the dynamic form for delay applies

$$X_4 = \frac{1}{D}\left[\alpha X_1 + (1-\alpha)X_2\right]$$

and which is an equivalent regression equation. Therefore, equating like factors yields values for D and α.

$D = 1/DR = 10$, so that longevity is 10 years. The shape parameter, $\alpha = w_1 = 0.516656$. Then since D and α are not equal to *one*, longevity for this population is an example of continuous delay.

2.12 INFORMATION FEEDBACK

Theoretically, systems are classified as *open* or *closed.* An open system is a system that interacts with elements outside of its circumscription. Conversely, a closed system does not interact with external elements. Systems are also described as *open loop* or *closed loop*. Both are open systems in that both experience input and output relative to the environment. In open loop systems, dependent variables have no influence on the level of independent variables. For example, $X_j = f(X_1, X_2, \ldots, X_i, \ldots, X_{j-1}, X_{j+1}, \ldots, X_n)$ represents a one-way function. On the other hand, closed loop systems experience *feedback*, and there exist independent variable(s) X_i in f such that $X_i = g(X_j)$. Feedback loops are either *positive* or *negative*. Response to positive feedback of information occurs as accelerated growth or decline: *Change reinforces change.*

In contrast, *goal seeking* requires negative feedback of information: Response is in proportion to the degree of failure in achieving the goal. Consequently, feedback loops in combination can produce many responses. Dynamic forms provide structure for closed-loop relationships. Dynamic forms introduce time-incremented *rates of change* (accretion or depletion) in *level of mass* of the system (subsystem), the rates themselves being dependent on levels (and possibly other rates). For additional discussion of feedback, see [13], [19], [38], [42], and [58].

2.13 DISTORTION REDUCTION

Mtm systems models are linear and readily solved by matrix inversion. System variables are partitioned into endogenous and exogenous variables:

$$\mathbf{X} = \begin{bmatrix} \mathbf{Y} \\ \mathbf{Z} \end{bmatrix}.$$

Y is the endogenous partition and **Z** is the exogenous partition. If **A** is the endogenous coefficient matrix, then solution $\mathbf{Y} = \mathbf{A}^{-1}\mathbf{b}$, where $\mathbf{A}^{-1}$ is the inverse of **A**, and **b** is a vector of constants that includes **Z** data. An inverse will produce distorted values for **Y** if **A** is an *ill-conditioned matrix* [25]. Although ill conditioning can be discovered in many different ways, examining the inverse of the matrix directly is very effective. Obtain the inverse by the method to be used in solving the model. If there are elements in the inverse that are many orders of magnitude larger than elements of the original matrix, then the matrix is probably ill conditioned. This condition can best be recognized if the rows in the original matrix are all scaled to have elements with a maximum magnitude of order 1. Another way is to formulate the matrix product $\mathbf{A}^{-1}\mathbf{A}$, comparing the result with the identity matrix **I**. Significant deviation from expected results indicates the presence of distortion. Role inversion, presented in Chapter 3, provides a way to detect distortion in specific applications of a model. Distortion is discovered by comparing model output to original database values.

A rule-of-thumb for visual inspection is as follows:

> *Examine all elements (coefficients) in* **A**. *If the elements are all of the same order of magnitude, inversion will produce little distortion. On the other hand, if elements span more than one order of magnitude, the greater the span, the greater will be the distortion.*

Y values are obtained iteratively, period by period. Linkage between periods is carried out as terminal levels are transferred to initial levels. Consequently in obtaining a computer solution, iterative, floating-point computations have the potential to interact with the resolution capability of the computer system—its combined hardware, software, and firmware—to induce distortion. For example, *roundoff*, *truncation*, and *limited word length* each induce some degree of distortion. Whether deployed as a computer program or input as data to a program, the model itself can be a propagator of distortion. However, distortion need not be a problem for a properly structured systems model. Awareness of how structure propagates distortion through positive feedback will help to avoid this effect. Both regression and balance equations will be examined as possible sources of propagation.

2.13.1 Regression Equations

Consider the linear regression of y on x as represented by Equation 2.26 and illustrated in Figure 2.20:

$$y = a + bx \tag{2.26}$$

Regression error always accompanies forward regression and is evidenced as deviation of the observed values, coordinate points $(x,\ y)$, from the regression line. However, regression-propagated distortion is not associated with data accuracy, but with the effect that regression-equation structure has on role inversion.

Consider the dependent/independent role inversion of x and y—that is, the regression of x on y:

$$x = c + dy. \tag{2.27}$$

Solve for y:

$$y = -\frac{c}{d} + \frac{1}{d} \cdot x. \tag{2.28}$$

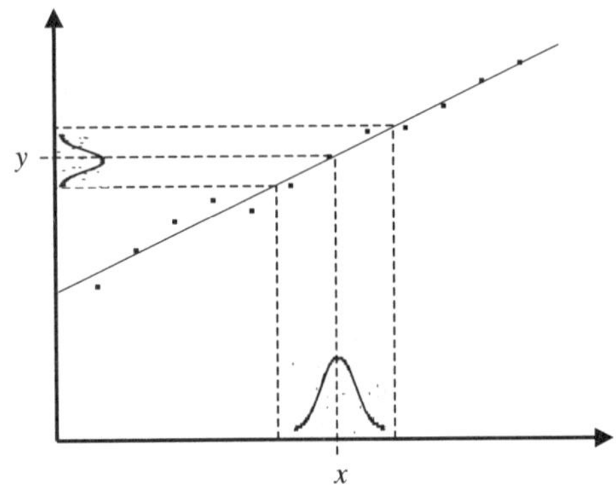

Figure 2.20 Regression of y on x

Question: Are Equations 2.26 and 2.28 the same? Does $a = -(c/d)$ and $b = 1/d$? Algebraically, *yes*, but in general, *no*, because of distortion effects. In Figure 2.20, examine the bell-shaped error distribution centered on x, exaggerated to emphasize distortion associated with x. This distortion is projected onto y but is reduced in scale because the linear-regression slope, a, is less than 1.0. To illustrate, examine the following, two algebraically equivalent equations:

$$y = 20 + 0.5x \tag{2.29}$$

$$x = -40 + 2y. \tag{2.30}$$

The effect each has on propagating distortion is different as seen by examining differentials for y and x. First, $dy = 0.5\,dx$, that is, y is relatively insensitive to change in independent variable x. Any distortion being propagated by x values due to whatever cause will be diminished. In contrast, $dx = 2\,dy$, that is, x is relatively sensitive to changes in y. Any distortion associated with y from whatever source will be amplified as y values are transferred to x. Thus, distortion propagation will be affected by role inversion if the controlling coefficients are much different from unity.

2.13.2 Unity Coefficient Equations

Consider x and y in a balance equation

$$w + x - y + z = 0.$$

When solved for y: $y = w + x + z$ yields $(dy)/(dx) = 1$. Likewise, solution for x yields: $x = -w + y - z$, with $(dx)/(dy) = 1$.

Thus, unlike regression equations, unity coefficient equations neither amplify nor attenuate distortion.

2.13.3 Distortion Reduction Principle

Propagation of distortion by the model can be reduced by applying the following principle:

> *For any endogenous (dependent) variable that takes solution definition from a regression equation, the closer its coefficient is to unity (1.0), the lesser the potential to propagate distortion.*

When distortion occurs, carefully examine the regression equations. Reducing distortion may require only a simple modification of the model. Often, scaling an entire equation will rectify a distortion-amplifying coefficient. Given that for a particular application of a model, X_5 takes definition from the following regression equation:

$$X_3 = \mathbf{0.01}\, X_5 + 0.075\, X_8.$$

Comparing model-generated values for X_5 with data-base values for X_5, distortion was noticeable, unfortunately exaggerated by the solution procedure. Solving for X_5 required that arithmetic operations be performed internally to render a unity coefficient for X_5, thus needlessly introducing distortion amplification. However, making these arithmetic operations external to the model, by scaling the X_5 coefficient, neutralizes propagation of distortion by X_5. Correction is easily effected by multiplying the entire equation by 100 to produce a unity coefficient for X_5

$$100\, X_3 = \mathbf{1.0}\, X_5 + 7.5\, X_8.$$

Distortion introduced by the software and hardware combination is becoming less and less of a problem with the passage of time. The precision capability of state-of-the-art equipment is phenomenal. For older equipment, specifying maximum precision, although slowing computations, is well worth the resulting reduction of distortion. Nonetheless, introduction of some distortion by outdated equipment is inevitable, especially in the deployment of floating-point arithmetic. Thus, this source of distortion can only be reduced by making upgrades in the computer system.

EXERCISES

1. Select a system for systems analysis at the $L_{1,j}$ level of detail.

    a. What are the bounds of analysis as established through *circumscription*?

    b. Make a list of knowledge items that you believe to be relevant for this level.

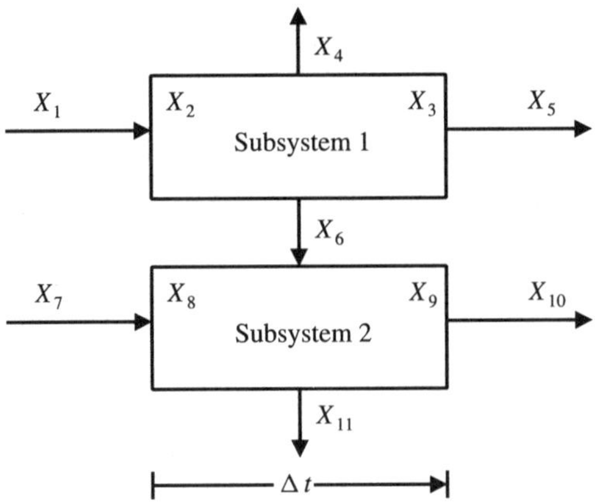

Figure 2.21 System Block Diagram

2. Using library resources, write a one-page discussion of the *conservation principle.*

3. How does the *correspondence principle* utilize the close parallel between the physical world and the model world?

4. Model variables perform three functions during the modeling process. Identify these functions and explain their purpose.

5. Using library resources, write a one- or two-page discussion of J. W. Forrester's classification of model variables as rates and levels.

6. Make and label a sketch to illustrate initialization of an IT pair for three consecutive periods.

7. Define the *principle of superposition. Hint*: Consult a text on introductory electrical circuit analysis.

8. Given the system block diagram in Figure 2.21:

   a. Write the subsystem balance equations.

   b. Write the system balance equation.

9. Which model equations and relationships require calibration and which do not?

10. With respect to variables in a regression equation, provide an example of

    a. physical correlation.

    b. statistical correlation.

11. In general, how can regression equations compensate for omission of external variables at a particular level of circumscription?

12. What is the Mtm triad of activities that are performed at each level of circumscription?

13. Contrast *forward* and *reverse* regression.

14. How are times series classified and how does calibration depend on classification?

15. Regarding calibration, what conditions indicate use of

    a. the average Δt period?

    b. a typical Δt period?

16. Explain the makeup and use of calibration equations.

17. Using library resources, write a one- or two-page paper on the role of feedback in system design and modeling.

18. Discuss the building block principle.

19. What is a dynamic form?

20. List and illustrate the standard dynamic forms.

21. Considering only Equations 2.4 and 2.10, derive the following auxiliary equations for calibrating the module of Figure 2.9 for both exponential growth and decline:

$$c_1 + c_2 = GR$$

$$c_3 + c_4 = DR$$

$$c_1 = \frac{GR}{DR} c_3 \text{ and } c_2 = \frac{GR}{DR} c_4.$$

22. Considering only Equation 2.4, verify that

$$w_1 = \frac{c_1}{c_1 + c_2} \text{ and } w_2 = \frac{c_2}{c_1 + c_2}.$$

23. Under what conditions is the power function $X_t = ab^t$ equivalent to the exponential function $X_t = X_0 e^{-DRt}$?

24. Verify Equations 2.19, 2.20, and 2.21 associated with the Malthusian form.

25. Distinguish between *continuous* delay and *transportation* delay.

26. Given the four-variable subsystem with exogenous X_3 in Figure 2.22.

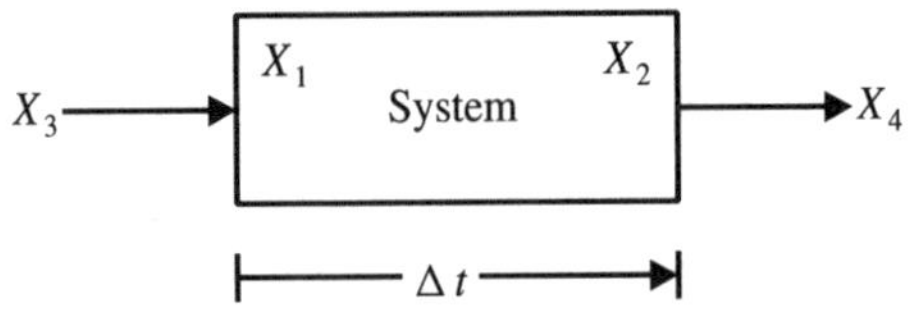

Figure 2.22 Subsystem

Period	1	2	3	4	5	6	7	8	9	10	11
X_3	0	3	6	0	0	0	0	0	0	0	0

Construct the vector table and write the linear model for the subsystem. Solve the model using the dynamic form for delay

$$X_4 = \frac{1}{D}[\alpha X_1 + (1-\alpha)X_2]$$

under the following conditions:

a. $X_1^0 = 0$; $D = 1$; $\alpha = 0, 1/2, 1$. Plot X_4 for each combination on the same graph.

b. Repeat (a) for $D = 2$; $\alpha = 0, 1/2, 1$.

c. Discuss the difference in responses relative to D and α.

27. Apply the following conditions to the linear model of Exercise 25:

Period	1	2	3	4	5	6	7	8	9	10	11
X_3	4	4	4	4	4	4	4	4	4	4	4

a. $X_1^0 = 0$; $D = 2$; $\alpha = 0, 1/2, 1$. Plot X_4 for each combination on the same graph.

b. Repeat (a) for $X_1^0 = 8$.

c. Discuss the difference in responses relative to X_1^0 and α.

28. For Exercise 25 (a), compute the average delay of individual input units (X_3) for $D = 1$ when $\alpha = 1/2$.

29. Given

$$\mathbf{A} = \begin{bmatrix} -1 & 120 & 0.005 \\ 50 & -1 & 0.075 \\ 0.15 & 250 & -1 \end{bmatrix}$$

Evaluate the product of **A** and its inverse using at least six-decimal place accuracy and two different methods (different matrix algebra software and/or computers having different precision capability). Compare values of the product with those of identity matrix **I**. Discuss results with respect to distortion.

30. What causes distortion in computer solutions of systems models?

31. How can distortion be reduced?

CHAPTER 3

Population Model: Calibration and Validation

3.1 INTRODUCTION

To introduce systems modeling, a four-variable population model was synthesized in Chapter 1 by each of three analytic approaches. Revisiting the population model is advantageous for two reasons: reader familiarity with the example and its simplicity. Four alternative calibration schemes are illustrated with the population model, and three general types of validation tests are introduced and illustrated. A two-phase approach, which readily accommodates identification and control of errors, is also presented.

3.1.1 Phase One and Phase Two Activities

Generally, the purpose of a systems model is transformation: transformation of system knowledge into data, and data into information. Activities associated with model synthesis are divided into two main phases.

- Phase One activities include the following:

  Systems analysis and knowledge acquisition

  Model structuring

  Calibration

  Generation and validation of a complete data base

- Phase Two activities involve validation of the model for each intended application using the following:

  Turing type tests

  Simulation tests such as history matching and perturbation.

3.2 POPULATION KNOWLEDGE BASE

Knowledge, established through systems analysis at the first level of detail, L_1, is formulated into a knowledge base from which a population model is to be synthesized. Knowledge items include the following:

- Population system exhibiting net exponential growth

$$x_t = x_o e^{(GR-DR)t} \text{ for } t = 1, 2, \ldots, 10 \text{ years.}$$

- Fundamental units of analysis

  Space: Population domain

  Mass: One unit of the population

  Time: One year (Δt)

- Number of subsystems: one
- System variables

  X_1 = initial population level, $X_1^0 = 10.0$ million

  X_2 = terminal population level

  X_3 = number of births per year, a rate

  X_4 = number of deaths per year, a rate

- Schematic: See Figure 3.1.
- *Vector table* for L_1 model structure

X_j	IT Pair	Exo	Endo	Equation Type	Equation No.
X_1	I	*		—	—
X_2	T		*	balance	1
X_3			*	regression	2
X_4			*	regression	3

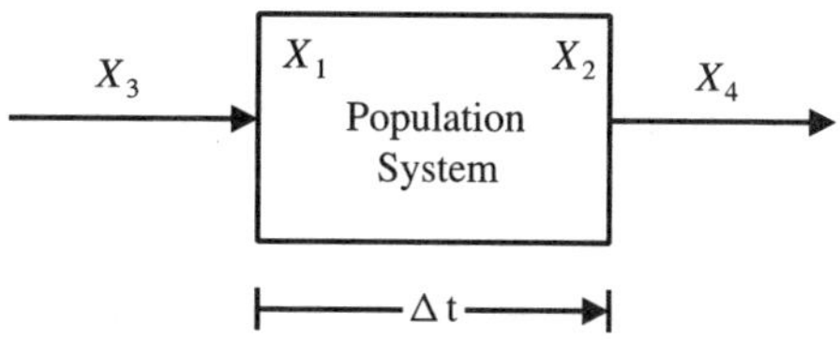

Figure 3.1 System Block Diagram

Structural specifications are listed in a *vector table* to provide a concise overview of the model. System variables are partitioned into *exogenous* (externally supplied) and *endogenous* (solution) variables in the vector table. Equation type is assigned to each solution variable. IT pair refers to a *level* variable having initial and terminal values separated by Δt.

- Dynamic form for exponential growth with $GR = 0.30$ birth per unit of the population. Provides values for X_3.
- Dynamic form for exponential decline with $DR = 0.10$ death per unit of the population. Provides values for X_4.
- Population data. Refer to Table 1.4 established by the theoretical model with calculations carrying eight-decimal-place accuracy. Population level at end of year one was found to be $X_2^1 = X_1^0 e^{(GR-DR)} = 12.214028$ million.

3.2.1 Population Model

Synthesis began by grouping a balance equation with the dynamic forms to represent a generic population model. From Figure 3.1, the balance equation was obtained by equating supply and demand variables

$$X_1 + X_3 = X_2 + X_4. \tag{3.1}$$

The dynamic form for exponential growth yielded the regression equation for X_3

$$X_3 = GR(w_1 X_1 + w_2 X_2) = c_1 X_1 + c_2 X_2 \tag{3.2}$$

where $w_1 + w_2 = 1.0$ and $c_1 + c_2 = GR$.

Similarly, the dynamic form for exponential decline gave the regression equation for X_4

$$X_4 = DR(w_1 X_1 + w_2 X_2) = c_3 X_1 + c_4 X_2 \tag{3.3}$$

where $w_1 + w_2 = 1.0$ and $c_3 + c_4 = DR$.

IT paired variables X_1 and X_2 were related by

$$X_1 = \begin{cases} X_1^0 & \text{initially,} \\ X_2 & \text{of previous year thereafter.} \end{cases}$$

3.3 CALIBRATION OF POPULATION MODEL

Calibration of the dynamic forms to represent a specific population is the next step in the synthesis process. Calibration is, perhaps, more of an art than a science, involving both choice of scheme and principle. Choice of calibration *scheme* varies with application, but most will fall into one of three categories: (1) simultaneous, (2) serial, or (3) a combination—partial serial-solution followed by simultaneous solution of the remainder. Several schemes are illustrated for the population model. Choice of calibration *principle* depends on classification of the time series. Because the population time series exhibits homogeneous nonstationarity within the 10-year span, any typical period, say, year one, is suitable for calibration.

3.3.1 Scheme 1

Two dynamic forms for number of births and number of deaths simultaneously impact population levels X_1 and X_2. Their regression equation equivalents contain four coefficients: c_1, c_2, c_3, and c_4. Thus, four auxiliary equations (used only for calibration) were required, two expressing individual impacts, and two expressing joint impacts:

1. $c_1 + c_2 = 0.30$ (birth impact)
2. $c_3 + c_4 = 0.10$ (death impact)
3. $c_1 = 3\,c_3$ or $c_2 = 3c_4$ (proportionality impact)
4. $(1 + c_1 - c_3)/(1 - c_2 + c_4) = 1.22140276$ (integrated impact)

Equation 3 was obtained by equating like coefficients between dynamic forms and regression equation forms within each of Equations 3.2 and 3.3 to obtain $GR = c_1/w_1$ and $DR = c_3/w_1$ and forming the proportion $GR/DR = (0.30)/(0.10) = 3$, expressed as

$$\frac{c_1}{w_1}/\frac{c_3}{w_1} = 3, \text{ or } c_1 = 3c_3.$$

Equation 4 was derived by a *serial* presolution of Equation 3.1 for X_2, replacing X_3 and X_4 by regression forms of Equations 3.2 and 3.3:

$$X_2 = \frac{1 + c_1 - c_3}{1 - c_2 + c_4} X_1.$$

From the knowledge base, $X_2 = X_1 e^{(GR-DR)}$, and equating coefficients lead to

$$\frac{1 + c_1 - c_3}{1 - c_2 + c_4} = e^{(GR-DR)} = 1.22140276.$$

Simultaneous solution of Equations 1 through 4 provided the following results:

$c_1 = 0.15499664 \equiv 0.154997$ (dropping to six decimal place accuracy)

$c_2 = 0.14500336 \equiv 0.145003$

$c_3 = 0.05166555 \equiv 0.051666$

$c_4 = 0.04833445 \equiv 0.048334$

3.3.2 Scheme 2

All four population variables (X_1, X_2, X_3, and X_4) are to be interacted with the four regression coefficients (c_1, c_2, c_3, and c_4). Because X_1 and X_2 have known values for year one, 10.0 and 12.214028, respectively, the remaining six unknowns required six calibration equations.

The model itself provided the following three equations:

1. $X_1 + X_3 = X_2 + X_4 \longrightarrow X_3 = X_4 + 2.214028$
2. $X_3 = c_1X_1 + c_2X_2 \longrightarrow X_3 = 10.0c_1 + 12.214028c_2$
3. $X_4 = c_3X_1 + c_4X_2 \longrightarrow X_4 = 10.0c_3 + 12.214028c_4$

Therefore, the following three auxiliary equations (used only for calibration) were required:

4. $c_1 + c_2 = 0.30$
5. $c_3 + c_4 = 0.10$
6. $c_1 = 3.0c_3$

Simultaneous solution of Equations 1 through 6 yielded the same values for c_1, c_2, c_3, and c_4 as did Scheme 1, plus providing "inspection" values for the two unknown variables for year 1:

$X_3 = 3.321041$

$X_4 = 1.107013$

3.3.3 Scheme 3

All four population variables (X_1, X_2,X_3, and X_4) are to be interacted with the two exponential weights of the dynamic forms, (w_1 and w_2) for year one. However, $X_1 = 10.0$ and $X_2 = 12.214028$ are known. Thus, four unknowns required a total of four calibration equations. Three were obtained from the model:

1. $X_1 + X_3 = X_2 + X_4 \rightarrow X_3 = X_4 + 2.214028$
2. $X_3 = 0.30(w_1X_1 + w_2X_2) \longrightarrow X_3 = 3.00w_1 + 3.664208w_2$
3. $X_4 = 0.10(w_1X_1 + w_2X_2) \longrightarrow X_4 = 1.0w_1 + 1.221403w_2$

The fourth equation (used only for calibration) was an auxiliary equation:

4. $w_1 + w_2 = 1.0$ (exponential weights must sum to 1.0)

To demonstrate *serial* solution, the equation system was solved manually. Equation 1 was substituted into Equation 2:

$$X_4 = 3.0w_1 + 3.664208w_2 - 2.214028$$

Equation 2 was substituted into Equation 3:

$$2w_1 + 2.442806w_2 = 2.214028$$

Equation 3 was substituted into Equation 4:

$$w_1 = 0.516655, \text{ and } w_2 = 0.483345,$$

leading to:

$$c_1 = 0.30w_1 = 0.154997, \quad c_2 = 0.30w_2 = 0.145003, \text{ and}$$
$$c_3 = 0.10w_1 = 0.051666, \quad c_4 = 0.10w_2 = 0.048334.$$

3.3.4 Scheme 4

Calibration can be simplified when *serial* presolution for the weights is possible. For example, $X_3 = 0.30(w_1X_1 + w_2X_2)$ and $X_4 = 0.10(w_1X_1 + w_2X_2)$ were substituted into $X_1 + X_3 = X_2 + X_4$ to obtain:

$$\frac{X_2}{X_1} = \frac{1 + 0.2\,w_1}{1 - 0.2\,w_2}, \text{ equivalent to } \frac{X_2}{X_1} = e^{(GR-DR)} = 1.221403.$$

Values for w_1 and w_2 were obtained by *simultaneous* solution of the following two equations:

1. $w_1 + w_2 = 1.0$
2. $\frac{1+0.2\,w_1}{1-0.2w_2} = 1.221403$

Again, $w_1 = 0.516655$, and $w_2 = 0.483345$.

3.3.5 Tentative Model

Equations 3.1, 3.2, and 3.3 together with the initialization relationship defined a generic model for the population system. Substituting values for c_1, c_2,c_3, and c_4, as derived from the knowledge base by any one of the four calibration schemes, yielded a tentative model of the population with six-decimal-place accuracy:

1. $X_1 + X_3 = X_2 + X_4$

2. $X_3 = 0.154997X_1 + 0.145003X_2$

3. $X_4 = 0.051666X_1 + 0.048334X_2$

$$X_1 = \begin{cases} X_1^0 & \text{initially,} \\ X_2 & \text{of previous year thereafter.} \end{cases}$$

Regression equations (dynamic forms) provide a mechanism for transforming *exogenous* system knowledge into *endogenous* knowledge, thus to shape the model's response characteristics. Taken together, balance and regression equations constitute a tentative model, a hypothesis to be validated via practical tests.

3.4 VALIDATION

Phase One of model synthesis is concluded by generating and validating an initial data base. Thus, the first priority for application of a model is to generate a complete data base, utilizing system knowledge. Therefore, initial validation testing of the model involves validating the data base in terms of its *secondary* data composition. Quite typically, the data base consists of a mixture of primary and secondary data. Model variables are partitioned into exogenous and endogenous variables. Exogenous variables receive primary data as input, and endogenous variables generate secondary data as output. Secondary data are evaluated via a *Turing Type* test.

3.4.1 Turing Type Test

Most often, the initial data base is incomplete, and completion means supplementing existing primary data by generating secondary data. Secondary data are obtained by reverse regression. Restating the reverse regression principle gives us the following:

> *Given adequate system knowledge, regression coefficients can be found that generate secondary data consistent with that knowledge.*

Thus, reverse regression essentially means "given knowledge-derived coefficients, solve the model for data." Endogenous knowledge is interacted with primary data to generate secondary data. Error in secondary data may result from the following:

- Incorrect model structure
- Incomplete model structure
- Incorrect primary data

Error must be distinguished from *distortion* (discussed in Chapter 2). Secondary data are *judged* for validity via a Turing Type (TT) test involving one or more knowledge domain experts. In the early days of artificial intelligence research, a computer was considered to be thinking only when a human interviewer, conversing with both an unseen human being and an unseen computer, could not determine which was which [51]. Tests of this type carry the name of their originator, Alan Turing of the United Kingdom [52]. Application of the Turing concept to a systems model is stated as follows:

> *A systems model passes a Turing Type test if a systems domain expert can not determine if the data presented for inspection came from the model or from the physical system.*

Domain experts have formal training and/or experience in the systems domain. However, domain experts are not infallible and have been known to reverse or even contradict themselves! Frequently, participation in calibrating and validating a systems model will be a learning experience for them. Even though validated by knowledgeable experts, the model continues to be regarded as a test hypothesis.

3.4.2 Simulation Tests

Partial validation is obtained for a systems model in Phase One by generating secondary data. Operating from the completed data base of Phase One, Phase Two activities are devoted to validating the model for specific applications via *simulation tests*. When structural deficiencies appear, revisions are made, and the model is recycled back through Phase One. System knowledge resides in exogenous form within the data base and in endogenous form within the model. Transformation of system knowledge into information, the ultimate purpose of a systems model, is provided by various applications of the model in response to "what if" questions. Many of the "what if" applications require data-base perturbation.

Various techniques can be contrived for simulation testing of the model. Exchange of role between endogenous and exogenous variables may be required by some applications. Of all the techniques, *role inversion* is perhaps the most useful as a generator of various applications for a systems model.

3.4.3 Role Inversion

Role inversion forms the basis for many tests and involves the interchange of exogenous and endogenous roles between pairs of model variables. Role inversion is essential to *history matching* tests, and provides a means for testing the susceptibility of an iteratively solved systems model to *distortion* as it operates within historical bounds.

Inversion is specified by redefining the solution (endogenous) vector prior to the test run of the model. However, inversion triggered during runtime, with optional automatic reinversion, greatly enhances flexibility in defining test scenarios. In general, the greater the number of secondary variables, the greater the number of possible tests. Potentially, a solution vector can be specified for any combination of variables. However, some vectors are not mathematically feasible. Feasibility requires that the coefficient matrix associated with the endogenous variables not be singular (contain or reduce to an all-zero row or column). Two frequently used tests, *history matching* and *perturbation*, use role inversion and are defined next.

3.4.4 History Matching Test

History matching is a test that questions, "Can the model accurately simulate values for primary data?" Primary variables, filling exogenous variable roles in Phase One, provide the opportunity for history matching. By this test, one or more exogenous variables are selected to exchange roles with variables previously designated as endogenous but with the restriction that any new solution vector be feasible, as described above. Rerunning the model for the new solution vector allows examination of how accurately the model matches historical data values taken as the standard for comparison. Because the model is truly a *systems* model complete with balance equation(s), role inversion merely imposes a different set of solution variables on the same set of equations. Any deviation, introduced by the change, is attributed to distortion. Minor distortion is ignored. Major distortion is dealt with by applying the distortion reduction principle of Chapter 2. Close reproduction of primary data values is unique to Mtm systems modeling. Exact matches are prevented only by distortion introduced by the computer system.

Very rarely do systems models have a complete, primary data base, and even more rarely are data error free. For a complete data base, each primary variable that is selected for an endogenous role in the model is represented by a regression equation. In this instance, the history matching test can be performed without role inversion because each solution variable has a "standard" against which to compare its values. Although primary data provide a standard, their absolute accuracy is questionable. Furthermore, for a complete primary data base, all regression equation coefficients are obtained by forward (standard) regression procedures. Consequently, the system of equations constitutes a multidependent, multivariate regression model. History matching by the model is *judged* by statistics domain experts on the basis of residual error sums (*the smaller, the better*) and the proportion of the variation in the dependent variables that is explained by their regression on the independent variables.

3.4.5 Perturbation Test

Systems models may produce valid dynamic response to times series inputs from a Phase One data base but may be found to generate obviously invalid responses to other inputs. This suggests further tests: perturbation of one or more input series. Various scenarios might be posed, asking "what if" questions of the model: what might have been, or what might occur in the future, and observing the model's response.

Secondary data generated in this manner are subjected to Turing Type examination as described for Phase One. Cause-effect reasoning may be validly applied to trace any nonfeasible response to its structural source.

3.5 VALIDATION OF POPULATION MODEL

Calibration to six decimal places yielded a tentative population model, a linear equation system. As a first application of the model, 10 years of secondary data were generated by simultaneous solution of the equation system, taken one year at a time.

3.5.1 Data Base Generation

Any software that features standard matrix algebra and data-base manipulation, and that can interface with user subroutines for special applications, is adequate for solving systems models. Examples presented in this text were solved using Fortran programs specifically designed by the author for synthesizing systems models. Table 3.1 displays 10 years of secondary data for the example population. This Phase One data base must yet be validated before conducting simulation testing of the model.

For a review of simultaneous solution of linear equations by matrix algebra, a manual solution of the model is outlined in Table 3.2 of Appendix A.

3.5.1.1 Turing Type Test

All secondary data are normally subjected to a TT test. $X_1 = 10.0$ was supplied initially. Thereafter, X_2 values, obtained from the system balance equation, were off set one year by the exogenous initialization relationship to provide values for X_1. Consequently, only values for the endogenous secondary variables X_2, X_3, and X_4 were examined for validity. Domain knowledge residing in Table 1.4, the population system data base, were obtained by the theoretical approach using nonlinear Equations 1.3, 1.4, and 1.5. These data provided the basis for a TT test. Values obtained from the model's linear equations qualify for a match to

Table 3.1 Database for Population Model

Period	X_1	X_2	X_3	X_4
1	10.000000	12.214028	3.321041	1.107013
2	12.214028	14.918248	4.056329	1.352109
3	14.918248	18.221190	4.954411	1.651469
4	18.221190	22.255413	6.051332	2.017109
5	22.255413	27.182823	7.391113	2.463703
6	27.182823	33.201177	9.027527	3.009173
7	33.201177	40.552010	11.026246	3.675413
8	40.552010	49.530339	13.467488	4.489159
9	49.530339	60.496495	16.449228	5.483072
10	60.496495	73.890588	20.091133	6.697039

these values at the fourth, fifth and sixth decimal places, and therefore meet the accuracy required by the TT test.

Surprising, indeed, is the discovery that a linear model can produce such a match of values for nonlinear trajectories of levels and rates corresponding to exponential growth. Amazingly few knowledge items were required for model calibration and data-base generation, which include the following:

- Dynamic form for exponential growth with $GR = 0.30$ birth per unit of the population
- Dynamic form for exponential decline with $DR = 0.10$ death per unit of the population
- Initial population level $X_1^0 = 10.0$ million
- Population level at end of year one: $X_2^1 = X_1^0 e^{(GR-DR)} = 12.214028$ million

3.5.2 Simulation Testing

Systems models usually do not have a complete primary data base to provide a standard of comparison for model output. In the absence of a complete standard, *simulation testing* of secondary data and of the model provides an alternative as illustrated by the following two tests.

3.5.2.1 History Matching Test

Secondary data have been generated by the model in Phase One, valid only to the extent of having passed a TT test. Can these secondary data and the same model equations be used to generate output that matches primary data values? History matching is used in validating any model ranging from the most simple to the most hypothetical and complex. Matching of historical data for a systems model is an expected outcome, and the history matching test would appear unnecessary were it not for the distortion detection capability of this test.

Consider the following naive illustration: Given that $x = 0.05y + 0.05z$, with "historical data" $y = 5$ and $z = 10$, find the value of x. Simple! $x = (0.05)(5) + (0.05)(10) = 0.75$. Apply history matching to y: Given the model, with $x = 0.75$ and $z = 10$, match the historical value of y. Again simple! $y = 1/(0.05)\ x - z = (0.75)/(0.05) - 10 = 5$. However, unaccounted for in this simple example is distortion—computation error that may be introduced by the computer system (hardware, software, and firmware) in iteratively solving a systems model. Nonetheless, the example illustrates how role inversion is used in performing a history matching test.

History matching for $X_1 = 10.000000$, the only primary data value, was carried out for the population model. X_1 and X_4 were selected for role inversion: X_1 Exo → X_1 Endo, X_4 Endo → X_4 Exo. Here is the vector table corresponding to this role inversion:

X_j	Exo	Endo	Equation Type	Equation No.
X_1		*	regression	3
X_2		*	balance	1
X_3		*	regression	2
X_4	*		—	—

Correspondingly, the model equations (repeated below) were solved for the *box-distinguished* variables. Initialization of X_1 was not required because the solution vector does not involve a terminal member of an *IT solution pair*. Some distortion was expected because X_1, in taking definition from a regression equation, has a relatively small, nonunity coefficient.

1. $X_1 + X_3 = \boxed{X_2} + X_4$

2. $\boxed{X_3} = 0.154997X_1 + 0.145003X_2$

3. $X_4 = 0.051666\boxed{X_1} + 0.048334X_2$

Test Results

Table 3.3 (see Appendix A) displays the computer solution for 10 yearly periods and was compared to Table 3.1. History matching for X_1 in year 1 produced a value that was up two population units—that is, $X_1 = 10.000002$ million—missing the exact primary value in the sixth decimal place. Considering all 10 years, the *cumulative* algebraic change was −0.000003. For year 10, X_1 was down 4 population units from the Phase One, data-base standard of 60,496,495. Values for

X_2 and X_3 remained close to their Phase One standards. Comparing values for all three variables, maximum *cumulative* algebraic change was 0.000008, occurring for X_2 at the end of year 5. But at the end of year 10, it was down only 5 population units from the data-base standard of 73,890,588 units. In conclusion, minor distortion was introduced by the computer system but not to the extent of disqualifying the model's accuracy.

3.5.2.2 Perturbation Test

Systems models are synthesized for more than history matching. For example, a question such as "What would be the resulting population levels due to a treatment that increased longevity during the ten years covered by the data base?" might be asked. To illustrate, suppose the number of deaths per year, X_4, is to be estimated by $f(t) \cdot X_4$, where $f(t)$ is the decimal equivalent of percent reduction in deaths as presented in the following table:

t	1	2	3	4	5	6	7	8	9	10
$f(t)$	.99	.99	.98	.98	.96	.96	.94	.94	.90	.90

Table 3.4 (see Appendix A) displays the perturbed X_4 values, perturbed relative to Table 3.1 (obtained previously in Phase One).

Model structure was modified to accommodate two exogenous variables as shown in the revised vector table.

X_j	IT Pair	Exo	Endo	Equation Type	Equation No.
X_1	I	*		—	—
X_2	T		*	balance	1
X_3			*	regression	2
X_4		*		—	—

Note the X_4 inversion of role from endogenous to exogenous. Also, Equation 3, associated with its endogenous role, was dropped. Presented here is the altered model.

1. $X_1 + X_3 = X_2 + X_4$
2. $X_3 = 0.154997X_1 + 0.145003X_2$

$$X_1 = \begin{cases} X_1^0 & \text{initially,} \\ X_2 & \text{of previous year thereafter.} \end{cases}$$

Discussion and Critique of Results

Dropping Equation 3 was not arbitrary. With increasing longevity, population level is expected to exceed data-base values. Output for this scenario, presented in Table 3.5 (see Appendix A), confirms this expectation. However, if Equation 1 is dropped instead of Equation 3, the model incorrectly projects a decline in population level, thus illustrating a basic principle:

> *Balance equations must not be dropped because they provide the skeletal structure of a systems model. Dropping the balance equation destroys the system's continuity. Loss of continuity allows output to drift.*

Data that satisfy the system balance equation are said to be *consistent*. The data base of Phase One, for which $X_{10} = 10.0$, $GR = 0.30$, and $DR = 0.10$, is both complete and consistent. Replacing X_4 by perturbed values, $X_4^{'} = f(t) \cdot X_4$, introduced discontinuities into the data base that do not correspond to $DR = 0.10$ for which the model was calibrated but rather to an undefined $DR^{'} < DR$. Thus, exogenous values for X_1 and $X_4^{'}$ supplied from the data base cannot both be consistent with respect to DR. Any *cumulative* effect introduced by increasing longevity is negated at the beginning of each year as X_1 reintroduces the unperturbed population level. Consequently, this solution represents a lower bound for expected population growth for this scenario.

Although the model behaved as expected (population levels increased), exact results (if required) are possible if the model is recalibrated with one regression equation for X_3 and with $X_1^0 = 10.0$ and $X_4^{'}$ as exogenous variables. Five sets of coefficients are required to accommodate the five different values of $f(t)$, that is, five two-period calibrations.

Rather than perturbing X_4 values, a better approach is to retain Equation 3 and to perturb its coefficient values during runtime

$$X_4^{'} = [f(t)(0.051666)]X_1 + [f(t)(0.048334)]X_2.$$

Values for $f(t)$ are input year by year: .99, .99, .98, .98, . . . , .90, .90.

3.6 ERROR CONTROL

In general, errors associated with mathematical modeling arise from two sources: (1) known and controllable, and (2) unknown and uncontrollable. Unknown errors derive from failing to recognize (and therefore failing to include) all variables that measurably influence model behavior. The presence of such errors is often acknowledged by including a random error term to explain the difference between observed data values and model output.

Mtm analysis offers a two-phase approach to mathematical modeling that affords many opportunities for error control by either elimination or reduction. Consequently, errors are further classified in terms of their Phase One or Phase Two occurrences.

3.6.1 Phase One Sources

3.6.1.1 Selection of Model Variables

Balance equation error for each system (or subsystem) is minimized by including all variables that significantly impact the balance equation. Whether to include a variable can be assessed with the assistance of a domain expert or with a statistical method such as experimental design through analysis of variance [27].

Calibration of Dynamic Forms

Each endogenous dependent variable is defined by a dynamic form composed of those independent variables that significantly impact its value. Regression coefficients must be selected so as to minimize regression error. When all variables that measurably influence model behavior are supported by primary data, standard linear regression techniques are applicable; otherwise, reverse regression techniques apply.

Primary Data

Primary data provide a standard by which model output is validated. However, the processes of measurement (or counting), recording, and compilation of primary data also have potential to introduce error. Error control measures include upgrading collection techniques: less manual processing and more use of automation and advanced technologies. If primary data exist for all variables of a system (subsystem), seldom, if ever, will the balance equation exhibit error-free balance. To compensate, a composite error term can be included in the system (subsystem) balance equation to account for imbalances. The size of the error term provides a relative measure of accuracy for the primary data.

Secondary Data

Estimation errors are introduced when secondary data are substituted for missing primary data. Exact error magnitudes cannot be determined in the absence of historical data. Because secondary data must pass TT tests, the process of structuring and calibrating the model keeps error magnitudes in check. Furthermore, estimation error tends to be distributed over all secondary data, making minimal impact on those variables having relatively great magnitudes.

3.6.2 Phase Two Sources

Error sources inherited from Phase One must be detected by simulation tests. Sources are (A) data base, (B) calibrated model, and (C) computerized system of solution. Errors introduced by A, B, and C are confounded (interacting and combined) and therefore difficult to isolate. Phase Two tests are mainly directed toward finding errors introduced by B and BC interactions. Upon detection, error is traced to its source, and model structure is revised.

3.6.2.1 Distortion

Perturbation applications of the model requiring role inversion may cause distortion (BC interaction errors). If role inversion involves one or more primary variables, distortion may be detected by history matching, that is, comparing computed values with primary values of the data base. Distortion reduction techniques of Chapter 2 must then be applied before performing perturbation tests.

3.6.2.2 Structural Deficiency

Model structure is further tested by perturbing exogenous time series of the data base, one at a time, and subjecting model responses to TT tests. An unsatisfactory response is traced to its cause in the knowledge base, and any item contributing to faulty structure is adjusted and the structure is revised. The domain expert(s) who participated in the knowledge acquisition process and TT tests will likely be required to assist in the revision.

EXERCISES

1. Describe the validation activities associated with
    a. Phase One.
    b. Phase Two.
2. Construct a generalized list of knowledge items appropriate to a system knowledge base.
3. Distinguish between exo*genous* and endo*genous* variables based on the Greek prefixes used in their names.
4. Theoretically, all Mtm systems models are macro extensions of micro forms. However, model synthesis and validation does not depend on "proof" of their connection or equivalence to underlying differential equations. In fact, only for relatively simple models, such as the population model, are the underlying differential equations readily identified. Derive Equation 1.2, the differential equation that underlies the population model. Start by considering any finite interval of time, say $\Delta t = T$, as in Figure 3.2
5. What factors determined the choice of calibration principle for the population model?

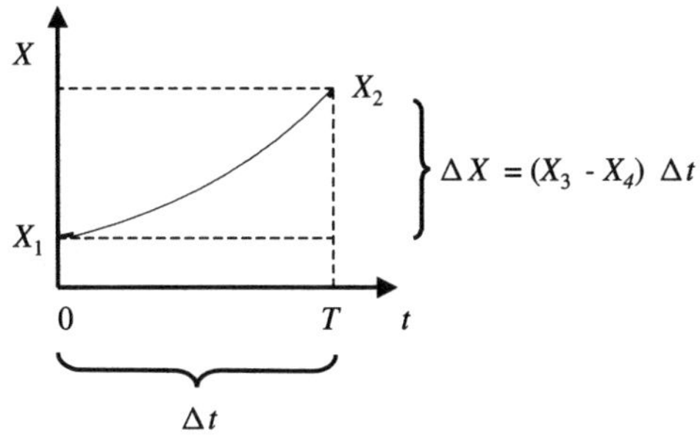

Figure 3.2 Growth Over a Finite Interval

6. Identify the category of each of the four calibration schemes presented for the population model.

7. Verify calibration Equation 4 of Scheme 1.

8. Show that if dynamic forms for X_3 and X_4 of the population model are calibrated independently, then Equations 3 and 4 of Scheme 1 are not satisfied.

9. Show that for the population model only one set of weights, w_1 and w_2, are required to calibrate the dynamic forms for births and deaths.

10. Describe a *Turing Type* test.

11. Test the following coefficient matrices for feasibility, and if not feasible, find the cause:

a. $\begin{bmatrix} -1 & 1 & -2 \\ 2 & 1 & 4 \\ .5 & 1 & 1 \end{bmatrix}$ b. $\begin{bmatrix} 1 & -1 & 1 \\ .15 & -1 & 0 \\ .05 & 0 & -1 \end{bmatrix}$ c. $\begin{bmatrix} -1 & 1 & 1 \\ 5 & 1 & 2 \\ 1 & .2 & .4 \end{bmatrix}$

12. Provide a short working definition for each of the following techniques:

    a. Role inversion

    b. History matching test

    c. Perturbation test

13. How does history matching apply to systems models whose data bases consist of

    a. some primary and some secondary data?

    b. all primary data?

14. How are the results of perturbation tests evaluated?

15. What opportunities does Mtm analysis afford for error reduction or elimination?

16. The population system model synthesized via the systems analysis approach produced data that matched that of the theoretical model as displayed in Table 1.4. Coefficients of the "systems analysis model" were accurate to eight decimal places (see Calibration 1). Table 3.1 data were computed using coefficients obtained by dropping to six decimal places. What maximum percent difference occurs in X_2 as a result of carrying six as opposed to eight decimal places?

17. To gain familiarity with the solution process, perform manual computations to verify the solution presented in Table 3.2 of Appendix A for the population model for $t = 2$.

18. Select computer software with which to calibrate and solve systems models. To gain familiarity with the software and model synthesis, verify the outcome for Phase One and Phase Two activities for the population model of this chapter. Produce complete documentation that someone else might follow to also verify the outcome. Activities include

    a. calibration

    b. completion of data base

    c. history matching test

    d. perturbation test.

19. Generate a data base for the population model using an *arithmetic* average of X_1 and X_2 in computing X_3 and X_4; that is, set $w_1 = w_2 = 1/2$. How does this data base compare to the data in Table 3.1 that were obtained by using a *weighted* average of X_1 and X_2?

20. Using $x = ae^{bt}$, a generic equation of exponential growth as an *explicit* function of time, derive an equation expressing exponential growth as an *implicit* function of time valid for any Δt and IT pair.

21. The population of a certain species of dolphin that inhabits tuna-fishing grounds is rapidly being decimated by tuna fisherman. Its population statistics indicate that this specie is facing extinction if present trends continue. Reliable estimates show the present population level to be 500,000 with a birth rate of 2 percent and a death rate from all causes of 12 percent.

    Synthesize and calibrate a systems model to obtain a 50-year projection of this species of dolphins. For convenience, code the data in thousands.

    Prepare the following materials:

- List of knowledge base items
    - System
    - Fundamental units of analysis
    - Number of subsystems
    - System variables
    - Schematic
    - Vector table
    - Dynamic forms
    - Data
- Model equations
- Calibration equations
- Phase One model
- Complete data base
- Plots of secondary data
- Specifically, find the following:

a. In what year will population drop below five thousand?

b. In what year will the number of births drop below one hundred?

c. What is the expected longevity of a dolphin?

d. What assumptions must be made to support your projections?

APPENDIX A

Table 3.2 Manual Solution of the Population Model

Formulate the augmented matrix for $\mathbf{A}\,\mathbf{X} = \mathbf{B}$, where $\mathbf{X}$ = vector of model variables:

$$
\begin{array}{ccccc}
\multicolumn{3}{c}{\text{Secondary (Endo)}} & \text{Primary (Exo)} & \text{RHS}\\
X_2 & X_3 & X_4 & X_1 & b_i
\end{array}
$$

$$
\left[\underbrace{\begin{matrix}1 & -1\\ .145003 & -1\\ .048334 & 0\end{matrix}}_{\mathbf{A}}\;\vdots\;\begin{matrix}1 & -1\\ 0 & .154997\\ -1 & .051666\end{matrix}\;\vdots\;\begin{matrix}0\\ 0\\ 0\end{matrix}\right]
$$

$\mathbf{A}$ = *Endogenous coefficient matrix* to be inverted by the Gauss-Jordan pivot method [49]:

$$
\mathbf{Y} = \begin{bmatrix} X_2 \\ X_3 \\ X_4 \end{bmatrix} = \textit{Endogenous vector}, \quad \mathbf{Y} = \mathbf{A}^{-1}\mathbf{B},
$$

where $\mathbf{B} = -\sum_{i=1}^{3}(c_{i1}X_1 + b_i) = \textit{ Exogenous vector}.$

$$
\left[\begin{array}{cccccc}
\boxed{1} & -1 & 1 & 1 & 0 & 0\\
.145003 & -1 & 0 & 0 & 1 & 0\\
.048334 & 0 & -1 & 0 & 0 & 1
\end{array}\right]
\begin{array}{l}
\text{divide by 1}\\
\text{add -.145003 x Row 1}\\
\text{add -.048334 x Row 1}
\end{array}
$$

$$
\left[\begin{array}{cccccc}
1 & -1 & 1 & 1 & 0 & 0\\
0 & \boxed{-.854997} & -.145003 & -.145003 & 1 & 0\\
0 & .048334 & -1.048334 & -.048334 & 0 & 1
\end{array}\right]
\begin{array}{l}
\text{add -1.169595 x Row 2}\\
\text{divide by -.854997}\\
\text{add -.056531 x Row 2}
\end{array}
$$

$$
\left[\begin{array}{cccccc}
1 & 0 & 1.169595 & 1.169595 & -1.169595 & 0\\
0 & 1 & .169595 & -1.169595 & -1.169595 & 0\\
0 & 0 & \boxed{-1.056531} & -.056531 & .056531 & 1
\end{array}\right]
\begin{array}{l}
\text{add 1.107014 x Row 3}\\
\text{add .160522 x Row 3}\\
\text{divide by -1.056531}
\end{array}
$$

$$
\left[\begin{matrix}1 & 0 & 0\\ 0 & 1 & 0\\ 0 & 0 & 1\end{matrix}\;\;\underbrace{\begin{matrix}1.107014 & -1.107014 & 1.107014\\ .160521 & -1.160521 & .160521\\ .053506 & -.053506 & -.946494\end{matrix}}_{\mathbf{A}^{-1} = \text{Inverse of } \mathbf{A}}\right]
$$

$$\mathbf{Y} = \mathbf{A}^{-1}\mathbf{B}$$

$$\mathbf{A}^{-1} = \begin{bmatrix} 1.107014 & -1.107014 & 1.107014 \\ 0.160521 & -1.160521 & 0.160521 \\ 0.053506 & -0.053506 & -0.946494 \end{bmatrix}$$

$$\mathbf{B} = \begin{bmatrix} -c_{11}X_1 - b_1 \\ -c_{21}X_1 - b_2 \\ -c_{31}X_1 - b_3 \end{bmatrix}$$

- Iteration for $t = 1$ with $X_1 = 10.000000$:

$$\mathbf{B} = \begin{bmatrix} -(-1)(10.000000) \quad - 0 \\ -(.154997)(10.000000) - 0 \\ -(.051666)(10.000000) - 0 \end{bmatrix} = \begin{bmatrix} 10.000000 \\ -1.549970 \\ -0.516660 \end{bmatrix}$$

$$\mathbf{Y} = \mathbf{A}^{-1}\mathbf{B} = \begin{bmatrix} (10.0)(1.107014) + (1.54997)(1.107014) - (.51666)(1.07014) \\ (10.0)(0.160521) + (1.54997)(1.160521) - (.51666)(0.160521) \\ (10.0)(0.053506) + (1.54997)(0.053506) + (.51666)(0.946494) \end{bmatrix}$$

$$\mathbf{Y} = \begin{bmatrix} 12.214028 \\ 3.321048 \\ 1.107008 \end{bmatrix} = \begin{bmatrix} X_2 \\ X_3 \\ X_4 \end{bmatrix}$$

- Iteration for $t = 2$ with $X_1 = 12.214028$:

$$\mathbf{B} = \begin{bmatrix} 12.214028 \\ -1.893138 \\ -0.631050 \end{bmatrix} \quad \mathbf{Y} = \begin{bmatrix} 14.918249 \\ 4.056338 \\ 1.352103 \end{bmatrix} = \begin{bmatrix} X_2 \\ X_3 \\ X_4 \end{bmatrix}$$

- Iterations for $t = 3$ to 9 are computed in similar manner.

⋮

- Finally, iterate for $t = 10$ with $X_1 = 60.496485$:

$$\mathbf{B} = \begin{bmatrix} 60.496485 \\ -9.376774 \\ -3.125611 \end{bmatrix} \quad \mathbf{Y} = \begin{bmatrix} 73.890581 \\ 20.091173 \\ 6.697011 \end{bmatrix} = \begin{bmatrix} X_2 \\ X_3 \\ X_4 \end{bmatrix}$$

As might be expected, manual solution did not match the superior accuracy of the Fortran software. Referring to Table 3.1, note that differences appear in the fifth and sixth decimal places.

Table 3.3 History Matching Test

Key: X_1 Initial Population Level
X_2 Terminal Population Level
X_3 Number of Births

SOLUTION FOR PERIOD 1

VARIABLE	VALUE	CHANGE	SUM*
X_1	10.000002	.000002	.000002
X_2	12.214030	.000002	.000002
X_2	3.321041	.000000	.000000

SOLUTION FOR PERIOD 2

VARIABLE	VALUE	CHANGE	SUM*
X_1	12.214032	.000004	.000006
X_2	14.918253	.000005	.000007
X_3	4.056330	.000001	.000001

SOLUTION FOR PERIOD 3

VARIABLE	VALUE	CHANGE	SUM*
X_1	14.918247	−.000001	.000004
X_2	18.221188	−.000002	.000005
X_3	4.954411	.000000	.000001

SOLUTION FOR PERIOD 4

VARIABLE	VALUE	CHANGE	SUM*
X_1	18.221189	−.000001	.000004
X_2	22.255412	−.000001	.000004
X_3	6.051331	−.000001	.000000

SOLUTION FOR PERIOD 5

VARIABLE	VALUE	CHANGE	SUM*
X_1	22.255416	.000003	.000006
X_2	27.182827	.000004	.000008
X_3	7.391114	.000001	.000002

SOLUTION FOR PERIOD 6

VARIABLE	VALUE	CHANGE	SUM*
X_1	27.182820	−.000003	.000004
X_2	33.201173	−.000004	.000004
X_3	9.027526	−.000001	.000000

SOLUTION FOR PERIOD 7

VARIABLE	VALUE	CHANGE	SUM*
X_1	33.201179	.000002	.000005
X_2	40.552013	.000003	.000007
X_3	11.026247	.000001	.000001

SOLUTION FOR PERIOD 8

VARIABLE	VALUE	CHANGE	SUM*
X_1	40.552006	−.000004	.000001
X_2	49.530334	−.000005	.000002
X_3	13.467487	−.000001	.000000

SOLUTION FOR PERIOD 9

VARIABLE	VALUE	CHANGE	SUM*
X_1	49.530339	.000000	.000002
X_2	60.496495	.000000	.000002
X_3	16.449228	.000000	.000000

SOLUTION FOR PERIOD 10

VARIABLE	VALUE	CHANGE	SUM*
X_1	60.496491	−.000004	−.000003
X_2	73.890583	−.000005	−.000003
X_3	20.091132	−.000001	−.000001

*SUM = Cumulative algebraic value of change

Table 3.4 Population Database after Perturbation

Period	X_1	X_2	X_3	X_4
1	10.000000	12.214028	3.321041	*1.095943*
2	12.214028	14.918248	4.056329	1.338588
3	14.918248	18.221190	4.954411	1.618440
4	18.221190	22.255413	6.051332	1.976767
5	22.255413	27.182823	7.391113	2.365155
6	27.182823	33.201177	9.027527	2.888806
7	33.201177	40.552010	11.026246	3.454888
8	40.552010	49.530339	13.467488	4.219809
9	49.530339	60.496495	16.449228	4.934765
10	60.496495	73.890588	20.091133	6.027335

Perturbations of X_4 values were obtained by multiplying original values by $f(t)$.
Example: $(.99)(1.107013) = 1.095943$.

Table 3.5 Perturbation Test

Key: X_2 Terminal Population Level
X_3 Number of Births

SOLUTION FOR PERIOD 1

VARIABLE	VALUE	CHANGE	SUM*
X_2	12.226975	.012947	.012947
X_3	3.322918	.001877	.001877

SOLUTION FOR PERIOD 2

VARIABLE	VALUE	CHANGE	SUM*
X_2	14.951552	.033304	.046251
X_3	4.063164	.006835	.008713

SOLUTION FOR PERIOD 3

VARIABLE	VALUE	CHANGE	SUM*
X_2	18.304810	.083620	.129871
X_3	4.971698	.017287	.026000

SOLUTION FOR PERIOD 4

VARIABLE	VALUE	CHANGE	SUM*
X_2	22.415557	.160144	.290015
X_3	6.087514	.036182	.062182

SOLUTION FOR PERIOD 5

VARIABLE	VALUE	CHANGE	SUM*
X_2	27.514419	.331596	.621611
X_3	7.464018	.072905	.135086

SOLUTION FOR PERIOD 6

VARIABLE	VALUE	CHANGE	SUM*
X_2	33.789904	.588727	1.210338
X_3	9.164290	.136763	.271850

SOLUTION FOR PERIOD 7

VARIABLE	VALUE	CHANGE	SUM*
X_2	41.605234	1.053224	2.263561
X_3	11.270218	.243972	.515822

SOLUTION FOR PERIOD 8

VARIABLE	VALUE	CHANGE	SUM*
X_2	51.268147	1.737808	4.001369
X_3	13.882722	.415234	.931056

SOLUTION FOR PERIOD 9

VARIABLE	VALUE	CHANGE	SUM*
X_2	63.485359	2.988864	6.990232
X_3	17.151977	.702749	1.633805

SOLUTION FOR PERIOD 10

VARIABLE	VALUE	CHANGE	SUM*
X_2	78.711463	4.820875	11.811110
X_3	21.253439	1.162306	2.796111

*SUM = Cumulative algebraic value of change

Part II

Deterministic Models

CHAPTER 4

Modeling with Dynamic Forms

Systems models are synthesized by applying items from a knowledge base that may or may not include data. Dynamic forms can be calibrated with or without data. This chapter illustrates both cases, first with a systems model for *harmonic oscillation*, without data, followed by synthesis of *polynomial* models and a data-fitting example. Application of dynamic forms to these two cases and others in following chapters, is highly suggestive that additional dynamic forms and their applications to systems modeling await discovery.

4.1 HARMONIC OSCILLATION MODEL

Sinusoidal oscillation is a phenomenon that frequently occurs in system dynamics. Common types of oscillation are the following:

- *Divergent* growth forced by positive feedback
- *Convergent* decline forced by negative feedback
- *Harmonic* free oscillation, as for a pendulum
- *Stochastic* random perturbations, as for the hydrologic cycle

Knowledge acquired through analysis of a freely oscillating pendulum, together with the exponential dynamic form, provide the basis for synthesizing a systems model for *harmonic oscillation.* Harmonic oscillation is a phenomenon well known to mathematical domain experts. For example in Figure 4.1, the point P, at radius distance A from the origin of the xy plane, exhibits angular rotation, θ. Instantaneous velocities in the x and y directions are given by

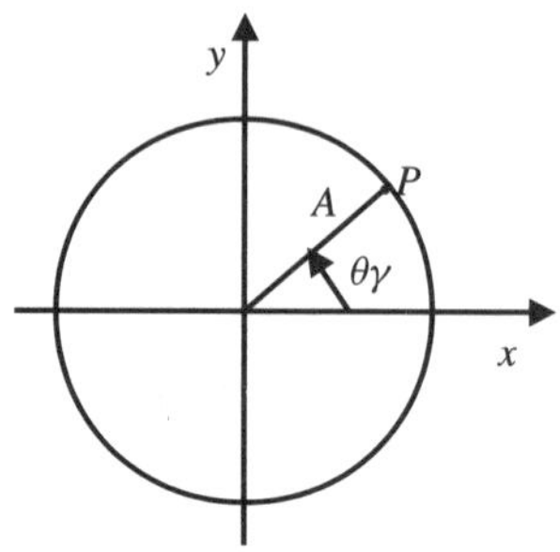

Figure 4.1 Angular Rotation of a Point

$$\frac{dx}{dt} = -A\sin\theta\frac{d\theta}{dt}, \text{ and}$$
$$\frac{dy}{dt} = A\cos\theta\frac{d\theta}{dt}.$$

Coordinates of the point P, $x = A\cos\theta$ and $y = A\sin\theta$, are orthogonal functions as evidenced by the derivatives of $\cos\theta$ and $\sin\theta$ with respect to θ:

- $\frac{d}{d\theta}\cos\theta = -\sin\theta$
- $\frac{d}{d\theta}\sin\theta = \cos\theta$

Furthermore, x and y are obtained by solving second-order differential equations:

- $x = A\cos\theta$ is defined as the solution of $(d^2x)/(d\theta^2) + x = 0$
- $y = A\sin\theta$ is defined as the solution of $(d^2y)/(d\theta^2) + y = 0$

4.1.1 Pendulum Systems Analysis

Consider the dynamics of a freely oscillating pendulum, as represented by its motion diagram in Figure 4.2. At the rest position between regions I and II, the pendulum possesses zero potential energy. At its zenith in either region, the pendulum possesses maximum potential energy. Motion of the pendulum transforms potential energy into kinetic energy. In each of the two regions, the pendulum alternately experiences growth and decline of potential energy. Thus, *harmonic oscillation* is exhibited in the alternating growth and decline in potential energy between its two zeniths.

In Figure 4.3, these regions of growth and decline are represented by $A\sin mt$, where A = constant amplitude of oscillation, $m = (2\pi)/p$ = angular velocity (growth rate of angular displacement), and p =fixed period of oscillation. Figure 4.3 is equivalent to Figure 4.1 in that $\theta = mt$ represents the angular displacement

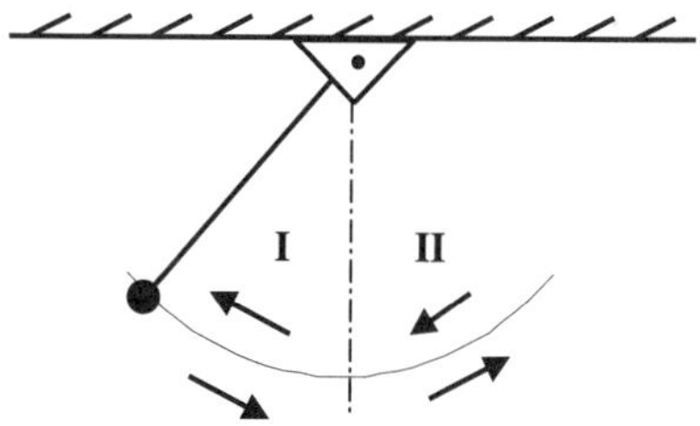

Figure 4.2 Oscillation of a Pendulum

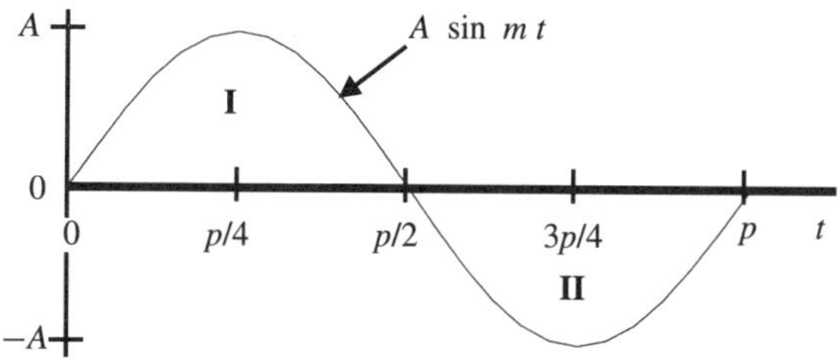

Figure 4.3 Harmonic Oscillation

Table 4.1 Pendulum Dynamics

Region I	Region II	Dynamic Form
$0 - p/4$	$p/2 - 3p/4$	growth
$p/4 - p/2$	$3p/4 - p$	decline

over time in radians of both the point P and the pendulum. Alternating regions of *growth* and *decline* in potential energy are identified in Table 4.1.

In the theory of *small* oscillations of a pendulum, $\sin\theta$ is defined by a Maclaurin series expansion of the form:

$$\sin\theta = \theta - \frac{\theta^3}{3!} + \frac{\theta^5}{5!} - \frac{\theta^7}{7!} + \cdots.$$

Although harmonic oscillation is highly nonlinear, it is possible to synthesize a *linear* systems model that defines "exact" values for $\sin\theta$ by calibrating a pair of dynamic forms for exponential growth.

In general, an exponential growth factor has the following form:

$$z = e^{\frac{1}{T}t} \tag{4.1}$$

where $1/T$ is the *growth rate* and T is the *time constant* of exponential change. For any point (t, z), the slope (derivative) is such that if the slope were to be continued beyond the point, z will double its value in time T. Mathematically, *sine* function values can be generated by subtracting time-phased exponential forms, where $i = \sqrt{-1}$ (see Exercise 4)

$$\sin mt = \frac{e^{i\,mt} - e^{-i\,mt}}{2\,i}. \tag{4.2}$$

Equating the growth rate in Equation 4.1 with its equivalent form in Equation 4.2 yields $1/T = m$, angular velocity of rotation in the complex plane.

The exponential $e^{i\,mt}$ defines a rotating vector reaching from the origin to the circumference of a unit circle (see Figure 4.4). By convention, $e^{i\,mt}$ evokes counter-clockwise rotation, and $e^{-i\,mt}$ evokes clockwise rotation. Subtracting these two counter-rotating vectors produces cycles of alternate reinforcement and cancellation referred to as oscillation.

4.1.2 Pendulum Knowledge Base

Knowledge obtained by analyzing an oscillating pendulum provides the basis for calibrating an $L_{1,2}$ systems model for harmonic oscillation without need of a primary data base. Knowledge items essential to the synthesis are as follows:

- Pendulum system exhibiting harmonic oscillation in potential energy:

$$A \sin mt = A\frac{e^{i\,mt} - e^{-i\,mt}}{2i} \qquad \text{At } t = 0, \sin mt = 0.$$

- Fundamental units of analysis:

Space:	Pendulum domain
Mass:	One pendulum bob, attribute: potential energy normalized to a range of ± 1
Time:	$\Delta t = 1$ unit of time

- Two subsystems:

SS_1:	Exponential growth, e^{mt}
SS_2:	Delayed exponential growth

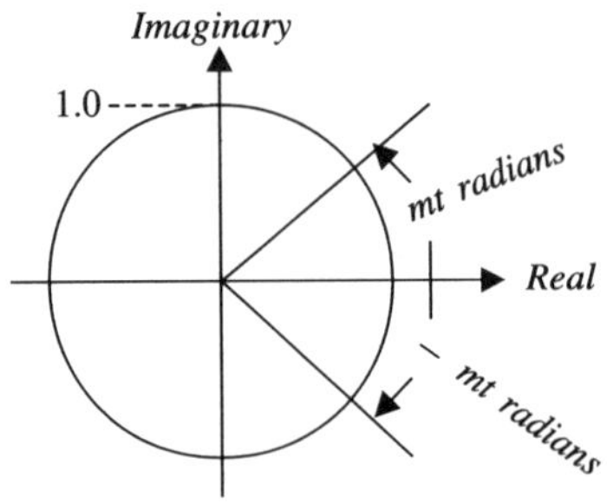

Figure 4.4 Counter Rotating Unit Vectors

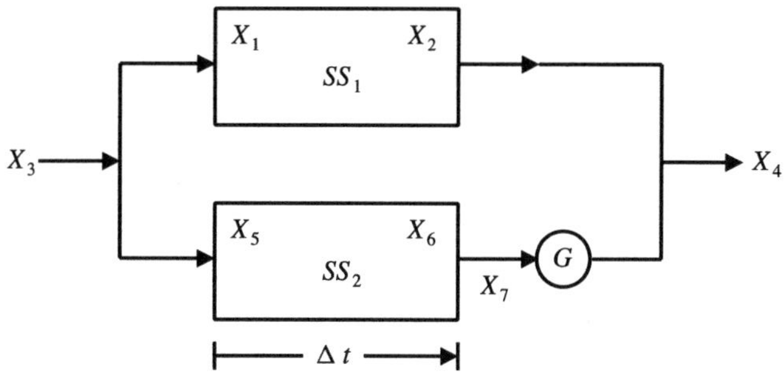

Figure 4.5 Pendulum System

- Schematic: See Figure 4.5.
- Mechanism for oscillation:

  In SS_1, exponential growth is introduced by X_3. Output from SS_2, X_7, is a delayed version of X_3. SS_1 leads off with a "head start." Although having the same exponential growth rate, SS_2 "overtakes" SS_1 via a gain, G, applied to output from SS_2 to produce X_4. X_4 is a negative component in the balance equation of SS_1, retarding growth and eventually driving X_1 negative, thereby reversing the roles of SS_1 and SS_2. First positive, then negative growth dominates, as feedback in the form of alternating exponential change flows from SS_2 to SS_1. Consequently, oscillation enfolds the entire system.

- System variables:

  SS_1: X_1 = initial level of potential energy
  X_2 = terminal level of potential energy
  X_3 = exponential growth, an input rate
  X_4 = delayed and modified exponential growth, an output rate

  SS_2: X_3 = exponential growth, an input rate
  X_5 = initial delayed level of potential energy
  X_6 = terminal delayed level of potential energy
  X_7 = delayed exponential growth, an output rate

- Vector table for model structure:

X_j	IT Pair	Exo	Endo	Equation Type	Equation No.
X_1	I	*		—	—
X_2	T		*	balance	1
X_3			*	regression	2
X_4			*	modification	3
X_5	I	*		—	—
X_6	T		*	balance	4
X_7			*	regression	5

- Dynamic form for exponential growth (see Equation 2.4):

  Growth rate $= m = (2\pi)/p =$ angular velocity (growth rate of angular displacement)

  Provides values for X_3

- Dynamic form for delayed exponential growth (a variant of Equation 2.4):

  Provides values for X_7

4.1.3 Calibration of Pendulum Model for Harmonic Oscillation

Five equations, as identified by *type* in the vector table, define the systems model for harmonic oscillation:

1. $X_2 = X_1 + X_3 - X_4$
2. $X_3 = c_1 X_1 + c_2 X_2$
3. $X_4 = G X_7$
4. $X_6 = X_3 + X_5 - X_7$
5. $X_7 = (1/D)[c_1 X_5 + c_2 X_6]$

$$X_1 = \begin{cases} X_1^0 & \text{initially,} \\ X_2 & \text{of previous } \Delta t \text{ thereafter;} \end{cases}$$

$$X_5 = \begin{cases} X_5^0 & \text{initially,} \\ X_6 & \text{of previous } \Delta t \text{ thereafter.} \end{cases}$$

X_2, *terminal level of oscillation* of SS_1, is selected to yield $A \sin mt$, with amplitude $A = 1.000000$. Selecting period length $p = 24\Delta t$'s, yields growth rate $m = (2\pi)/p = [(2)(3.1415926535)]/24 = 0.2617993878$ radians per Δt, equivalent to 15° per Δt. SS_1 is driven by X_3 of Equation 2, the dynamic form for exponential growth. By Equations 2.7 and 2.9,

$$c_1 + c_2 = m = 0.2617993878, \text{ and}$$
$$\frac{1 + c_1}{1 - c_2} = e^m = 1.299265864,$$

whence $c_1 = 0.136604752$ and $c_2 = 0.125194635$.

SS_2 introduces negative feedback, with gain (G) via Equation 3, into SS_1 as a demand component (X_4) of Equation 1. In this macro model, $G > 1$ roughly corresponds to energy replacement in compensation for entropic loss. A phase shift (delay) in exponential growth (X_3) is produced by X_7 of Equation 5. Equation 5 is a variant of Equation 2.4 in that a delay parameter D is included for fine-tuning the phase shift relative to X_3 of SS_1.

There are nine remaining unknowns—X_1^0, X_2, X_3, X_4, X_5^0, X_6, X_7, G and D—but only five equations. Therefore, four of these unknowns must be assigned to auxiliary equations or given tentative values. Choosing the latter, tentative values were supplied as follows:

- Set $X_1^0 = 0$, the point of origin of the sine curve.
- Pick a tentative value for X_5^0, delayed relative to X_1 that is starting out at $X_1^0 = 0$, say, $X_5^0 = -2.0$.
- Start D at 1.0, so that initially there is no variation from the dynamic form of SS_1.
- Set $G = 2.0$ to provide gain in the negative feedback loop.

4.1.3.1 Calibration Procedure

Using double-precision software and hardware capable of carrying 10 significant-digits accuracy, the five-equation model was calibrated using a time span of 50 consecutive Δt's via the following procedure:

- Make iterative runs of the model, holding X_1^0 and X_5^0 constant but adjusting initial values of D and G between runs until (period) $p = 24\Delta t$'s.
- Calibrate X_5^0 to obtain the target amplitude of $A = 1.000000$ for X_2. Let A' be the amplitude that results from $X_5^0 = -2.0$. Then by direct proportion, $A/A' = X_5^0/-2.0$, so that

$$X_5^0 = (-2.0)\frac{A}{A'}. \tag{4.3}$$

4.1.3.2 Heuristics

- D and G in combination control period length and divergence/convergence: $G \uparrow \Rightarrow p \downarrow$ and $D \uparrow \Rightarrow p \uparrow$. (Arrows indicate direction of change.)
- Sensitivity impinges upon both G and D.
- Adjust D to obtain *uniform* amplitude A' at the extremes of X_2 for the 50 time units (no divergence/convergence).
- Adjust G to obtain zero-valued axis crossings to six decimal places (0.000000).
- Display a complete database output for each run to facilitate inspection of results and subsequent adjustment of D and G.
- Carry 10 significant digits in D and G to effect six-decimal-place accuracy in $X_1 = \sin mt$. (Calibration required only slight adjustment in D and G although carried out over many iterations.) Final values were

$$D = 0.9888590330, \quad G = 1.999960839, \quad \text{and } A' = 4.023083.$$

- Calibrate X_5^0 to $A = 1.00000$ by Equation 4.3:

$$X_5^0 = -\frac{2.0}{4.023083} = -0.4971312$$

4.1.3.3 The Calibrated Model

1. $X_2 = X_1 + X_3 - X_4$
2. $X_3 = 0.136604752X_1 + 0.125194635X_2$
3. $X_4 = 1.999960839X_7$
4. $X_6 = X_3 + X_5 - X_7$
5. $0.9888590330X_7 = 0.136604752X_5 + 0.125194635X_6$

$$X_1 = \begin{cases} 0.0000000 & \text{initially,} \\ X_2 & \text{of previous } \Delta t \text{ thereafter;} \end{cases}$$

$$X_5 = \begin{cases} -0.4971312 & \text{initially,} \\ X_6 & \text{of previous } \Delta t \text{ thereafter.} \end{cases}$$

4.1.3.4 Results and Discussion

Figure 4.6 provides confirmation of the mechanism for oscillation that was presented in the knowledge base. Delayed X_4 "pursues" after X_3 in the positive direction, catches up with, and then overshoots as X_3 turns downward. X_4 turns downward, and although delayed by its overshoot, again catches up with X_3 in the negative direction. But then, negative overshoot delays X_4 and the "chase" renews in the positive direction, thus completing an entire cycle of oscillation.

Data were generated by the harmonic oscillation model with X_2 calibrated to yield sin $0.2617993878\,t$, as demonstrated in Figure 4.7. All other variables of the model exhibited sinusoidal oscillation with $p = 24$.

Table 4.2 lists model-generated X_2 data, exact to six decimal places, except for $t = 8$ where the value dipped by 0.000001. In implementing the given procedure, accuracy at the sixth decimal place was dependent on carrying 10 significant digits for D and G. Thus, to improve accuracy requires extending the number of significant digits, which ultimately depends on the software/hardware combination.

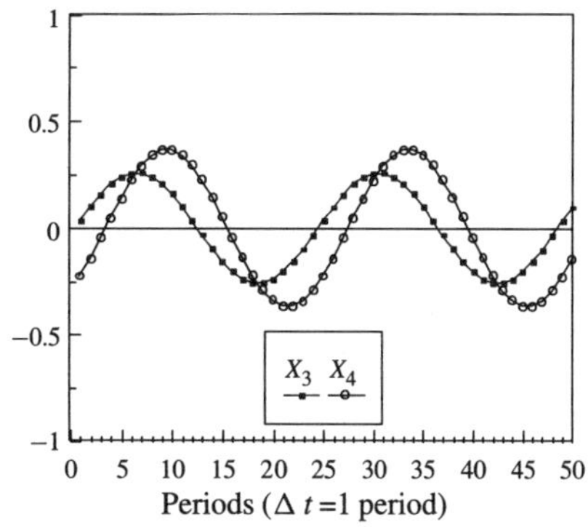

Figure 4.6 Confirmation of Oscillation Mechanism

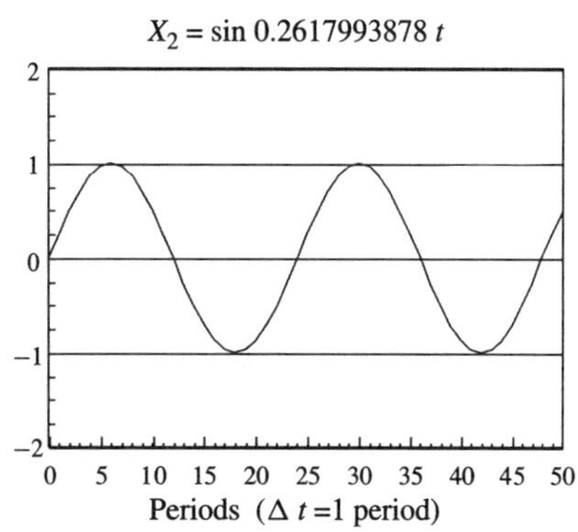

Figure 4.7 Harmonic Oscillation

Table 4.2 Data for Sin mt in 15 Degree Intervals

Time	X_2	Time	X_2
0	0.000000	24	0.000000
1	0.258819	25	0.258819
2	0.500000	26	0.500000
3	0.707107	27	0.707107
4	0.866025	28	0.866025
5	0.965926	29	0.965926
6	1.000000	30	1.000000
7	0.965926	31	0.965926
8	0.866025	32	0.866026
9	0.707107	33	0.707107
10	0.500000	34	0.500000
11	0.258819	35	0.258819
12	0.000000	36	0.000000
13	−0.258819	37	−0.258819
14	−0.500000	38	−0.500000
15	−0.707107	39	−0.707107
16	−0.866025	40	−0.866025
17	−0.965926	41	−0.965926
18	−1.000000	42	−1.000000
19	−0.965926	43	−0.965926
20	−0.866026	44	−0.866026
21	−0.707107	45	−0.707101
22	−0.500000	46	−0.500000
23	−0.258819	47	−0.258819
		48	0.000000
		49	0.258819
		50	0.500000

Distortion introduced by the computer system was discussed in Chapter 2.

4.1.3.5 More Detailed Analysis

The $L_{1,2}$ version of harmonic motion modeled only the potential energy *effect* of a pendulum bob in motion. However, a more detailed $L_{2,j}$ version can include both *cause* and *effect*. *Cause* is attributed to gravity acting upon the pendulum bob and modeled by variables representing acceleration, velocity, elevation, and energy input required to sustain oscillation. Refer to the ballistics model in Chapter 5.

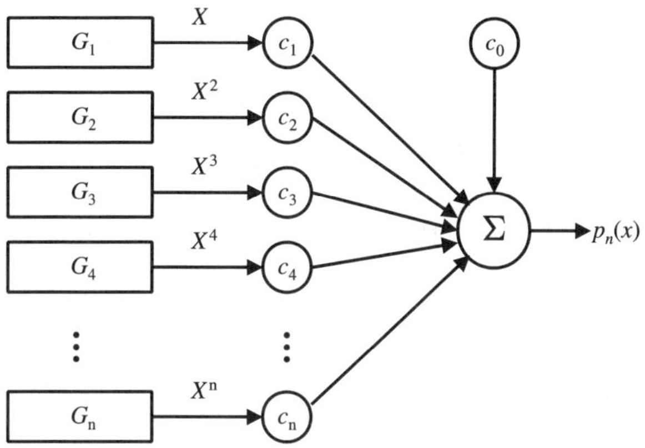

Figure 4.8 Polynomial Synthesis

4.2 POLYNOMIAL MODELS

Approximation of a *nonlinear* function $y = f(x)$ by an nth-*order collocation polynomial*

$$p_n(x) = c_0 + c_1 x + c_2 x^2 + \cdots + c_n x^n$$

was among the early accomplishments in the field of *numerical analysis*, and continues to hold its place as a popular numerical method. A *collocation* polynomial coincides with $f(x)$ at certain specified points. Synthesis of a systems model for an nth-order polynomial includes synthesis of n *groups* of modules, one *group (G)* for each power of x, as illustrated in Figure 4.8. Furthermore, each nth-order group is an nth-order *accumulator* obtained by combining n modules in series. Each module is a *linear accumulator* comprised of a *constant form* and a *balance equation.*

4.2.1 Calibration of Systems Models for Polynomials

In a systems model, dependent variable y represents traversal of *space* or accretion/depletion of *mass* (or an attribute of mass) as a function of *time*: $y = f(t)$. Thus, $x = t$, and $p_n(t) = c_0 + c_1 t + c_2 t^2 + \cdots + c_n t^n$ is to be modeled by linear equations with time *implicitly* represented by equally-spaced Δt increments. To demonstrate the ease with which polynomial systems models can be synthesized from standard dynamic forms, groups of modules are defined for $n = 1$ (linear), $n = 2$ (quadratic), $n = 3$ (cubic), and $n = 4$ (quartic). Although calibration is presented for each of the four module groups, sample output is included for only the cubic group.

Knowledge items, the product of systems analysis, provide the essential raw materials for the synthesis.

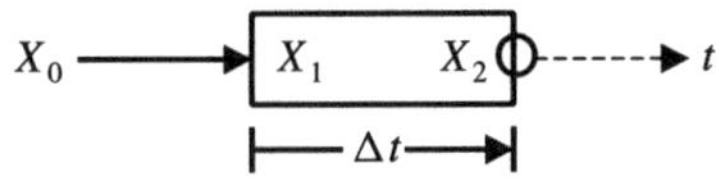

Figure 4.9 Linear Accumulator

4.2.1.1 *Linear Accumulator—synthesizes first-order growth*

Knowledge Items

- System: $y = t$
- Fundamental units of analysis:

  Space: Generic domain

  Mass: Generic unit (or attribute)

  Time: $\Delta t = 1$, for $t = \ldots - 2,\ -1,\ 0,\ 1,\ 2, \ldots$
- Schematic: Examine Figure 4.9, and note the absence of an output rate. X_2 is *tapped* for *information*, indicated by a small circle. *Information flow* is indicated by the dashed-line arrow for t.
- System variables:

  X_0 = input rate

  X_1 = initial level of accumulation

  X_2 = terminal level of accumulation
- Vector table:

X_j	IT Pair	Exo	Endo	Equation Type	Equation No.
X_0				constant form	—
X_1	I	*		—	—
X_2	T		*	balance	1

- Dynamic form for constant rate of change (see Equation 2.3): $X_0 = K > 0$ for linear accumulation.

Calibration

Tabular computations were performed for these four, lower-order polynomial models in order to discover programmable relationships to be used with computer calibration of higher-order polynomial models. Because there are no regression equations, no coefficients are involved. Proceeding from X_2 to X_1 to X_0, calibration was carried out with a *backward-looking* algorithm for *forward differencing* of IT pairs, (X_I, X_T). X_2 was supplied known values from the first-order function of time, t, incremented in Δt steps. Unique values were determined for X_1^0 (bolded) and the constant form $X_0 = K$.

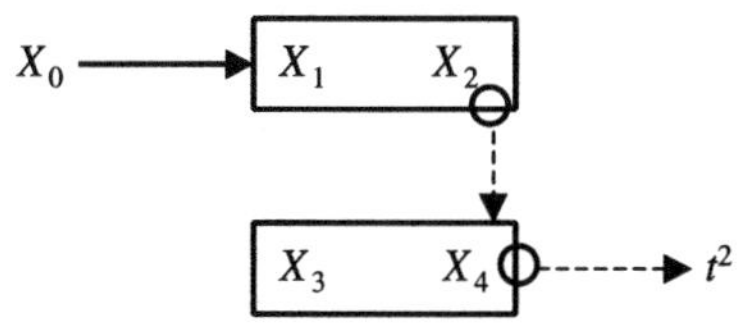

Figure 4.10 Quadratic Accumulator

t	X_2	X_1	X_0
−1	−1		
0	0	**−1**	1
1	1	0	1

Tabular entries begin in the X_2 column, with $X_2 = t$. Entries in the X_1 column are offset values of X_2. Entries in the X_0 column are forward differences: $\Delta X_1 = X_2 - X_1$.

New Knowledge Items

- Linear Accumulator Model:

$$X_2 = X_1 + X_0 \qquad [t]$$
$$X_I = \begin{cases} X_I^0 & \text{at } t = 0, \\ X_T & \text{of previous } \Delta t \text{ thereafter.} \end{cases}$$

- $X_0 = 1$, $X_1^0 = -1$

4.2.1.2 Quadratic Accumulator—synthesizes second-order growth

Additional Knowledge Items

- System: $y = t^2$, spanning two Δt
- Subsystem in *addition* to SS_1 (linear accumulator):

  SS_2: a *second* linear accumulator

- Schematic (see Figure 4.10).
- SS_2 variables:

  X_3 = initial level of accumulation

  X_4 = terminal level of accumulation

- Vector table:

X_j	IT Pair	Exo	Endo	Equation Type	Equation No.
X_0				constant form	
X_1	I	*		—	—
X_2	T		*	balance	1
X_3	I	*		—	—
X_4	T		*	balance	2

Calibration

X_4 coincides with values from the second-order function of time, t^2, where t is varied in Δt steps. Again, calibration only involved finding unique values for X_3^0, X_1^0 (bolded), and the constant form $X_0 = K$. Tabular computations are presented below:

t	X_4	X_3	X_2	X_1	X_0
−2	4				
−1	1	4	−3		
0	0	**1**	−1	**−3**	2
1	1	0	1	−1	2
2	4	1	3	1	2

Tabular entries begin in the X_4 column with $X_4 = t^2$. Entries in the X_3 column are offset values of X_4. Entries in the X_2 column are obtained from $\Delta X_3 = X_4 - X_3$. Entries in the X_1 column are offset values of X_2. Entries in the X_0 column are obtained from $\Delta X_1 = X_2 - X_1$.

New Knowledge Items

- Quadratic Accumulator Model

  1. $X_2 = X_1 + X_0$
  2. $X_4 = X_2 + X_3 \quad [t^2]$

$$X_I = \begin{cases} X_I^0 & \text{at } t = 0, \\ X_T & \text{of previous } \Delta t \text{ thereafter} \end{cases}$$

- $X_0 = 2$, $RX_1^0 = -3$, $X_3^0 = 1$

4.2.1.3 Cubic Accumulator—synthesizes third-order growth

Additional Knowledge Items

- System: $y = t^3$, spanning three Δt

- Subsystems:

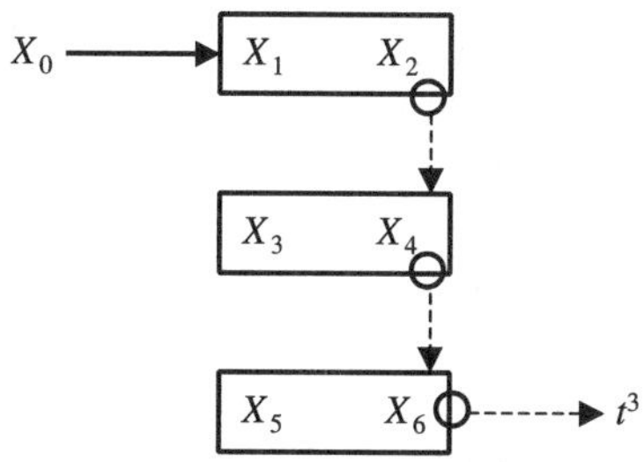

Figure 4.11 Cubic Accumulator

SS3: a *third* linear accumulator

- Schematic: See Figure 4.11.
- SS3 variables:

X_5 = initial level of accumulation

X_6 = terminal level of accumulation

- Vector table:

X_j	IT Pair	Exo	Endo	Equation Type	Equation No.
X_0				constant form	—
X_1	I	*		—	—
X_2	T		*	balance	1
X_3	I	*		—	—
X_4	T		*	balance	2
X_5	I	*		—	—
X_6	T		*	balance	3

Calibration

X_6 yields values for the third-order function of time, t^3, where t is varied in Δt steps. Calibration involves finding values for X_5^0, X_3^0, X_1^0 (bolded), and the constant form $X_0 = K$. Tabular computations are presented next:

t	X_6	X_5	X_4	X_3	X_2	X_1	X_0
−3	−27						
−2	−8	−27	19				
−1	−1	−8	7	19	−12		
0	0	**−1**	1	**7**	−6	**−12**	6
1	1	0	1	1	0	−6	6
2	8	1	7	1	6	0	6
3	27	8	19	7	12	6	6

Tabular entries begin in the X_6 column with $X_6 = t^3$. Entries in the X_5 column are offset values of X_6. Entries in the X_4 column are obtained from $\Delta X_5 = X_6 - X_5$. Entries in the X_3 column are offset values of X_4. Entries in the X_2 column are obtained from $\Delta X_3 = X_4 - X_3$. Entries in the X_1 column are offset values of X_2. Entries in the X_0 column are obtained by from $\Delta X_1 = X_2 - X_1$.

New Knowledge Items

- Cubic Accumulator Model

1. $X_2 = X_1 + X_0$
2. $X_4 = X_2 + X_3$
3. $X_6 = X_4 + X_5 \quad [t^3]$

$$X_I = \begin{cases} X_I^0 & \text{at } t = 0, \\ X_T & \text{of previous } \Delta t \text{ thereafter} \end{cases}$$

- $X_0 = 6$, $X_1^0 = -12$, $X_3^0 = 7$, $X_5^0 = -1$

Sample Output $(X_6 = t^3)$

t	X_1	X_2	X_3	X_4	X_5	X_6
0	−12	−6	7	1	−1	**0**
1	−6	0	1	1	0	**1**
2	0	6	1	7	1	**8**
3	6	12	7	19	8	**27**
4	12	18	19	37	27	**64**
5	18	24	37	61	64	**125**
6	24	30	61	91	125	**216**
7	30	36	91	127	216	**343**
8	36	42	127	169	343	**512**
9	42	48	169	217	512	**729**
10	48	54	217	271	729	**1000**

4.2.1.4 Quartic Accumulator—synthesizes fourth-order growth

Additional Knowledge Items

- System: $y = t^4$, spanning four Δt

- Subsystems:

SS_4: a *fourth* linear accumulator

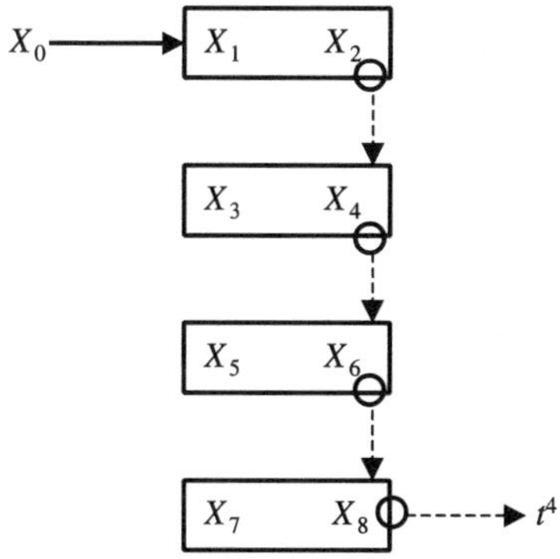

Figure 4.12 Quartic Accumulator

- Schematic: See Figure 4.12.
- SS_4 variables:

  X_7 = initial level of accumulation

  X_8 =terminal level of accumulation

- Vector table:

X_j	IT Pair	Exo	Endo	Equation Type	Equation No.
X_0				constant form	—
X_1	I	*		—	—
X_2	T		*	balance	1
X_3	I	*		—	—
X_4	T		*	balance	2
X_5	I	*		—	—
X_6	T		*	balance	3
X_7	I	*		—	—
X_8	T		*	balance	4

Calibration

X_8 yields values for the fourth-order function of time, t^4, where t is varied in Δt steps. Calibration involves finding values for X_7^0, X_5^0, X_3^0, X_1^0 (bolded), and the constant form $X_0 = K$. Tabular computations are presented next:

t	X_8	X_7	X_6	X_5	X_4	X_3	X_2	X_1	X_0
−4	256								
−3	81	256	−175						
−2	16	81	−65	−175	110				
−1	1	16	−15	−65	50	110	−60		
0	0	**1**	−1	**−15**	14	**50**	−36	**−60**	24
1	1	0	1	−1	2	14	−12	−36	24
2	16	1	15	1	14	2	12	−12	24
3	81	16	65	15	50	14	36	12	24
4	256	81	175	65	110	50	60	36	24

Tabular entries begin in the X_8 column with $X_8 = t^4$. Entries in the X_7 column are offset values of X_8. Entries in the X_6 column are obtained from $\Delta X_7 = X_8 - X_7$. Entries in the X_5 column are offset values of X_6. Entries in the X_4 column are obtained from $\Delta X_5 = X_6 - X_5$. Entries in the X_3 column are offset values of X_4. Entries in the X_2 column are obtained from $\Delta X_3 = X_4 - X_3$. Entries in the X_1 column are offset values of X_2. Entries in the X_0 column are obtained from $\Delta X_1 = X_2 - X_1$.

New Knowledge Items

- Quartic Accumulator Model

  1. $X_2 = X_1 + X_0$
  2. $X_4 = X_2 + X_3$
  3. $X_6 = X_4 + X_5$
  4. $X_8 = X_6 + X_7 \quad [t^4]$

$$X_I = \begin{cases} X_I^0 & \text{at } t = 0, \\ X_T & \text{of previous } \Delta t \text{ thereafter} \end{cases}$$

- $X_0 = 24$, $X_1^0 = -60$, $X_3^0 = 50$, $X_5^0 = -15$, $X_7^0 = 1$

4.2.2 Generic Model for Polynomial Terms

To summarize, systems models have been synthesized for the first through fourth polynomial terms.

Successive *first forward differences* were used in a *backward-looking* algorithm to obtain values for initial conditions and constant dynamic forms.

4.2.2.1 Backward-Looking Algorithm

Each group G_j $(j = 1, 2, \ldots, n)$ of Figure 4.8 consists of one or more linear accumulators. Starting with output from the last linear accumulator, $x^n = t^n$, first forward differencing was used to obtain an initial starting value. In backing through the group of linear accumulators, the forward differencing operation was repeated for each linear accumulator to find initial starting values, and finally, to determine the input value of the constant form K.

4.2.2.2 First Forward Difference

Let $y = f(t)$ be any function of time. If continuous, the derivative operator can be applied successively: $dy/dt, d/dt\ dy/dt = d^2y/dt^2, d/dt\ d^2y/dt^2 = d^3y/dt^3$, and so on. Similarly, although continuity is not required, the forward difference operator, Δ, can be applied successively: $\Delta = \Delta^1, \Delta\,\Delta^1 = \Delta^2, \Delta\,\Delta^2 = \Delta^3$, and so on. The first forward difference of $f(t)$ is defined as $\Delta f_1 = f_{t+1} - f_t$. For example, in Equation 4 of the quartic model, X_6 is a first forward difference of X_7: $X_6 = \Delta X_7 = X_8 - X_7$. Note that X_7 and X_8 are the IT pair of the module. All calculations essential to calibrating the quartic (fourth-order) accumulator are reproduced in the following matrix:

$$\begin{bmatrix} \underline{\text{period}} & \underline{t} & \underline{t^4} & \underline{\Delta^1} & \underline{\Delta^2} & \underline{\Delta^3} \\ (-4,-3) & -4 & 256 & & & \\ (-3,-2) & -3 & 81 & -175 & & \\ (-2,-1) & -2 & 16 & -65 & 110 & \\ (-1,\ 0) & -1 & \mathbf{1} & \mathbf{-15} & \mathbf{50} & \mathbf{-60} \end{bmatrix}$$

For each period in column 1, time t of column 2 is incremented in Δt steps. Three successive first forward differences are generated. Initial values for the IT pairs, X_7^0, X_5^0, X_3^0, and X_1^0, are highlighted in the last row. Instead of differencing one last time to obtain the constant form K, differentiating $y = (t \cdot \Delta t)^4$ four times with respect to t yields the same result: $(d^4y/dt^4) = 4! \cdot (\Delta t)^4 = 4 \cdot 3 \cdot 2 \cdot 1 \cdot (1)^4 = 24$.

To generalize, an nth-order polynomial model for $y = t^n$ will have $2n$ variables, n linear accumulators with n initial conditions, constant form $K = n! \cdot (\Delta t)^n$, and consist of n equations:

1. $X_2 = X_1 + n! \cdot (\Delta t)^n$
2. $X_4 = X_2 + X_3$

$\vdots$

n. $X_{2n} = X_{2n-2} + X_{2n-1}$

The initial condition algorithm follows:

$$\begin{bmatrix} -n\Delta t & (-n)^n(\Delta t)^n & & & & & \\ (-n+1)\Delta t & (-n+1)^n(\Delta t)^n & \Delta^1 & & & & \\ (-n+2)\Delta t & (-n+2)^n(\Delta t)^n & \Delta^1 & \Delta^2 & & & \\ (-n+3)\Delta t & (-n+3)^n(\Delta t)^n & \Delta^1 & \Delta^2 & & & \\ \vdots & & & & & & \\ -2\Delta t & (-2\Delta t)^n & \Delta^1 & \Delta^2 & \cdots & \Delta^{n-2} & \\ -\Delta t & \mathbf{(-\Delta t)^n} & \mathbf{\Delta^1} & \mathbf{\Delta^2} & & \mathbf{\Delta^{n-2}} & \mathbf{\Delta^{n-1}} \end{bmatrix}$$

Column 1 of the matrix contains values of t from $-n\Delta t$ to $-\Delta t$. Initial conditions, generated in reverse order, are bolded in the last row.

Appendix B contains *Fortran language source code* to generate *nontruncated* initial conditions for any specified order up through $n = 20$.

4.2.3 Fourth-order Example

Calibrate a collocation polynomial to fit $y = f(t)$ at the five data points of Table 4.3 each carrying six digits.

Table 4.3 Raw Data Points

t	0	10	20	30	40
$f(t)$	0.00000	4.00000	−72.0000	**6.00000**	160.000

Because $f(0) = 0$, a fourth-order polynomial is sufficient:

$$p_4(t) = c_1 t + c_2 t^2 + c_3 t^3 + c_4 t^4 \tag{4.4}$$

Substituting the data values into Equation 4.4 yields four calibration equations:

1. $10c_1 + 100c_2 + 1000c_3 + 10000c_4 = 4$
2. $20c_1 + 400c_2 + 8000c_3 + 160000c_4 = -72$
3. $30c_1 + 900c_2 + 27000c_3 + 810000c_4 = 6$
4. $40c_1 + 1600c_2 + 64000c_3 + 2560000c_4 = 160$

Simultaneous solution for the four coefficients yields $c_1 = 20$, $c_2 = -3$, $c_3 = 0.117$, and $c_4 = -0.0013$, resulting in a fourth-order polynomial model for

$$y = f(t) : p_4(t) = 20t - 3t^2 + 0.117t^3 - 0.0013t^4 \tag{4.5}$$

Because of its nonlinearity, Equation 4.5 cannot be linked to a systems model for simultaneous solution by matrix inversion. However, precise values for $p_4(t)$ can be obtained from its equivalent, linear systems model by applying knowledge items from the preceding systems analyses.

4.2.3.1 Knowledge Items

- System: $p_4(t) = 20t - 3t^2 + 0.117t^3 - 0.0013t^4$
- Fundamental units of analysis

  Space: Generic domain

Mass: Arbitrary unit of an attribute

Time: $\Delta t = 1$

- Groups of polynomial accumulators

G_1: linear with $K = 1$

G_2: quadratic with $K = 2$

G_3: cubic with $K = 6$

G_4: quartic with $K = 24$

- Schematic: See Figure 4.8 with $n = 4$.
- System variables:

G_1: IT pair (1,2) with $X_1^0 = -1$

G_2: IT pairs (3,4), (5,6) with $X_3^0 = -3$, $X_5^0 = 1$

G_3: IT pairs (7,8), (9,10), (11,12), with $X_7^0 = -12$, $X_9^0 = 7$, $X_{11}^0 = -1$

G_4: IT pairs (13,14), (15,16), (17,18), (19,20) with $X_{13}^0 = -60$, $X_{15}^0 = 50$, $X_{17}^0 = -15$, $X_{19}^0 = 1$, $X_{21} \equiv p_4(t)$

- Vector table for polynomial model structure:

X_j	IT Pair	Exo	Endo	Equation Type	Equation No.
X_1	I	*		—	—
X_2	T		*	balance	1
X_3	I	*		—	—
X_4	T		*	balance	2
X_5	I	*		—	—
X_6	T		*	balance	3
X_7	I	*		—	—
X_8	T		*	balance	4
X_9	I	*		—	—
X_{10}	T		*	balance	5
X_{11}	I	*		—	—
X_{12}	T		*	balance	6
X_{13}	I	*		—	—
X_{14}	T		*	balance	7

Continued on next page

Continued

X_{15}	I	*		—	—
X_{16}	T		*	balance	8
X_{17}	I	*		—	—
X_{18}	T		*	balance	9
X_{19}	I	*		—	—
X_{20}	T		*	balance	10
X_{21}			*	regression	11

4.2.3.2 Model Equations

G_1:

1. $X_2 = X_1 + 1 \quad [t]$

G_2:

2. $X_4 = X_3 + 2$
3. $X_6 = X_4 + X_5 \quad [t^2]$

G_3:

4. $X_8 = X_7 + 6$
5. $X_{10} = X_8 + X_9$
6. $X_{12} = X_{10} + X_{11} \quad [t^3]$

G_4:

7. $X_{14} = X_{13} + 24$
8. $X_{16} = X_{14} + X_{15}$
9. $X_{18} = X_{16} + X_{17}$
10. $X_{20} = X_{18} + X_{19} \quad [t^4]$

$p_4(t)$:

11. $X_{21} = 20X_2 - 3X_6 + 0.117X_{12} - 0.0013X_{20}$

IT pairs:

$$X_I = \begin{cases} X_I^0 & \text{initially,} \\ X_T & \text{of previous } \Delta t \text{ thereafter} \end{cases}$$

4.2.3.3 Computer Solution

Values for the X_{21} equivalent of $p(t)$ were computed for 50 Δt intervals (51 points) by simultaneous solution of Equations 1 through 11. Figure 4.13 presents a plot of these 51 point values.

4.2.3.4 Discussion

Validation of this linear systems model may be achieved by “history” matching with five raw data points, $f(t)$ at $t = 0$, 10, 20, 30, and 40. Reproduction of these five primary data values by the model as X_{21} was indeed quite accurate, considering that 51 iterations were employed in calculating these five data and 46 intermediate values. Compare values of $p(t)$ in Table 4.4 with values of $f(t)$ in

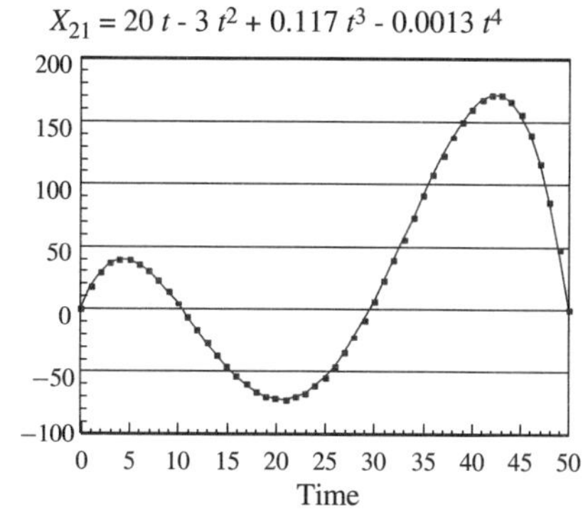

Figure 4.13 Fourth Order Polynomial

Table 4.4 Model Generated Data Points

t	0	10	20	30	40
$p(t)$	0.00000	4.00000	−72.0000	**6.0000**	160.000

Table 4.3. Note that $p(30)$ displays the only deviation in accuracy; the number of significant digits was reduced from six to five. Referring to terminology from *numerical analysis*, X_{21} is also an *interpolating polynomial*, providing a smooth curve at unit increments between the raw data points.

EXERCISES

1. What are the four common types of sinusoidal oscillation that occur in system dynamics?
2. Give at least one example for sinusoidal oscillation associated with
   a. negative feedback.
   b. positive feedback.
3. Show that $(e^{i\,mt} - e^{-i\,mt})/2i = \sin mt$.
4. Evaluate $\sin mt$ for an angular displacement of $mt = \pi/6$ radians $\equiv 30^0$ by
   a. the Maclaurin series to five-term accuracy, and
   b. Equation 4.2, using an appropriate sketch.

5. Explain in your own words how two dynamic forms for exponential growth can be combined to produce a complex dynamic form for harmonic oscillation.

6. Calibrate a model for harmonic oscillation to yield $A \sin mt$, for $A = 2.0$. and $p = 20\Delta t$'s. Give the linear equation model and its output for two full periods. Provide a tabulated comparison of $A \sin mt$ with values generated by computer software of your own choice (e.g., programming language, spreadsheet, etc.).

7. Experiment with the parameters in the model of Exercise 6, D, G, X_1^0, X_5^0, to explore

   a. convergent oscillation.

   b. divergent oscillation.

8. Equation 5 of the calibrated harmonic oscillation model, as a variant of Equation 2.4, introduces a phase shift (delay) in X_3. Rewrite Equation 5 in regression equation form, $X_7 = c_3 X_5 + c_4 X_6$, and equate its coefficients to the coefficients of Equation 2.24 to determine values for delay, D', and shaping parameter, α, for SS_2.

9. Why does calibration of a polynomial systems model require a backward-looking algorithm?

10. Verify that the dynamic form for constant rate of change that comprises input to the generic systems model for t^n, is given by $K = n!\,(\Delta t)^n$, allowing that Δt need not be equal to *one*.

11. (a) Fit a unique third-order polynomial to the following raw data points:

t	0	1	2	3
$f(t)$	0	1	16	81

    (b) Present the equivalent systems model for the third-order polynomial in (a), and give all initial conditions.

12. Derive the constant form and initial conditions for a quintic (fifth-order) accumulator with lag 0.

13. Of what potential use are standard dynamic forms to systems modeling?

APPENDIX B

Initial conditions were calculated manually using a tabular format for dynamic polynomial forms of order 1 through 4 by an algorithm that iteratively applies the difference operator, Δ. To obtain initial conditions for higher orders, the algorithm was coded in Microsoft Fortran. Using double precision, 20 is the maximum order that can be computed by the algorithm without significant digit loss. The algorithm makes three queries for user input:

- Order of polynomial: $1 \leq \text{ORDER} \leq 20$, input as *integer*
- $DT = \Delta t > 0$, input as *real*
- Initial conditions for Lag 0 or Lag 1: Lag 0 locates X_I^0 as the origin of the IT series (X_I, X_T) at $t = 0$. Lag 1 locates X_I^{-1} as the origin at $t = -\Delta t$.

Source Code:

```
      DOUBLE PRECISION  CONSTANT, X(21,20)
      INTEGER ORDER
      WRITE(*,'(1X,A)') 'Input integer order of Polynomial:'
      READ(*,'(I3)') ORDER
      WRITE(*,'(1X,A)') 'Input floating point delta t:'
      READ(*,'(F10.6)') DT
      WRITE(*,'(1X,A)') 'Initial conditions for Lag 0 or Lag 1?  [0,1]'
      READ(*,'(I2)') L
      CONSTANT = 1
      X(1,1) = 0.

      DO I = 1, ORDER
        C = I*DT
        CONSTANT = C*CONSTANT
        X(I+1,1) = (-C)**ORDER
      END DO

      K = ORDER + 1
      I = 0

      DO J = 2, ORDER
        K = K - 1
10      I = I + 1
        X(I,J) = X(I,J-1) - X(I+1,J-1)
        IF(I.LT.K) GO TO 10
```

```
    I = 0
  END DO

  WRITE(*,'(1X,A,F32.8)\') 'Constant Form = ',CONSTANT
  WRITE(*,'(1X,A,I3)\') 'Initial Conditions for Order = ',ORDER
  DO J = ORDER, 1, -1
     WRITE(*,'(1X,F32.8)') X(L+1,J)
  END DO
  END
```

Example Output: Order = 20, Lag = 1

```
Constant form = 2432902008176640000.000000
                Initial Conditions for Order = 20
                -2554547147137679000.000000
                 1234697768216771000.000000
                -3640229629900401000.000000
                 7313100694733609000.000000
               -10591436150757540000.000000
                11409426238686140000.000000
                -9297191415447228000.000000
                 5771534434353111000.000000
                -2727531721102075000.000000
                  973492794090470100000.000000
                 -258323865658578500000.000000
                   49703462510770270000.000000
                   -6678555173493034000.000000
                     592638891947625500.000000
                    -31902815355364800.000000
                       90990301641624.000000
                      -1089054420300.000000
                          3484687250.000000
                            -1048575.000000
                                   1.000000
```

CHAPTER 5

Small Arms Exterior Ballistics Model

Exterior ballistics is the study of the motion of a projectile fired from a gun. Physical forces that act on the projectile have been under study for more than a century [47]. Trajectories, the curved paths of projectiles, have been described by differential equations. Solution by integration has yielded further information about these physical forces.

Researchers have concentrated their studies of exterior ballistics mainly in the area of military applications. Earliest studies concerned effectiveness of bomb placement by developing equations to calculate terminal position and energy of the projectile. Later studies concentrated on gaining knowledge about large, fixed-position artillery [6]. However, in previous analyses, differential equation models have been loosely coordinated with observed facts. Attempts to apply this knowledge to smaller weapons have met with varying degrees of success. Thus, discovery of a more accurate approach to modeling would be a welcome improvement over traditional analytical methods.

Mtm systems analysis and modeling provide an alternative way to achieve accuracy in modeling exterior ballistics. To initiate systems analysis, ballistics data were sought from the test laboratories of two manufacturers, Remington and Winchester. Both, as a matter of policy, refused the request on the grounds of confidentiality and suggested that some data were already available in published form. However, despite refusal to provide test data, systems modeling becomes an even more attractive approach in that knowledge items provide a suitable substitute for data in building a small arms exterior ballistics model. The model can then be used to synthesize relevant data. Adequate knowledge for model calibration does indeed exist in gun magazines, reloading manuals, and handbooks. Furthermore, a ballistics model is simply described by a set of initial conditions and linear equations that account for changing levels and rates. Mathematical knowledge, obtained from applying the laws of physics to projectile flight for military applications, is readily transferred to appropriate dynamic forms.

Small arms exterior ballistic modeling applies specifically to the motion of a projectile fired from a handheld or shoulder-supported gun. These guns are typically used for target practice, hunting, and shooting matches. Ballistics performance information gained from a systems model is particularly useful to ballistics professionals or sportspersons desiring to improve their shooting effectiveness. Although systems models can be synthesized for any number of firearms, one popular firearm was selected for illustration.

5.1 PROJECTILE KNOWLEDGE BASE

Forces that act on a projectile during its flight are identified through systems analysis. For a horizontally fired projectile, the time span of exterior ballistics covers the flight of the projectile from the instant of departure from the muzzle of the barrel and continues until its motion ceases. By dividing the time span into Δt segments, velocity and distance can be represented by an L_1 linear-equation model that relates levels and rates.

5.1.1 Knowledge Items

- Projectile system:

  Gun: 270 Winchester rifle, 24 inch barrel

  Telescopic sight: line of aim is 0.75 inches above rifle bore center line

  Ammunition: 130 grain bullet, standard load

- Fundamental units of analysis:

  Space: xy coordinate axes

  Mass: Projectile (130 grain bullet)

  Time: $\Delta t \approx 0.02$ seconds (arbitrary)

- Subsystem groups: *Horizontal* and *vertical* flight components of a projectile require two subsystem groups.

  Range ballistics: *velocity* and *distance* along the horizontal x-axis

  Elevation ballistics: *velocity* and *distance* along the vertical y-axis

- Range ballistics: Air resistance during flight, produced by bullet shape and air density, provides a negative force acting on the projectile to slow its horizontal velocity [32], [54]. Thus, velocity in the x direction is subject to exponential decline:

$$v_t = v_o e^{-DRt} \tag{5.1}$$

where v_o is initial velocity in feet/second, v_t is velocity at time t, with $v_t < v_o$, and DR is the decline rate, applicable to $t = n\Delta t$, where n is the number of Δt's beyond $t = 0$.

Horizontal distance in feet traversed by the bullet during flight time t is obtained by integrating Equation 5.1:

$$s_t = \frac{v_0}{DR}(1 - e^{-DRt}) \tag{5.2}$$

Because of declining velocity, additional distance (Δs) traversed during each successive Δt undergoes an exponential decline:

$$\frac{\Delta s}{\Delta t} = (w_1 v_o + w_2 v_t) \tag{5.3}$$

where w_1 and w_2 are positive weights, summing to one, that shape exponential decline.

- Elevation ballistics: *Trajectory* refers to the vertical motion of a bullet in flight. The bullet rises from $y = 0$ (the x axis) but subsequently reverses direction and at some distance drops below the x axis. Vertical velocity is subject to gravitational force, which produces a constant acceleration in the negative y direction [39]. Air resistance also retards acceleration exponentially but is negligible for vertical velocities achieved in less than one second. Thus, vertical velocity accumulates linearly over time:

$$v_t = v_o + gt \tag{5.4}$$

where v_o is initial velocity in feet/second, v_t is velocity at time $t = n\Delta t$, and $g = -32.174$ feet/second^2.

Vertical distance, obtained by integration of Equation 5.4, is described by a second-order polynomial:

$$s_t = v_o t + \frac{1}{2} g t^2 \tag{5.5}$$

where s_t is distance in feet at time $t = n\Delta t$.

- Schematic: Subsystem (SS) groups for range and elevation ballistics are presented in Figure 5.1. Each IT pair is separated by Δt.

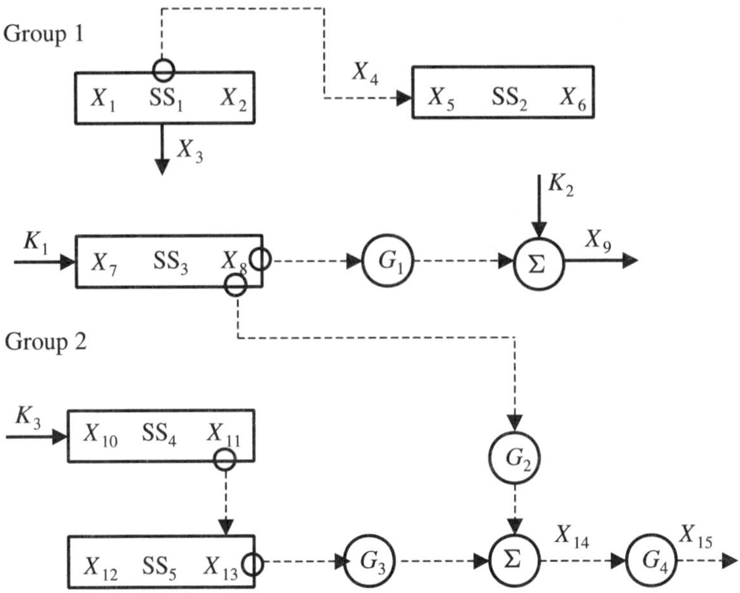

Figure 5.1 Block Diagram of Projectile System

- Variables by subsystem group:

  Group 1: Range ballistics

  X_1 = initial x velocity of bullet

  X_2 = terminal x velocity of bullet

  X_3 = deceleration in x velocity

  X_4 = incremental growth in x distance

  X_5 = initial x distance of bullet

  X_6 = terminal x distance of bullet (Malthusian growth)

  Group 2: Elevation ballistics:

  X_7 = initial level of SS_1 accumulation of t

  X_8 = terminal level of SS_1 accumulation of t

  X_9 = terminal y velocity of bullet

  X_{10} = initial level of SS_4 accumulation of t

  X_{11} = terminal level of SS_4 accumulation of t

  X_{12} = initial level of SS_5 accumulation of t^2

  X_{13} = terminal level of SS_5 accumulation of t^2

  X_{14} = terminal y distance of bullet in feet

  X_{15} = terminal y distance of bullet in inches

- Handbook data [56]:

Distance (yards)	Muzzle	100	200	300
Velocity (feet/second)	3,160	2,970	—	—
Mid Range Trajectory (inches)	—	0.50	2.00	4.50

- Flight time to (maximum) midrange rise for 200-yard sighting, although unknown, is greater than flight time to 100 yards because of diminishing velocity. Velocity at 100 yards and midrange rise are random variables with probability distributions. Consequently, these ballistics data are nominal values that estimate ballistics performance.

- Vector table for model structure:

X_j	IT Pair	Exo	Endo	Equation Type	Equation No.
X_1	I	*		—	—
X_2	T		*	balance	1
X_3			*	regression	2
X_4			*	regression	3
X_5	I	*		—	—
X_6	T		*	balance	4
X_7	I	*		—	—
X_8	T		*	balance	5
X_9			*	modification	6
X_{10}	I	*		—	—
X_{11}	T		*	balance	7
X_{12}	I	*		—	—
X_{13}	T		*	balance	8
X_{14}			*	modification	9
X_{15}			*	modification	10

- Dynamic form for exponential decline:

  Provides values for X_3

  Provides values for X_4

  Rate of decline in X_4 varies directly with rate of decline in X_3.

- Polynomial Accumulators:

  Linear

  Provides values for X_6

Provides values for X_8

Quadratic

Provides values for X_{13}

- Constant form:

  Provides value for K_1 input to linear accumulator, SS_3

  Provides value for K_2 to initialize y velocity

  Provides value for K_3 input to quadratic accumulator at SS_4

5.2 CALIBRATION OF BALLISTICS MODEL

Subsystem groups of the L_1 ballistics model are independent and can therefore be represented on independent axes. Consequently, equations of the model can be calibrated by group. *Group 1* contains only exponential decline forms, and *Group 2* involves only polynomial forms. Considering handbook data, the model can be calibrated for a bull's eye at 100, 200, or 300 yards. Bull's eye at 200 yards was arbitrarily selected. Calibration can be initiated by either of the following two schemes:

- Starting with velocity of 2,970 feet/second at 100 yards, find flight time to 200 yards and *check* midrange trajectory (maximum rise) of 2.00 inches against the model's predicted rise at midflight time.
- Starting with midrange rise of 2.00 inches, find the flight time to midrange and *check* velocity at 100 yards against the model's prediction.

In comparing sensitivities of the two schemes, scheme one was selected because less distortion was observed. Period one, a typical Δt period, was selected for calibration.

5.2.1 Range Ballistics Equations

Referring to Figure 5.1 and the vector table for model structure, equations for SS_1 and SS_2 are the following:

1. $X_1 = X_2 + X_3$
2. $X_3 = c_1 X_1 + c_2 X_2$

3. $X_4 = c_3 X_1 + c_4 X_2 = DR \cdot \Delta t(w_1 X_1 + w_2 X_2)$ for $\Delta t \neq 1.0$

4. $X_6 = X_4 + X_5$

$$X_1 = \begin{cases} 3,160 & \text{initially,} \\ X_2 & \text{of previous } \Delta t \text{ thereafter} \end{cases}$$

$$X_5 = \begin{cases} 0 & \text{initially,} \\ X_6 & \text{of previous } \Delta t \text{ thereafter} \end{cases}$$

SS_1 accounts for the horizontal deceleration of the bullet. In Equation 1, X_1 and X_2 (initial and terminal velocities) are *levels*, and X_3 (deceleration) is the decreasing *rate* by which velocity declines during Δt. Equation 2 defines exponential decline in x velocity (deceleration).

SS_2 accumulates horizontal distance traveled by the bullet. X_5 and X_6 (initial and terminal distances) are *levels*, and X_4 (incremental distance) is the decreasing *rate* by which distance grows during Δt. Equation 3 defines exponential decline in x distance. Both Equations 2 and 3 require calibration.

5.2.1.1 X Velocity

The dynamic form for X_3, exponential decline in x *velocity* during each successive Δt, is modeled by Equation 2 (SS_1). Calibration equations for exponential decline are

$$c_1 + c_2 = DR\ \Delta t$$
$$\frac{1 - c_1}{1 + c_2} = e^{-DR\ \Delta t}.$$

There are four unknowns: c_1, c_2, DR, and Δt. If DR and Δt can be synthesized from handbook data, then these two equations can be solved for the two coefficients.

Given v_t = 2,970 (velocity at 100 yards) and v_o = 3,160 (muzzle velocity), Equation 5.1 was solved for $DR\, t_{100}$:

$$3,160e^{-DR\,t} = 2,970.$$

$DR\, t_{100} = 0.062010075$. Rate of exponential decline in velocity, DR, was obtained from Equation 5.2:

$$\frac{3,160}{DR}(1 - e^{-0.062010075}) = 300 \text{ feet}$$

yielding $DR = 0.633333351$. Dividing former by latter yielded time to 100 yards:

$$t_{100} = \frac{DR\, t_{100}}{DR} = 0.09791064201 \text{ second.}$$

Again, Equation 5.2 was solved for time to 200 yards:

$$\frac{3,160}{0.633333351}(1 - e^{-0.633333351\,t_{200}}) = 600 \text{ feet}$$

with the result that $t_{200} = 0.2022964732 = n\,\Delta t$. Picking $n = 10$ (for $\Delta t \approx 0.02$),

$$\Delta t = 0.02022964732\text{second.}$$

New knowledge items are the following:

- $DR = 0.633333351$
- $\Delta t = 0.02022964732$ second
- Time to 100 yards, $t_{100} = 0.09791064201$ second
- Time to (max) midrange rise, $t_m = 5\Delta t = 0.10114823660$ second
- Time to 200 yards, $t_{200} = 10\Delta t = 0.2022964732$ second

Simultaneous solution of the two calibration equations completed the x-velocity calibration, and yielded

$$c_1 = 0.006392$$
$$c_2 = 0.006420.$$

5.2.1.2 X Distance

The dynamic form for X_4, exponential decline in x *distance* during each successive Δt, is introduced by Equation 3 (SS_2). Referring to Equation 5.3, X_4 is equivalent to Δs. The change in distance is

$$X_4 = \Delta t(w_1 v_o + w_2 v_t) = c_3 X_1 + c_4 X_2.$$

Equating coefficients, $c_3 = \Delta t w_1$, and $c_4 = \Delta t w_2$, with $w_1 + w_2 = 1.0$. Furthermore, because exponential decline in distance is directly proportional to exponential decline in velocity, the following relationship holds:

$$\frac{c_3}{c_4} = \frac{c_1}{c_2}.$$

Thus, the two calibration equations relating distance coefficients to velocity coefficients are

$$c_3 + c_4 = \Delta t(w_1 + w_2) = \Delta t$$
$$c_3 = \frac{c_1}{c_2} c_4.$$

Simultaneous solution yielded

$$c_3 = 0.010093$$
$$c_4 = 0.010137.$$

5.2.2 Elevation Ballistics Equations

Referring to Figure 5.1, the equations for SS_3, SS_4, and SS_5 are the following:

5. $X_8 = X_7 + K_1 \quad [\Delta t]$
6. $X_9 = K_2 + G_1 X_8$
7. $X_{11} = X_{10} + K_3$
8. $X_{13} = X_{11} + X_{12} \quad [(\Delta t)^2]$
9. $X_{14} = G_2 X_8 + G_3 X_{13}$
10. $X_{15} = G_4 X_{14}$

$$X_7 = \begin{cases} X_7^0 & \text{initially,} \\ X_8 & \text{of previous } \Delta t \text{ thereafter} \end{cases}$$

$$X_{10} = \begin{cases} X_{10}^0 & \text{initially,} \\ X_{11} & \text{of previous } \Delta t \text{ thereafter} \end{cases}$$

$$X_{12} = \begin{cases} X_{12}^0 & \text{initially,} \\ X_{13} & \text{of previous } \Delta t \text{ thereafter.} \end{cases}$$

Vertical distance is related to vertical velocity through Equation 5.5:

$$s_t = v_o t + \frac{1}{2} g\, t^2.$$

Two points on the trajectory were selected for application:

$$\frac{s_m}{12} = v_o(5\Delta t) + \frac{1}{2} g(5\,\Delta t)^2$$
$$0 = v_o(10\Delta t) + \frac{1}{2} g(10\Delta t)^2$$

where s_m is maximum rise at midrange in inches, and "0" indicates a bull's eye at 200 yards. With $g = -32.174$ feet/second^2, and $\Delta t = 0.02022964732$ second, simultaneous solution produced the following two new knowledge items:

- $s_m = 1.975$ inches, 1.25% low relative to the handbook datum, 2.00 inches. However, 1.975, is possibly closer to the population mean for s_m than is 2.00, and thus better expresses the central tendency of shot groups at midrange.
- $v_o = 3.254321488$ feet/second. Because v_o is so small relative to initial x velocity of 3,160 feet/second, correction of x velocity to account for the minuscule angle of trajectory amounts to less than one foot/second and was therefore neglected.

5.2.2.1 Y Velocity

The dynamic form for linear accumulation of $\dot{\Delta t}$ time units is modeled by Equation 5 (SS_3). K_1 and X_7^0 were obtained by running the initial condition program of Appendix B for *order* $= 1$, $\Delta t = 0.02022964732$ second, and *lag* $= 0$:

$$K_1 = \Delta t = 0.02022964732$$
$$X_7^0 = 0.0.$$

Equation 6 (SS_3) incorporates Equation 5.4 for y velocity: $v_t = v_o + g\,t$. The constant form is easily identified:

$$K_2 = v_o = 3.254321488 \text{ feet/second.}$$

Likewise, gain is

$$G_1 = g = -32.174 \text{ feet/second}^2$$

and $X_8 = t$, accumulated units of Δt.

5.2.2.2 Y Distance

The dynamic form for quadratic accumulation of Δt time units is made up of Equation 7 (SS_4) and Equation 8 (SS_5). Values for K_3, X_{10}^0, and X_{12}^0 were also obtained from the initial condition program for *order* $= 2$, $\Delta t = 0.02022964732$, and *lag* $= 0$:

$$K_3 = 2!\,(\Delta t)^2 = 0.00081848$$
$$X_{10}^0 = -.00040924$$
$$X_{12}^0 = 0.0.$$

Equation 9 (SS_5) brings Equation 5.5

$$s_t = v_o t + \frac{1}{2} g\,t^2$$

into the model, with gain

$$G_2 = v_o = 3.254321488 \text{ feet/second}$$

and gain

$$G_3 = -\frac{1}{2}\,g = -16.087 \text{ feet/second}^2$$

and $X_{13} = t^2$, in terms of Δt. Equation 10 of SS_5 scales feet to inches, with modifier

$$G_4 = 12 \text{ inches/foot.}$$

5.2.3 Tentative Model

Calibration of the small arms ballistics model has been carried out "piecemeal," group by group, subsystem by subsystem. Only *tentative status* can be claimed for the model prior to making a test run for performance accuracy. Equations of the tentative model are as follows:

1. $X_1 = X_2 + X_3$
2. $X_3 = 0.006392X_1 + 0.006420X_2$
3. $X_4 = 0.010093X_1 + 0.010137X_2$
4. $X_6 = X_4 + X_5$

$$X_1 = \begin{cases} 3,160 & \text{initially,} \\ X_2 & \text{of previous } \Delta t \text{ thereafter} \end{cases}$$

$$X_5 = \begin{cases} 0 & \text{initially,} \\ X_6 & \text{of previous } \Delta t \text{ thereafter} \end{cases}$$

5. $X_8 = X_7 + 0.02022964732 \quad [\Delta t]$
6. $X_9 = 3.254321488 - 32.174X_8$
7. $X_{11} = X_{10} + 0.00081848$
8. $X_{13} = X_{11} + X_{12} \quad [(\Delta t)^2]$
9. $X_{14} = 3.254321488X_8 - 16.087X_{13}$
10. $X_{15} = 12X_{14}$

$$X_7 = \begin{cases} X_7^0 & 0.0 \text{ initially,} \\ X_8 & \text{of previous } \Delta t \text{ thereafter} \end{cases}$$

$$X_{10} = \begin{cases} X_{10}^0 - 0.00040924 & \text{initially,} \\ X_{11} & \text{of previous } \Delta t \text{ thereafter} \end{cases}$$

$$X_{12} = \begin{cases} X_{12}^0 & 0.0 \text{ initially,} \\ X_{13} & \text{of previous } \Delta t \text{ thereafter.} \end{cases}$$

5.3 VERIFICATION AND VALIDATION OF MODEL

Using computer software, Equations 1 through 10 were solved in matrix form to produce test values for each of the 10 endogenous variables for 40 Δt intervals. The result was a complete 40×15 data base that provided ballistics data from the muzzle out to 2,000 feet for each Δt increment ($\Delta t \approx 0.02$ second). Appendix

C presents segments of the data base, consisting of values for velocities (X_2, X_9), distances (X_6, X_{15}), time (X_8), headed by their initial values. Data base values are to be checked at significant points in time in order to verify and validate the model. Graphs of these data are also presented for visual inspection.

5.3.1 Numerical Results

5.3.1.1 End of First Δt Period

Velocity predicted by the model at end of the first Δt period, $X_2 = 3,119.772342$, was verified by making an independent check by Equation 5.1 for X_2:

$$X_2 = 3,160e^{-(0.633333351)(0.02022964732)} = 3,119.771984 \text{ feet/second.}$$

In absolute comparison by rounding to identical values, $X_2 = 3,119.772$ displays three more digits than the handbook's four significant digits. In relative comparison, the model's value is $(1.1)\,10^{-5}\%$ higher than the check value. Differences of this magnitude are due to *distortion*, which was discussed in Chapter 2.

At Time to 100 Yards

From the new knowledge item list, $t_{100} = 0.097911$ sec. The two X_8 values that bracket t_{100} were used to linearly interpolate between the two corresponding X_2 values, yielding $\widehat{X}_2 = 2,970.0033580$. Note the close comparison to 2,970 feet/second, the handbook estimation. Again, interpolation between the two corresponding X_6 values, found $\widehat{X}_6 = 299.954803$ feet $\equiv$ 100 yards.

At Time to Maximum Rise

Another new knowledge item revealed $t_m = 5\Delta t = 0.101148$ second, where t_m is time to maximum rise. Checking the data base, maximum rise, $X_{15} =$ 1.974994 inches, does occur at $X_8 = 0.101148$ second. Rounded to three significant digits, $X_{15} = 1.98$ inches, is perhaps more accurate than the 2.00 (sample mean) value presented in the handbook. In further validation, vertical velocity at maximum rise must be zero. Except for distortion, the model does predict zero velocity: $X_9 = -0.000022$ feet/second. Note that $t_m > t_{100}$, as observed under knowledge item "handbook data," has been verified by the model . Of course, maximum rise cannot occur at 300 feet because of declining horizontal velocity. The model predicts this distance to be $X_6 = 309.611204$ feet.

At Time to 200 Yards

Again, another new knowledge item, $t_{200} = 10\Delta t = 0.2022964732$ second, provides a basis for evaluation. Checking at $X_8 = 0.202297$ second, the data base shows $X_6 = 600.010642$ feet $\equiv$ 200 yards. Next, check the data for a bull's eye at 200 yards. Trajectory starts from a vertical elevation of $X_{15} = 0.00$, rises

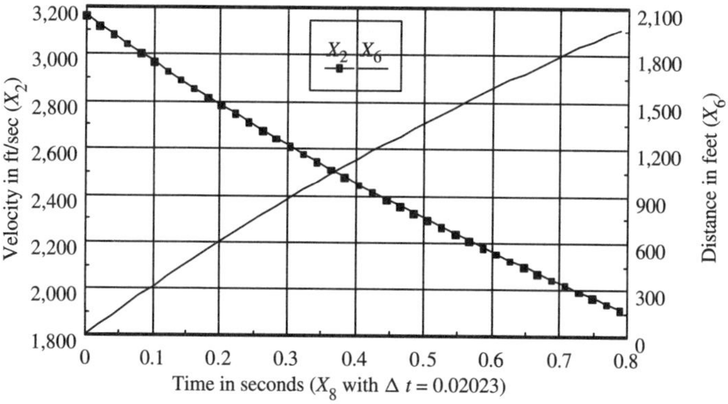

Figure 5.2 Horizontal Ballistics

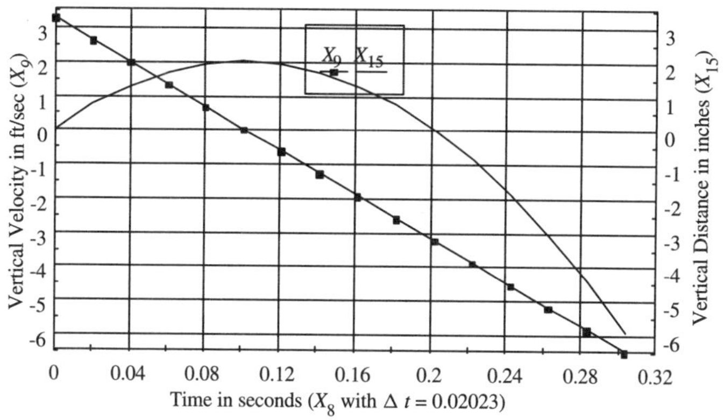

Figure 5.3 Vertical Ballistics

to maximum elevation of 1.98 inches at 103.20 yards, and drops through 0.00 at 200 yards. Except for distortion, the bullet's path includes a *perfect* bull's eye: $X_{15} = -0.000078$ inch at 600 feet.

5.3.2 Graphical Inspection of Trajectory

Figure 5.2 shows horizontal velocity X_2 exponentially declining from $X_2^0 = 3{,}160$ at $t = 0$ to 1,900 feet/second, 0.8 second later. Exponentially declining (Malthusian) growth in distance is evidenced in the "bending over" of the X_6 curve. Starting from $X_6^0 = 0$ at $t = 0$, distance covered declines with each Δt, and reaches 1,980 feet, 0.8 second later. Figure 5.2 can be used to scale velocity at any desired distance, or vice versa. For example, at 600 feet, velocity is approximately 2,780 feet/second.

In Figure 5.3, vertical velocity X_9 declines linearly, responding to the constant acceleration of gravity. Starting from $X_9^0 = 3.25$ feet/second at $t = 0$, velocity

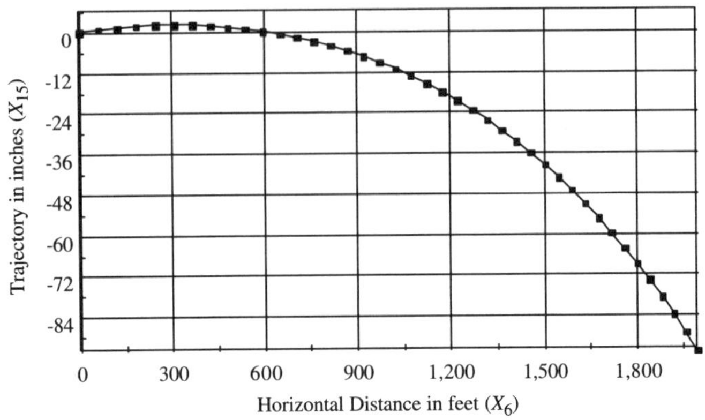

Figure 5.4 270 Winchester Trajectory

drops to zero at $t_m = 0.10$ second ($\approx 5\Delta t$), to reach -6.5 feet/second at $t = 0.30$ second ($\approx 15\Delta t$) later. Vertical distance X_{15} varies with time, experiencing a second-order polynomial trajectory. Starting from $X_{15}^0 = 0$ at $t = 0$, increasing to $X_{15} = 2.0$ inches at $t_m = 0.10$ second ($\approx 5\Delta t$), decreasing to 0.0 at $t_{600} = 0.20$ ($\approx 10\Delta t$) second, and completing the graph with a drop to -6.0 inches at $t = 0.30$ ($\approx 15\Delta t$) second. Variation within the physical realm, increasing distance with declining velocity, occurs with *continuity* (as opposed to step or discrete variation). Although variation simulated by the model occurs from point to point at Δt intervals, the smooth-curve interpolation between points greatly enhances the appearance of continuous response.

Trajectory for this particular 270 Winchester cartridge, sighted at 200 yards, was plotted from the data base and can be observed in Figure 5.4. Of interest to the sports person is the *drop* of the bullet at various ranges: at 600 feet, bull's eye; at 900 feet, about 7 inches low; at 1,200 feet, about 19 inches low; at 1,500 feet, about 40 inches low; at 1,800 feet, about 68 inches low; and at 2,000 feet, about 94 inches low. Thus, the range for shooting at medium-sized targets does not extend beyond 400 yards, unless drop can be expertly compensated for by *elevated aim*. Furthermore, trajectory data suggest an alternative method for sighting the rifle for this cartridge at 200 yards. Refer to Figure 5.5 where trajectory is plotted out to almost 300 yards. In addition to trajectory, *line of aim* has been plotted: a straight line extending through the telescopic sight, at a vertical distance of 0.75 inches above the *line of flight*, to the bull's eye, at 200 yards distant. By inspection, the intersection point of these two lines occurs at about 62 feet. This suggests that, alternatively, the target could be placed at 62 feet, and the telescope adjusted to group shots in the bull's eye. Sighting at 62 feet is advantageous in that bullet holes are easily visible from that distance without having to walk to the target. At 200 yards, a *spotting telescope* or a *spotter* (one who calls and records the shots) can be used in lieu of walking. Thus, another advantage is that neither are required at

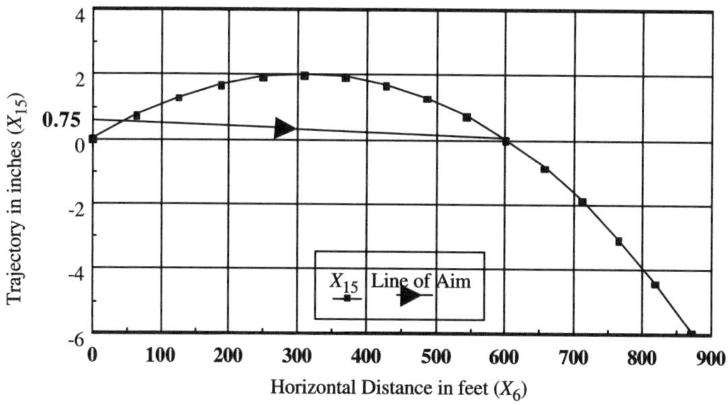

Figure 5.5 Alternative Method for Sighting

62 feet, an obvious safety feature for the spotter. In addition, the bullet encounters negligible wind drift at 62 feet, even if wind velocity is moderate to high! However, sighting should not be performed at 62 feet if *spiraling* is a problem. Spiraling is caused by the spin of an imbalanced bullet about its line of flight.

5.4 CONCLUSION AND DISCUSSION

Calibration of a small arms ballistics model provided an excellent demonstration of how knowledge about the system can be transformed into data by systems analysis and modeling. Model synthesis based on data alone would not have been possible. Characteristic systems knowledge and knowledge of velocity and distance relationships were valuable substitutes for actual data. A complete data base was the *reward* for having constructed a systems model.

Model validity was established by numerical results and by visual inspection of graphical plots of the data base, summarized as follows:

- Accuracy of the model was verified by "history matching" at three points in time out to 600 feet.
- Accurate interpolation between these three points was assured for each Δt because of the model's capability to simulate continuity in the variation of distance and velocity.
- Simulated continuity also implied validity for extrapolated values, although with declining accuracy, over a restricted range as from 600 feet to 2,000 feet.

5.4.1 Interpretation of Handbook Velocities

Handbooks give horizontal velocities at specific ranges in 100-yard increments. Nothing is provided to identify whether these data are *average* velocities with respect to distance, or *actual* velocities at specific distances. However, because velocity data must be consistent with other ballistics data and system knowledge items, the truth becomes known! Such was encountered in the foregoing calibration example.

First, velocity at 100 yards, 2,970 feet/second, was assumed to be average velocity, in that average velocity is higher than actual velocity because of exponential decline. (Reporting the higher of the two values has its advantages!) Thus, time to reach 100 yards was obtained by solving $s = \overline{v}\, t$ for t:

$$t_{100} = \frac{300}{2,970} = 0.1010101 \text{ second.}$$

Initializing the calibration procedure with this value for t_{100} resulted in $\Delta t =$ 0.02035706, leading to $DR = 0.69246842$. To check the consistency of DR and t, their values were substituted into Equation 5.1, with $v_0 = 3,160$:

$$v_{100} = 3,160e^{-DRt} = 2,946.65$$

Obviously, v_{100} is not the correct 100-yard velocity. Subsequently, by the same method, *2,970* was found to be *actual* velocity, much to the credit of the handbook publisher [56] for not publishing misleading data!

5.4.2 Possible Extension

Drift ballistics, accounting for *windage* and *spiraling*, were not essential for the level of detail articulated by the foregoing L_1 analysis. Windage and spiraling require finer detail and expansion to a third axis. Mtm analysis leads to a refined level, where range might be restricted to 1,200 feet, with Δt reduced to 0.005 seconds or less. Drift suggests at least two additional types of dynamic forms: (1) *stochastic,* to input a time series for windage, and (2) *harmonic oscillation*, to introduce spiraling. Thus, the resulting L_2 model would be comprised of three subsystem groups, one each for *range, elevation*, and *drift* ballistics. Readers who are interested in further refinement are challenged to synthesize an L_2 model as an advanced exercise.

EXERCISES

1. Using library resources, write a one- or two-page paper describing the traditional differential equation models of exterior ballistics.

2. Verify Equation 5.2 by integrating Equation 5.1.

3. Perform *dimensional analysis* to show that the weights of Equation 5.3 are dimensionless.

4. Verify Equation 5.5 by integrating Equation 5.4.

5. Based on handbook data from a knowledge item, two schemes for initiating calibration of the ballistics model were presented. Make and label trajectory sketches to illustrate each. Can you suggest an additional initiation scheme based on this knowledge item, again for bull's eye at 200 yards?

6. Calibration equation $c_1 + c_2 = DR \cdot \Delta t$ is used to calibrate model Equation 2, the exponential form for declining velocity. Verify that this calibration equation is dimensionally correct. *Hint*: Show that X_3 must be proportioned to Δt because $\Delta t \neq 1.0$ second.

7. Verify through *dimensional analysis* that calibration of model Equation 3, the exponential form for declining distance, requires that $c_3 + c_4 = \Delta t$.

8. Show that DR is the *constant of proportionality* that relates to the knowledge item "Rate of decline in X_4 varies directly with rate of decline in X_3."

9. Verify that the initial vertical component of velocity, $v_y = 3.254$ *fps*, decreases the initial horizontal component of velocity, $v_x = 3,160$ *fps*, by a negligible amount.

10. Calibrate, verify, and validate the ballistics model for bull's eye at 100 yards or 300 yards.

11. Are model Equations 1 through 4 (SS_1 and SS_2) an equivalent expression of a Malthusian form for growth in x distance even though maximum distance, X_{max} was not utilized? Preliminary to your analysis, solve the submodel corresponding to Group 1 for 40 periods, with $X_1^0 = 3,160$ and $X_5^0 = 0$, to obtain a reference data base. To answer the question:

    a. Solve Equation 2.16, with $c_0 = DR \cdot \Delta t \cdot X_{max}$, for X_{max} by inserting values for period one.

    b. Verify your result in a. by Equation 5.2 for $t \to \infty$.

    c. Substitute the value for c_0 from part a. and values for c_1 and c_2 from the x-velocity calibration into the standard Malthusian form:

- $X_6 = X_4 + X_5$
- $X_4 = c_0 - c_1 X_5 - c_2 X_6$,

With initial distance $X_5^0 = 0.0$ feet/second, solve these two equations for 40 periods, and compare the resulting X_6 values to X_6 values in your reference data base.

12. Calibrate a ballistics model for a rifle, ammunition, and handbook data of your own choosing.

APPENDIX C

X_2	X_6	$X_8 = n\Delta t$	X_9	X_{15}
3160.000000	000000	.000000	3.254321	.000000
3119.772342	63.519012	.020230	2.603453	.711004
3080.056792	126.229410	.040459	1.952584	1.264006
3040.846832	188.141488	.060689	1.301715	1.659004
3002.136026	249.265408	.080919	.650846	1.896000
2963.918019	309.611204	.101148	−.000022	1.974994
2926.186538	369.188781	.121378	−.650891	1.895985
2888.935388	428.007920	.141608	−1.301760	1.658973
2852.158456	486.078275	.161837	−1.952629	1.263958
2815.849704	543.409379	.182067	−2.603497	.710941
2780.003172	600.010642	.202297	−3.254366	−.000078
2744.612976	655.891356	.222526	−3.905235	−.869101
2709.673308	711.060693	.242756	−4.556104	−1.896126
2675.178430	765.527709	.262985	−5.206972	−3.081154
2641.122682	819.301346	.283215	−5.857841	−4.424184
2607.500473	872.390429	.303445	−6.508710	−5.925217
2574.306284	924.803674	.323674	−7.159579	−7.584253
2541.534665	976.549685	.343904	−7.810447	−9.401291
2509.180239	1027.636954	.364134	−8.461316	−11.376332
2477.237693	1078.073869	.384363	−9.112185	−13.509376
2445.701784	1127.868708	.404593	−9.763054	−15.800422
2414.567336	1177.029645	.424823	−10.413922	−18.249471
2383.829238	1225.564750	.445052	−11.064791	−20.856523
2353.482444	1273.481990	.465282	−11.715660	−23.621577
2323.521973	1320.789231	.485512	−12.366529	−26.544634
2293.942907	1367.494237	.505741	−13.017397	−29.625694
2264.740391	1413.604676	.525971	−13.668266	−32.864756
2235.909630	1459.128117	.546201	−14.319135	−36.261821
2207.445893	1504.072032	.566430	−14.970004	−39.816888
2179.344507	1548.443798	.586660	−15.620873	−43.529959
2151.600860	1592.250700	.606889	−16.271741	−47.401031
2124.210396	1635.499929	.627119	−16.922610	−51.430107
2097.168621	1678.198583	.647349	−17.573479	−55.617185
2070.471095	1720.353671	.667578	−18.224348	−59.962266
2044.113435	1761.972114	.687808	−18.875216	−64.465349
2018.091316	1803.060742	.708038	−19.526085	−69.126435
1992.400465	1843.626301	.728267	−20.176954	−73.945524
1967.036666	1883.675450	.748497	−20.827823	−78.922616

1941.995755	1923.214762	.768727	−21.478691	−84.057710
1917.273621	1962.250728	.788956	−22.129560	−89.350806
1892.866207	2000.789755	.809186	−22.780429	−94.801906

CHAPTER 6

Inventory Systems Models: Shaping Dynamic Response

The time series response of inventory systems results from the attempt to match supply to demand. The modeling of inventory systems is one of the major focuses of *Operations Research* [57] (see Chapters 16 through 18). Inventory levels are controlled for the purpose of cost containment. In this chapter, a series of seven inventory systems models is synthesized, with emphasis on the shaping of dynamic response. Analysis and modeling are circumscribed at the $L_{1,j}$ level ($j = 1, 2, 3$ subsystems). Depth of detail is confined to $i = 1$; that is, vertical expansion does not occur. First, dynamic response is explored by synthesizing three, single-subsystem $L_{1,1}$ models. The first $L_{1,1}$ model accumulates inventory through a process of diminishing, incremental accretion (growth). The second $L_{1,1}$ model reduces inventory through a process of diminishing, incremental depletion (decline). The third $L_{1,1}$ model integrates accretion and depletion into an *ideal* or *baseline* inventory system. For the remaining systems models, analysis and modeling are extended horizontally from the ideal model by increasing the number of subsystems, beginning with a two-subsystem model with order delay, and ending with a three-subsystem, three-policy model with random demand. As the scope or breadth of coverage of the model is extended, the knowledge base must also be extended.

6.1 INVENTORY ACCUMULATION MODEL

To initiate the series of models, analysis is focused on accumulation of a single inventory item by its manufacturer. A contract calls for delivery in 20 months of a larger number of items than are currently in stock. Because the contract specifies a

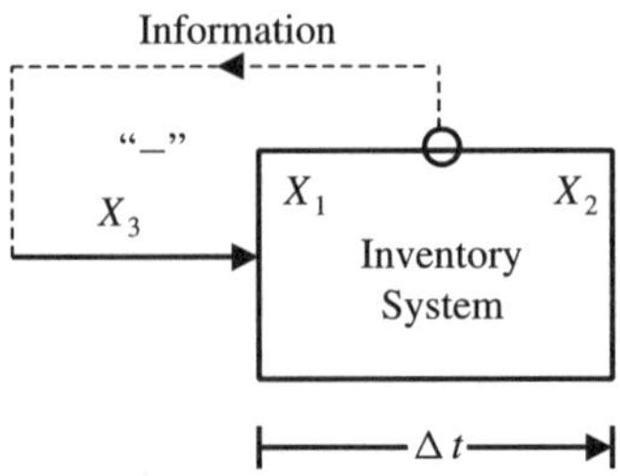

Figure 6.1 Inventory Accumulation System

penalty cost for shortfall or late delivery, this item is to receive high priority at the start. However, as prospects of shortfall diminish with inventory accumulation, manufacturing capacity is released to other products. An $L_{1,1}$ model capable of goal seeking (exhibiting negative feedback) and of generating alternative inventory profiles will be sufficient.

6.1.1 Knowledge Base

To synthesize an $L_{1,1}$ model, only knowledge items commensurate with that level need be presented. Initial knowledge items are the following:

- Inventory system exhibiting Malthusian growth to a target level:

$$x_t = X_{\max} - (X_{\max} - x_o)e^{-DRt}$$

- Fundamental units of analysis:

  Space: Warehouse domain

  Mass: One stock keeping unit (SKU)

  Time: Δt = one period (month)

- Number of subsystems: one
- Schematic: See Figure 6.1.
- System variables

  X_1 = initial inventory level

  X_2 = terminal inventory level

  X_3 = inventory accretion (supply rate)

Figure 6.1 illustrates the relationship between these variables. Knowledge of the system will be used to derive a regression equation for X_3, the system's only rate variable. Negative feedback of information is represented by a *dashed line*, an optional addition to the block diagram when emphasis is desired.

- Vector table for $L_{1,1}$ model:

X_j	IT Pair	Exo	Endo	Equation Type	Equation No.
X_1	I	*		—	—
X_2	T		*	balance	1
X_3			*	regression	2

- Malthusian dynamic form to sensitize declining supply rate, X_3, to accumulation of inventory.

- Specifications for inventory accumulation system:

Decline rate: $DR = 0.20$

Initial inventory position: $X_1^0 = 1,000$

Fixed goal: $X_{\max} = 6,000$

Planning horizon: 20 months

6.1.2 Calibration

6.1.2.1 Model Equations

1. $X_2 = X_1 + X_3$

2. $X_3 = c_0 - c_1 X_1 - c_2 X_2$

$$X_1 = \begin{cases} 1,000 & \text{at } t = 0, \\ X_2 & \text{of previous } \Delta t \text{ thereafter.} \end{cases}$$

Calibration equations for the Malthusian form yield

$$c_0 = DR \times X_{\max} = (0.20)(6,000) = 1,200$$

$$\left. \begin{array}{l} c_1 + c_2 = DR \\ \frac{1-c_1}{1+c_2} = e^{-DR} \end{array} \right\} \; c_1 = 0.096669, \text{ and } c_2 = 0.103331$$

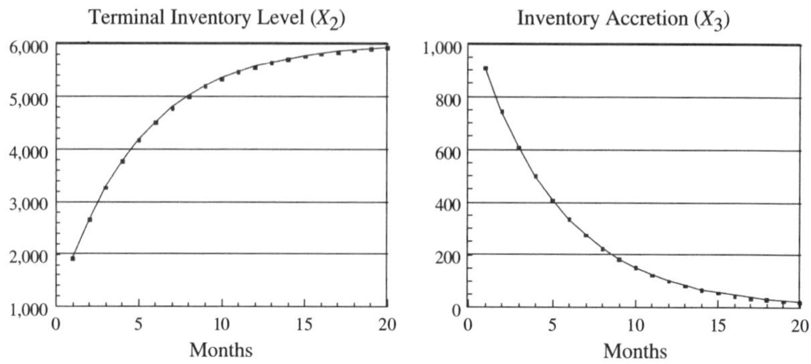

Figure 6.2 Inventory Accumulation System Response

6.1.3 Data Base Generation

Coefficient values were inserted into the model for computer processing. A complete data base for 20 periods was generated. Inventory level X_2 had accumulated to 5,908.4 SKUs, only 1.53% below the goal of $X_{\max} = 6,000$ SKUs. Plots of X_2 and X_3 are displayed in Figure 6.2. Experimenting with this inventory accumulating system by modifying values of X_1^0, DR, and $X_{\max}$ is informative.

6.2 INVENTORY REDUCTION MODEL

A manufacturer has ceased production of a repair part. Initially, the inventory of this part is to be depleted as rapidly as possible to relieve warehouse congestion as a new product is phased in. However, as inventory level is substantially reduced, the goal of reduction to zero in 20 months has less priority, and delaying total stockout becomes desirable. A goal-seeking (negative feedback) $L_{1,1}$ model to effect inventory reduction can be synthesized from a list of essential knowledge items.

6.2.1 Knowledge Base

- Inventory system exhibiting exponential reduction to zero level:

$$x_t = X_{\max} e^{-DR\,t}$$

- Fundamental units of analysis:

Space: Warehouse domain

Mass: One stock keeping unit (SKU)

Time: Δt = one period (month)

- Number of subsystems: one
- Schematic: See Figure 6.3.
- System variables

X_1 = initial inventory level

X_2 = terminal inventory level

X_4 = inventory depletion

- Vector table for $L_{1,1}$ model:

X_j	IT Pair	Exo	Endo	Equation Type	Equation No.
X_1	I	*		—	—
X_2	T		*	balance	1
X_4			*	regression	2

- Dynamic form for exponential decline.
- Specifications for inventory reduction system:

Decline rate: $DR = 0.20$

Initial inventory position: $X_1^0 = 6000$

Fixed goal: $X_{final} = 0$

Planning horizon: 20 months

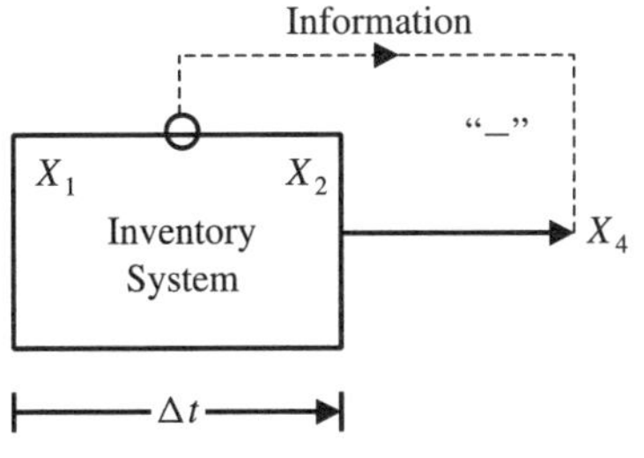

Figure 6.3 Inventory Reduction System

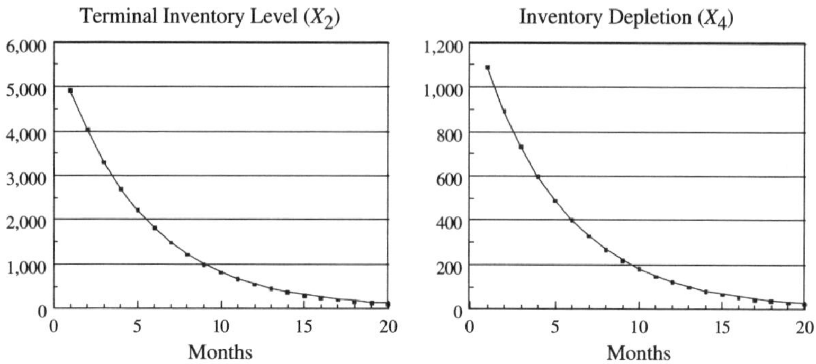

Figure 6.4 Inventory Reduction System Response

6.2.2 Calibration

6.2.2.1 Model Equations

1. $X_2 = X_1 - X_4$

2. $X_4 = c_1 X_1 + c_2 X_2$

$$X_1 = \begin{cases} 6,000 & \text{at } t = 0, \\ X_2 & \text{of previous } \Delta t \text{ thereafter.} \end{cases}$$

Calibration equations for exponential decline yield:

$$\left. \begin{array}{l} c_1 + c_2 = DR \\ \frac{1-c_1}{1+c_2} = e^{-DR} \end{array} \right\} \; c_1 = 0.096669 \text{ and } c_2 = 0.103331$$

6.2.3 Data Base Generation

A data base was generated by iterative solution of the model's equations. Inventory level X_2 has been reduced from $X_1^0 = 6,000$ SKUs to 109.9 SKUs in 20 periods, on its way to $X_{\text{final}} = 0$. Figure 6.4 presents plots of X_2 and X_4. Further experimentation could explore modifying the values of X_1^0 and DR.

6.3 MODEL OF AN IDEAL INVENTORY SYSTEM

Inventories deal with maintaining sufficient stocks of goods (for example, parts and raw materials) that will ensure a smooth operation of a production system or business activity. Traditionally, inventory has been viewed by business and industry as a necessary evil: Too little of it may cause costly interruptions in the operation of the system, and too much of it can ruin the competitive edge and profitability of the business. From that standpoint the only effective way of coping with inventory is to minimize its adverse impact by striking a "happy medium" between the two extreme cases [49].

6.3.1 Ideal Profile

Systems analysis and modeling provide the means for establishing an *ideal* system or *happy medium* as a standard of performance. However, in practice, actual system performance often fails to conform to the ideal. Constraints restrict performance, resulting in a *displaced ideal*. Consider an ideal profile: The ideal system maintains a steady-state inventory level that just matches a given steady-state demand rate. Furthermore, the system must also be able to achieve a steady-state response that yields supply rate equal to steady-state inventory level.

To model such a system, automatic, transient response is required to counteract any positive or negative perturbation in inventory level from its steady-state value. Targeted steady-state values can be achieved via an $L_{1,1}$ model that contains two negative-feedback, goal-seeking loops. One loop makes accretion responsive to any perturbation in inventory level from steady state. The other loop makes depletion responsive to inventory level.

6.3.2 Knowledge Base

- Ideal inventory system exhibiting the following properties:

  Supply is always available to meet demand.

  Items are warehoused for a minimum of one time period.

  Supply rate, demand rate, and initial and terminal inventory levels are equal.

- Fundamental units of analysis:

  Space: Warehouse domain

  Mass: One stock keeping unit (SKU)

  Time: Δt = one period (month)

- Number of subsystems: one

- Schematic: See Figure 6.5.

- System variables:

  X_1 = initial inventory level

  X_2 = terminal inventory level

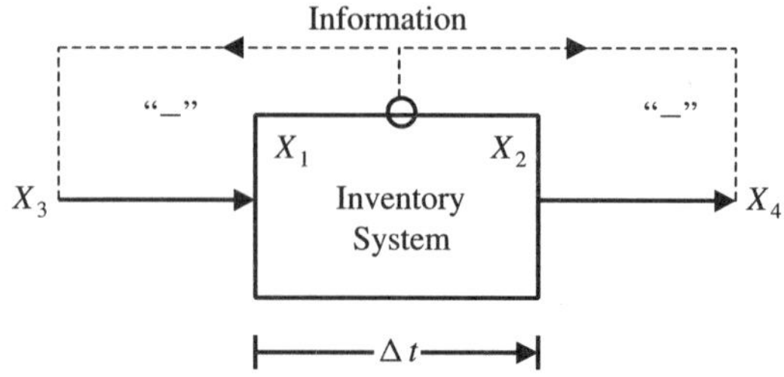

Figure 6.5 Ideal Inventory System

X_3 = inventory accretion (supply rate)

X_4 = inventory depletion (demand rate)

- Vector table for $L_{1,1}$ model:

X_j	IT Pair	Exo	Endo	Equation Type	Equation No.
X_1	I	*		—	—
X_2	T		*	balance	1
X_3			*	regression	2
X_4			*	regression	3

- Malthusian dynamic form to sensitize supply rate, X_3, to positive or negative displacements from steady-state inventory level.
- Dynamic form for continuous delay for X_4.
- Specifications for ideal inventory system:

Delay: $D = 1.0$ period

Steady-state inventory position: $X_{\text{ss}} = 3,000$

Initial inventory position: $X_1^0 = 5,000$ (to illustrate dynamics)

Planning horizon: 20 months

6.3.3 Calibration

6.3.3.1 *Model Equations*

1. $X_1 + X_3 = X_2 + X_4$
2. $X_3 = c_0 - c_1 X_1 - c_2 X_2$
3. $X_4 = c_1 X_1 + c_2 X_2$

$$X_1 = \begin{cases} 5,000 & \text{at } t = 0, \\ X_2 & \text{of previous } \Delta t \text{ thereafter} \end{cases}$$

6.3.3.2 Derivation of Calibration Equations

Equation 2 expresses supply as a Malthusian form composed of two dynamic forms:

$$X_3 = \underbrace{c_0}_{T_1} - \underbrace{(c_1 X_1 + c_2 X_2)}_{T_2}$$

- $T_1 = c_0$ is a constant form.
- $T_2 = c_1 X_1 + c_2 X_2$ is an exponential decline form.

Equation 3 expresses demand and is the realization of either of two equivalent dynamic forms, *delay* or *exponential decline*:

$$X_4 = c_1 X_1 + c_2 X_2 = \frac{1}{D}[\alpha X_1 + (1-\alpha) X_2] \equiv \beta(w_1 X_1 + w_2 X_2),$$

whose weights must sum to 1.0. Since for the ideal system, $D = 1.0$, $\beta = 1/D$. Therefore, the regression coefficients must satisfy:

$$c_1 + c_2 = \beta = 1.0. \tag{6.1}$$

$X_1^0 = 5{,}000$ introduces a perturbation of 2,000 from steady state, thereby initiating a transient response as sketched in Figure 6.6. Negative feedback loops drive inventory level, the IT pair (X_1, X_2), to $X_{ss} = 3,000$. Inventory level changes whenever X_3 is not equal to X_4. From model Equations 2 and 3

$$\frac{\Delta X}{\Delta t} = X_3 - X_4 = c_0 - 2(c_1 X_1 + c_2 X_2). \tag{6.2}$$

The *constant form*, $T_1 = c_0$, is the system's *forcing function*. Analysis of Equation 6.2 in steady state reveals the forcing function. Consider Figure 6.6 as X goes from X_1^0 to X_{ss}. Transient response dies out as X approaches steady-state response. Thus, steady state is marked by the following:

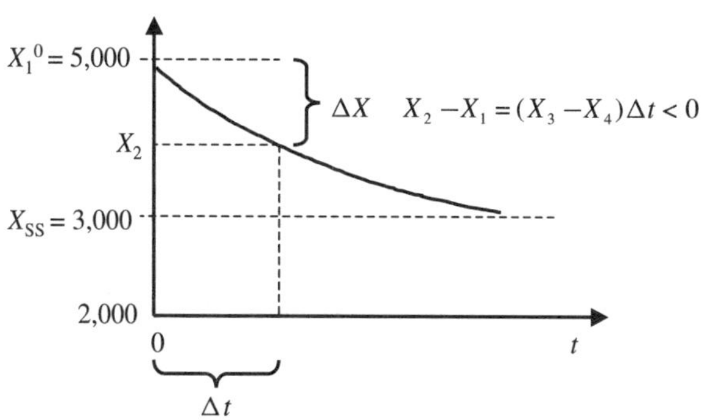

Figure 6.6 Transient Response

- $\Delta X/\Delta t = 0$
- $X_1 = X_2 = X_3 = X_4 = X_{ss}$
- $c_0 = 2(c_1 + c_2)X_{ss}$

From Equation 6.1,

$$c_0 = 2\beta X_{ss} \tag{6.3}$$

To discover the effect of the dynamic forms (X_3, X_4) on inventory level, substitute Equations 2 and 3 into Equation 1 and refine by Equations 6.1 and 6.3 to obtain

$$X_2 = \frac{1 - 2c_1}{1 + 2c_2} X_1 + \frac{2c_1 + 2c_2}{1 + 2c_2} X_{ss}$$

By analogy to the calibration equations for the Malthusian form

$$\frac{1 - 2c_1}{1 + 2c_2} = e^{-2\beta}. \tag{6.4}$$

Solution of Equations 6.1, 6.3, and 6.4 yields the model coefficients

$$c_0 = 6,000$$
$$c_1 = 0.343482$$
$$c_2 = 0.656518.$$

6.3.4 Data Base Generation

Knowledge items that describe an ideal inventory system were used to synthesize a three-equation model. The model was run for 20 periods to synthesize a data base. Plots of X_2, X_3, and X_4 are presented in Figure 6.7. Further analysis might probe the effects that the X_1^0 perturbation and β have on transient response.

6.3.5 Efficiency of Inventory Systems

Denoting efficiency by E, then $E = 100\% \times \beta$, where β represents the steady-state *turnover rate* of a unit of stock. Substitute steady-state conditions for the ideal inventory system, $(X_1)_{ss} = (X_2)_{ss} = (X_4)_{ss}$, into Equation 3

$$(X_4)_{ss} = c_1(X_2)_{ss} + c_2(X_2)_{ss} = (c_1 + c_2)(X_2)_{ss} = \beta(X_2)_{ss}$$

Thus, $\beta^* = (X_4)_{ss}/(X_2)_{ss} = 1.0$ is the *ideal* turnover rate for the ideal inventory system and corresponds to $E = 100\%$. Viewed in terms of delay, $D = 1/\beta^* = 1.0$, each unit of stock is warehoused *one Δt period.*

The value of β can be determined empirically for any inventory system

$$\beta = \frac{\textit{Steady-State Demand}}{\textit{Total Steady-State Inventory Level}}. \tag{6.5}$$

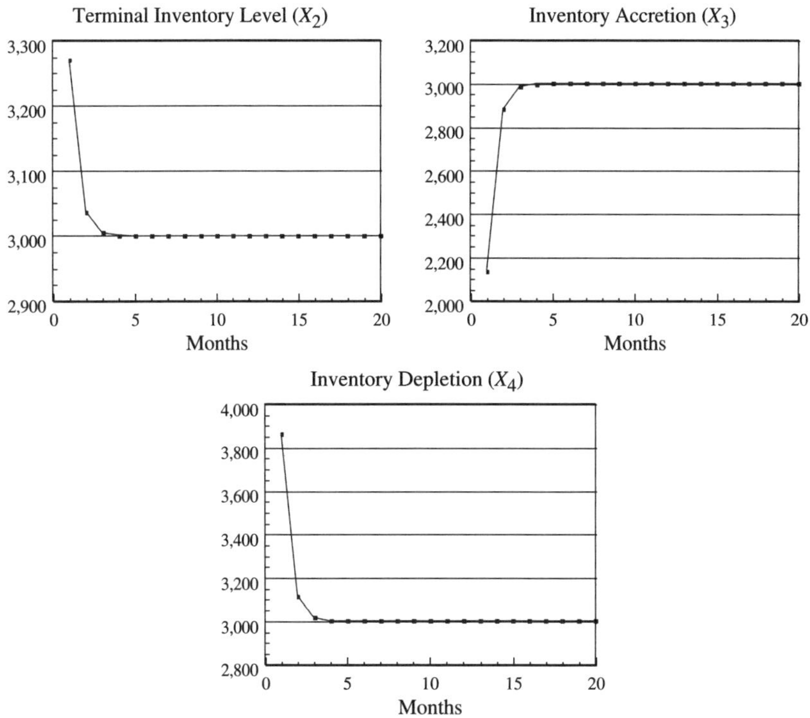

Figure 6.7 Ideal Inventory System Response

Inventory systems with $\beta < \beta^* = 1.0$ exhibit a *displaced ideal.* For example, if a system's steady-state demand is 1,200 and steady-state inventory level is 3,000, then $\beta = 0.40$ and $D = 1/\beta = 2.5$ periods. Consequently, efficiency can be raised from 40% to 100% if steady-state inventory level can be reduced to 1,200. However, rapid-turnover inventory systems exhibiting $D < 1.0$, corresponding to $\beta > 1.0$ and $E > 100\%$, may need to adopt a shorter inventory review period, $\Delta t \leq D$, to guard against stockout. If steady-state averages are used in Equation 6.5, β and E can be calculated for any stochastic inventory system.

6.4 ORDER DELAY MODEL

An inventory system that experiences *order delay* will consequently experience displacement from the ideal system. In general, an inventory system may experience a combination of *transportation* delay and *continuous* delay. Order lead time is a transportation delay. Order delay may occur at a receiving department as *continuous* delay: unloading, inspecting, report preparation, and dispatching the items to storage. Any form of delay other than pure transportation delay is modeled as continuous delay.

An order delay model is obtained by horizontal expansion of the ideal inventory model. An order delay subsystem will be introduced by including an additional negative feedback loop. The complete knowledge base for this $L_{1,2}$ model follows.

6.4.1 Knowledge Base

- Inventory system exhibiting transportation order-delay.
- Fundamental units of analysis:

  Space: Warehouse domain

  Mass: One stock keeping unit (SKU)

  Time: Δt = one period (month)

- Two subsystems:

  SS_1: Storage

  SS_2: Order Delay

- Schematic: See Figure 6.8.
- System variables:

  SS_1: X_1 = initial inventory on hand
  X_2 = terminal inventory on hand
  X_3 = supply rate (replenishment)
  X_4 = demand rate (depletion)

  SS_2: X_3 = supply rate (replenishment)
  X_5 = order quantity (accretion)
  X_6 = initial inventory on order
  X_7 = terminal inventory on order

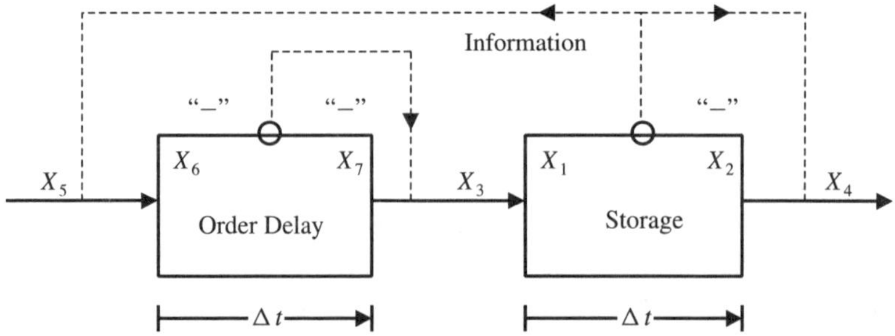

Figure 6.8 Order Delay Inventory System

- Vector table for $L_{1,2}$ model:

X_j	IT Pair	Exo	Endo	Equation Type	Equation No.
X_1	I	*		—	—
X_2	T		*	balance	1
X_3			*	regression	2
X_4			*	regression	3
X_5			*	regression	4
X_6	I	*		—	—
X_7	T		*	balance	5

- Dynamic form for transportation delay for X_3.
- Dynamic form for exponential decline for X_4.
- Malthusian dynamic form to sensitize order quantity, X_5, to displacements from steady-state inventory level.
- Specifications for order delay system:

  Transportation order-delay: $D = 1$ period, $\alpha = 1$

  Storage turnover rate: $\beta = 0.2$

  Initial inventory position: $X_1^0 = 10$

  Initial order position: $X_6^0 = 0$

  Steady-state inventory position: $X_{ss} = 30$

  Planning horizon: 20 months

6.4.2 Calibration

6.4.2.1 Model Equations

1. $X_1 + X_3 = X_2 + X_4$
2. $X_5 + X_6 = X_3 + X_7$
3. $X_3 = \frac{1}{D}\left[\alpha X_6 + (1-\alpha)X_7\right]$
4. $X_4 = c_1 X_1 + c_2 X_2$
5. $X_5 = c_0 - c_1 X_1 - c_2 X_2$

$$X_1 = \begin{cases} 10 & \text{at } t = 0, \\ X_2 & \text{of previous } \Delta t \text{ thereafter} \end{cases}$$

$$X_6 = \begin{cases} 0 & \text{at } t = 0, \\ X_7 & \text{of previous } \Delta t \text{ thereafter} \end{cases}$$

6.4.2.2 Derivation of Calibration Equations

Equation 3 reduces to $X_3 = X_6$, pure transportation delay with $D = \alpha = 1$.

Although Equations 4 and 5 contain identical dynamic forms for exponential decline, Equation 4 can be calibrated independently of Equation 5. The calibration equations for exponential decline yield

$$\left.\begin{array}{l} c_1 + c_2 = \beta \\ \frac{1-c_1}{1+c_2} = e^{-\beta} \end{array}\right\} c_1 = 0.096669 \text{ and } c_2 = 0.103331$$

However, to define c_0, all five equations must be interacted. Let x and y be defined as *inventory on hand* and *inventory on order*, respectively. Then from Equations 1, 3, and 4,

$$\frac{\Delta x}{\Delta t} = (X_3 - X_4) = X_6 - c_1 X_1 - c_2 X_2.$$

Similarly, from Equations 2, 3, and 5,

$$\frac{\Delta y}{\Delta t} = (X_5 - X_3) = c_0 - c_1 X_1 - c_2 X_2 - X_6.$$

At steady state, $\Delta x / \Delta t = \Delta y / \Delta t = 0$, leading to

$$(X_6)_{\text{ss}} = c_1 X_{\text{ss}} + c_2 X_{\text{ss}}$$
$$c_0 = c_1 X_{\text{ss}} + c_2 X_{\text{ss}} + (X_6)_{\text{ss}}.$$

Substituting β for $c_1 + c_2$, and combining these two expressions yields

$$c_0 = 2\beta X_{\text{ss}} = 12.$$

6.4.3 Data Base Generation

For the given initial values and derived coefficients, the order delay model was run to generate a complete data base for 20 periods. Figure 6.9 presents system response. Applying Equation 6.5 to the data of Figure 6.9 yields $\beta = (X_4)_{\text{ss}}/(X_2)_{\text{ss}} = 6.0/30.0 = 0.20$, thus matching the turnover rate specified in the knowledge base. Relative to the ideal system, the storage subsystem is 20% efficient. Further understanding of system response can be obtained by perturbing D, α, and β values.

6.5 RANDOM DEMAND MODEL

Inventory systems can be driven by *internal* goals or by *external* demand. The three previous examples were deterministic, and the inventory systems were driven by

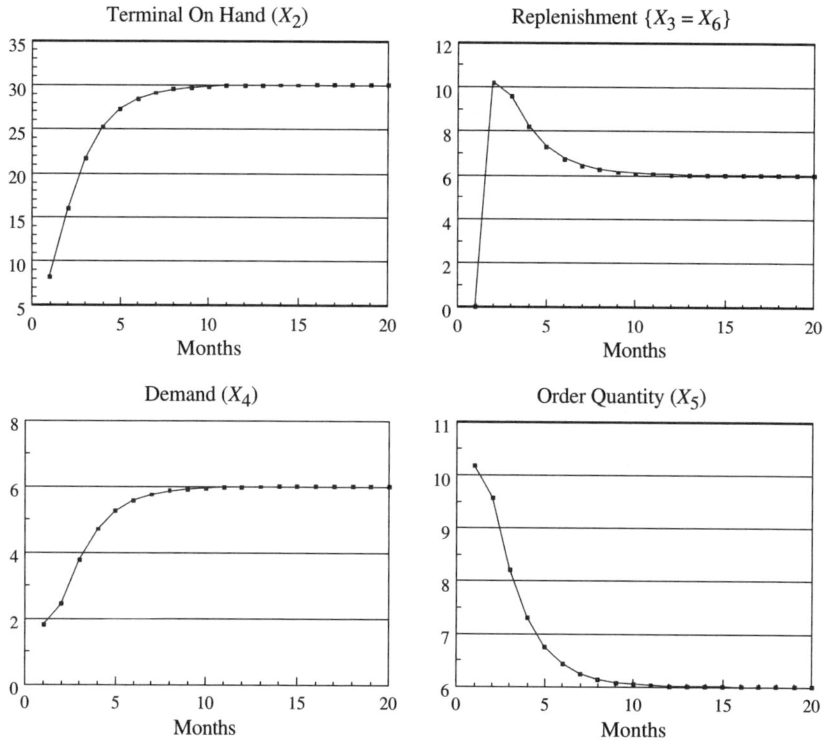

Figure 6.9 Order Delay Inventory System Response

internal goals—for example, that a specific steady-state storage level be achieved. Demand can be *deterministic* or *stochastic* (random). Most often, inventory systems must contend with one or more random variables.

To illustrate, a random demand model is obtained by horizontal expansion of the order delay model. To arrive at an order quantity, external demand data are to be accumulated and processed, thereby introducing another source of delay. Therefore, a third subsystem for *demand delay* will be included to "translate" demand into order quantity, thus introducing an additional negative feedback loop. The complete knowledge base for this $L_{1,3}$ model follows.

6.5.1 Knowledge Base

- Inventory system exhibiting order delay and delayed random demand.
- Fundamental units of analysis:

Space: Warehouse domain

Mass: One stock keeping unit (SKU)

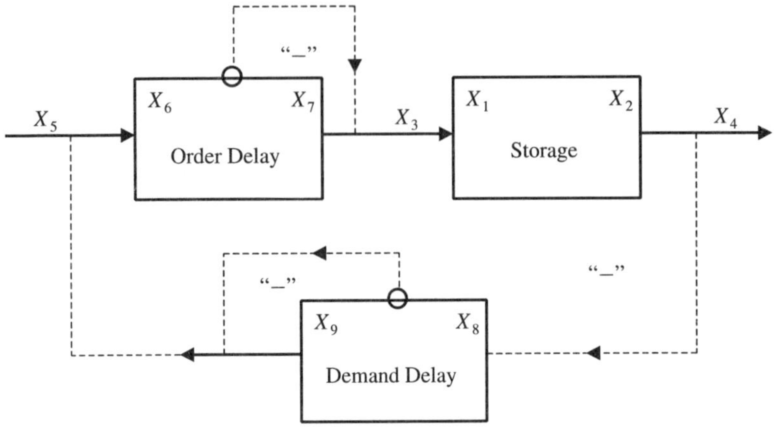

Figure 6.10 Random Demand Inventory System

Time: Δt = one period (month)

- Schematic: See Figure 6.10. Each IT pair is separated by Δt.
- Three subsystems:

  SS_1: Storage

  SS_2: Order Delay

  SS_3: Demand Delay

- System variables:

  SS_1: X_1 = initial inventory on hand
  X_2 = terminal inventory on hand
  X_3 = supply rate (replenishment)
  X_4 = demand rate (depletion)

  SS_2: X_3 = supply rate (replenishment)
  X_5 = order quantity (accretion)
  X_6 = initial inventory on order
  X_7 = terminal inventory on order

  SS_3: X_4 = demand rate (depletion)
  X_5 = order quantity (accretion)
  X_8 = initial delayed demand
  X_9 = terminal delayed demand

- Vector table for $L_{1,3}$ model:

X_j	IT Pair	Exo	Endo	Equation Type	Equation No.
X_1	I	*		—	—
X_2	T		*	balance	1
X_3			*	regression	2
X_4		*		—	—
X_5			*	regression	3
X_6	I	*		—	—
X_7	T		*	balance	4
X_8	I	*		—	—
X_9	T		*	balance	5

- Dynamic form for continuous delay for X_3.
- Dynamic form for continuous delay for X_5.
- Specifications for random demand system:

  Order delay: $D_1 = 1$ period, $\alpha_1 = 0.1$ (relatively high inertia)

  Demand delay: $D_2 = 1$ period, $\alpha_2 = 0.5$ (moderate response to delayed demand)

  Initial inventory position: $X_1^0 = 10$

  Initial order position: $X_6^0 = 5$

  Initial delayed demand: $X_8^0 = 8$

  Planning horizon: 30 months

- Primary data: Random demand, X_4, is defined by an *exogenous stochastic form.* A 30-month sequence was obtained by *Monte Carlo simulated sampling* (see Chapter 8) from a Poisson distribution [27] with mean $\lambda = 6$ SKUs. Because λ remains constant, the series fluctuates about $X_4 = 6.0$—that is, it exhibits *zero trend.* These primary data appear by column in Table 6.1 and in Figure 6.11 as a *stochastic* time series.

6.5.2 Calibration

6.5.2.1 Model Equations

1. $X_1 + X_3 = X_2 + X_4$
2. $X_3 = 1/D_1[\alpha_1 X_6 + (1 - \alpha_1)X_7]$
3. $X_5 = 1/D_2[\alpha_2 X_8 + (1 - \alpha_2)X_9]$

Table 6.1 Random Demand in SKUs

3	6	14	4	5	3
2	4	6	5	3	7
7	3	8	6	8	9
8	5	9	3	9	9
8	8	3	3	9	7

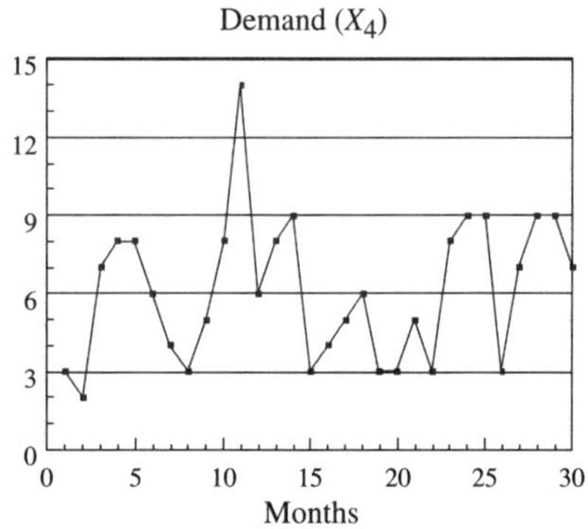

Figure 6.11 Primary Data for Random Demand Inventory System

4. $X_5 + X_6 = X_3 + X_7$
5. $X_4 + X_8 = X_5 + X_9$

$$X_1 = \begin{cases} X_1^0 & \text{at } t = 0, \\ X_2 & \text{of previous } \Delta t \text{ thereafter} \end{cases}$$

$$X_6 = \begin{cases} X_6^0 & \text{at } t = 0, \\ X_7 & \text{of previous } \Delta t \text{ thereafter} \end{cases}$$

$$X_8 = \begin{cases} X_8^0 & \text{at } t = 0, \\ X_9 & \text{of previous } \Delta t \text{ thereafter} \end{cases}$$

6.5.2.2 Calibration of Model

Because parameter values were specified for order delay and demand delay, no calibration equations are required, and Equations 2 and 3 become

$$X_3 = 0.1X_6 + 0.9X_7$$
$$X_5 = 0.5X_8 + 0.5X_9.$$

Values for initial conditions were specified but were arbitrary in that their only purpose was to initialize the first iteration. Arbitrary initial conditions introduce *nonzero trends*. These are transient response components that can be removed by successively solving the model. Zero trend was established for the three IT pairs by successive replacement of X_I^0 values by X_T^{30} values until $X_I^0 = X_T^{30}$ (see Chapter 8).

6.5.3 Data Base Generation

A complete data base was generated for the random demand model. Initial values that introduced zero trend were found to be

$$X_1^0 = 7.539913$$
$$X_6^0 = 7.893333$$
$$X_8^0 = 7.566755.$$

Figure 6.12 presents the secondary time series for order quantity. Comparison with Figure 6.11 clearly shows the one-period delay and the smoothing effect introduced by the demand delay subsystem. The nine-variable data base is displayed as Table 6.2 in Appendix D. Plots of the remaining secondary time series are continued as Figure 6.16 following Table 6.2. The zero-trend, steady-state response of the system is shaped by "transients," introduced by Poisson demand and system delays in processing that demand. Further analysis of transient response is left to Exercise 14.

Performance efficiency of the system over the 30 months was compared to that of the ideal inventory system. The average total delay of the inventory system is $D_t = D_1 + D_2 + 1/\beta_3$. From the data of Table 6.2 for the storage subsystem, $\overline{X}_2 = 10.73$ and $\overline{X}_4 = 6.13$, so that $\beta_3 = \overline{X}_4/\overline{X}_2 = 0.57$. Thus, $D_t = 1 + 1 + 1/0.57 = 3.75$ months. The *turnover rate* for this random demand inventory model is equivalent to that of an $L_{1,1}$ model with $\beta = 1/D_t = 0.27$, or $E = 27\%$ efficiency.

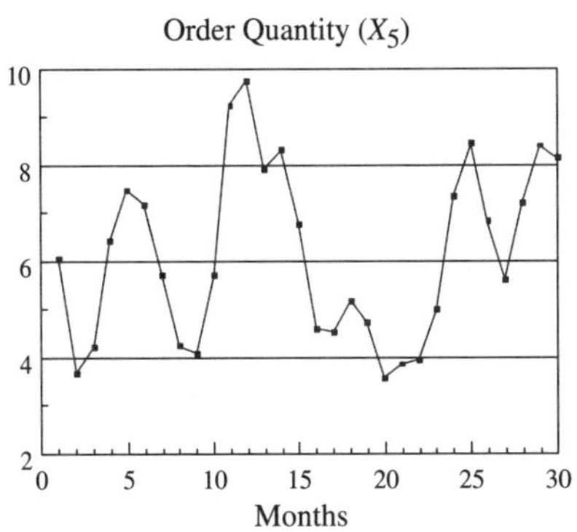

Figure 6.12 Order Quantity for Random Demand Inventory System

6.5.4 Inventory Costs

Although this chapter focuses on stock dynamics, cost dynamics are easy to include. For example, consider ordering and storage costs. Ordering cost is composed of a fixed and a variable component

$$\textit{Ordering Cost} = c_0 + c_1 X_5,$$

where *constant form* c_0 is the cost of processing an order and *Gain* c_1 is the processing cost associated with each SKU of the order. Storage cost is defined via an *arithmetic average form*:

$$\textit{Storage Cost} = c_3 \left(\frac{X_1 + X_2}{2} \right),$$

where c_3 is the cost of storing one SKU for one month and is assessed against the average inventory on hand. Exercise 15 explores one new variable, X_{10} = inventory cost, which is added to the model as Equation 6.6:

$$X_{10} = c_0 + c_1 X_5 + c_3 \left(\frac{X_1 + X_2}{2} \right). \tag{6.6}$$

6.6 ORDER POLICY MODEL

Inventory systems act as buffers between transformation systems in the *pipeline* that connects raw material sources to finished product markets. Inventory levels are controlled by an explicit order policy when more than trivial investment is involved. Order policies are often derived from an optimization objective based on monetary data.

Policies govern the operation of inventory systems. Ideally, policies have the potential to effect systems integration within the total span of control of the policymaker. But unfortunately, control most often fails to focus on systems integration. For example, when control focuses on the supply or raw material end of the pipeline, order policy reflects a *push* philosophy. Economic order quantities are excessively high, minimized only to the extent that carrying costs balance ordering costs. Raw materials are pushed through the system to achieve target inventory levels "just in case" raw material supplies might dwindle. On the other hand, when control focuses on the demand or market end of the pipeline, inventory policy reflects a *pull* philosophy. Raw materials are pulled through the system "just in time" to satisfy accumulated demand, with the result that extremely low inventory levels are required. Although carrying costs are reduced, they are offset by increased equipment costs.

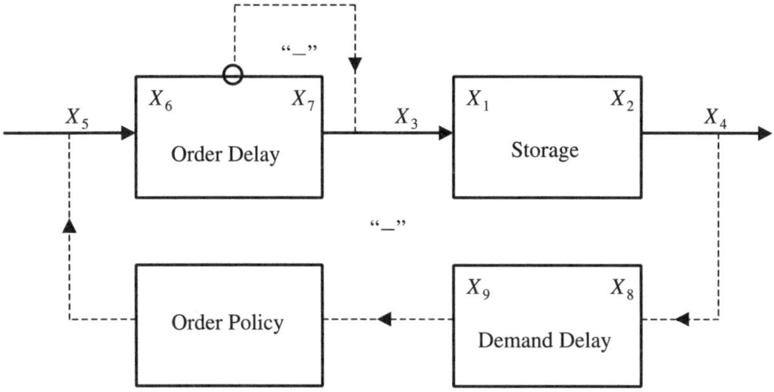

Figure 6.13 Order Policy Inventory System

6.6.1 Knowledge Base Addition

An order policy model is constructed as a variant of the random demand model by inserting an order policy between the demand delay subsystem and the order delay subsystem. Equation 3 is replaced by an exogenous *logical form* that expresses the order policy. The previous knowledge base is updated by adding one new knowledge item and modifying the system schematic and vector table:

- Order policy: Order $X_5 = \begin{cases} 14 & \text{if } X_8 > 11 \text{ SKUs,} \\ 0 & \text{if } X_8 \leq 11 \text{ SKUs} \end{cases}$
- Schematic: See Figure 6.13.
- Vector table for $L_{1,3}$ model:

X_j	IT Pair	Exo	Endo	Equation Type	Equation No.
X_1	I	*		—	—
X_2	T		*	balance	1
X_3			*	regression	2
X_4		*		—	—
X_5		*		—	—
X_6	I	*		—	—
X_7	T		*	balance	3
X_8	I	*		—	—
X_9	T		*	balance	4

6.6.2 Order Policy Simulation

6.6.2.1 Model Equations

1. $X_1 + X_3 = X_2 + X_4$

2. $X_3 = 0.1X_6 + 0.9X_7$ with $D_1 = 1$ period, $\alpha_1 = 0.1$.

$$X_5 = \begin{cases} 14 & \text{if } X_8 > 11 \text{ SKUs,} \\ 0 & \text{if } X_8 \leq 11 \text{ SKUs} \end{cases}$$

3. $X_5 + X_6 = X_3 + X_7$

4. $X_4 + X_8 = X_5 + X_9$

$$X_1 = \begin{cases} 7.539913 & \text{at } t = 0, \\ X_2 & \text{of previous } \Delta t \text{ thereafter} \end{cases}$$

$$X_6 = \begin{cases} 7.893333 & \text{at } t = 0, \\ X_7 & \text{of previous } \Delta t \text{ thereafter} \end{cases}$$

$$X_8 = \begin{cases} 7.566755 & \text{at } t = 0, \\ X_9 & \text{of previous } \Delta t \text{ thereafter} \end{cases}$$

6.6.2.2 Performance Test

The order policy model is a variant of the random demand model. Performance of the order policy was tested by the following procedure:

- Generate a *standard* data base using the random demand model (see Table 6.2) of Appendix D.
- Generate a *test* data base for the order policy model.
- Compare the two data bases.

An abbreviated 30-month comparison of the two data bases is presented in Table 6.3 of Appendix D. Data are given in SKUs.

6.6.2.3 Analysis of Results

- Average terminal inventory level, $\overline{X}_2 = 212.347/30 = 7.08$, was 3.65 lower (109.56/30).
- Average replenishment, $\overline{X}_3$, showed a 0.12 reduction (3.60/30).
- Demand, X_4, was fully met except for month 11, when a maximum demand of 14 created a backlog of $X_2 = -0.53$. Average demand, $\overline{X}_4 = 184/30 = 6.13$.
- An order quantity of $X_5 = 14$ was placed only 13 months out of the 30-month period, resulting in 17 fewer orders. See Figure 6.14.
- Average order quantity, $\overline{X}_5 = (13)(14)/30 = 6.07$, a 0.07 decrease (1.99/30).
- Average terminal on order, $\overline{X}_7$, was 0.11 lower (3.44/30).

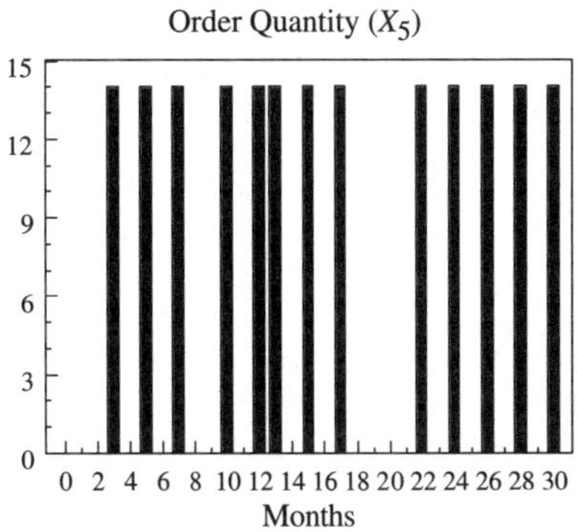

Figure 6.14 Order Placement

- Average terminal delayed demand, $\overline{X}_9 = 296.995/30 = 9.90$, was 3.77 higher (113/30).
- For the demand delay subsystem, $\beta_2 = \overline{X}_5/\overline{X}_9 = 6.07/9.90 = 0.61$.
- For the storage subsystem, $\beta_3 = \overline{X}_4/\overline{X}_2 = 6.13/7.08 = 0.87$.
- Average total delay of the order-policy model:

$$D_t = D_1 + \frac{1}{\beta_2} + \frac{1}{\beta_3} = 1 + 1.64 + 1.15 = 3.79,$$

  and thus is quite comparable to 3.75 of the random demand model.
- The turnover rate for an equivalent $L_{1,1}$ model is $\beta_e = 1/D_t = 0.26$.

6.7 THREE-POLICY MODEL

Policies govern many of the variables associated with inventory systems, as for example, the previous order policy that has been retained for use in a three-policy model:

$$X_5 = \begin{cases} 14 & \text{if } X_8 > 11 \text{ SKUs}, \\ 0 & \text{if } X_8 \leq 11 \text{ SKUs} \end{cases}$$

Other policies cover purchase decisions, service level, backlogging of demand, safety stock, materials handling, stock rotation, obsolescence, and so forth. Two more policies will be added to the order policy model: a *backlog* policy, and a *pallet* policy.

6.7.1 Knowledge Base Addition

Orders unloaded at the dock of a receiving department may not be sized (number of SKUs) to readily accommodate materials handling but rather may require makeshift storage until they can be grouped into pallet-sized lots for transfer into materials storage.

- Pallet policy: Two SKUs make up one pallet. Storage is replenished at the rate of four pallets per month only if four are available. Otherwise, the product is delayed at the receiving department for another month. Thus, the equation for X_3 is replaced by an exogenous *logical form*:

$$\text{Transfer } X_3 = \begin{cases} 8 & \text{if } X_6 \geq 8 \text{ SKUs}, \\ 0 & \text{if } X_6 < 8 \text{ SKUs} \end{cases}$$

  If customer loyalty exists, demand can be backlogged after on-hand inventory is exhausted—that is, when $X_2 < 0$. In the order policy model, backlogging could not be prevented in month 11 when X_2 became negative. Consequently, backlogged demand had to be made up from inventory on hand in month 12.

- Backlog policy: Backlogging is controlled by a simple "yes-no" policy that replaces no equation. *Backlogging "yes"* automatically occurs if X_2 is allowed to become negative. *Backlogging "no"* requires that X_2 be prevented from becoming negative. Backlogging is prevented by inverting the roles of X_2 and X_4 during runtime whenever X_2 becomes negative:

$$\text{If } endo\ X_2 < 0, \text{ then invert, set } exo\ X_2 = 0, \text{ and make } X_4\ endo.$$

  The roles of X_2 and X_4 are then automatically reinverted for the next month.

- Vector table with schematic following in Figure 6.15:

X_j	*IT* Pair	Exo	Endo	Equation Type	Equation No.
X_1	*I*	*		—	—
X_2	*T*		*	balance	1
X_3			*	—	—
X_4		*		—	—
X_5		*		—	—
X_6	*I*	*		—	—
X_7	*T*		*	balance	2
X_8	*I*	*		—	—
X_9	*T*		*	balance	3

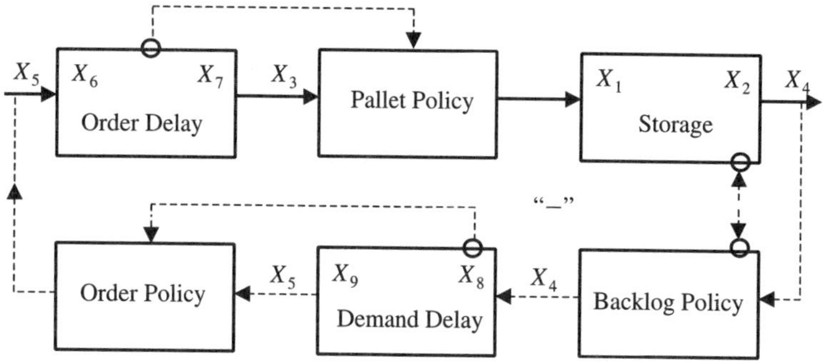

Figure 6.15 Three-Policy Inventory System

6.7.2 Three-Policy Simulation

6.7.2.1 Model Equations

1. $X_1 + X_3 = X_2 + X_4$

$$X_3 = \begin{cases} 8 & \text{if } X_6 \geq 8 \text{ SKUs}, \\ 0 & \text{if } X_6 < 8 \text{ SKUs} \end{cases}$$

$$X_5 = \begin{cases} 14 & \text{if } X_8 > 11 \text{ SKUs}, \\ 0 & \text{if } X_8 \leq 11 \text{ SKUs} \end{cases}$$

2. $X_5 + X_6 = X_3 + X_7$

3. $X_4 + X_8 = X_5 + X_9$

$$X_1 = \begin{cases} 7.539913 & \text{at } t = 0, \\ X_2 & \text{of previous } \Delta t \text{ thereafter} \end{cases}$$

$$X_6 = \begin{cases} 7.893333 & \text{at } t = 0 \\ X_7 & \text{of previous } \Delta t \text{ thereafter} \end{cases}$$

$$X_8 = \begin{cases} 7.566755 & \text{at } t = 0 \\ , X_9 & \text{of previous } \Delta t \text{ thereafter} \end{cases}$$

6.7.2.2 Performance Tests

Dynamic response was compared to the previous standard, the data base of the random demand model (Table 6.2 in Appendix D). Performance of the three-policy model for 30 months was tested under two scenarios.

Scenario 1 Demand was allowed to backlog in the event of stockout. An abbreviated 30-month comparison of the two data bases is presented in Table 6.4 of Appendix D. Data are given in SKUs. Analysis of results follows:

- Backlogged demand, $X_2 < 0$ occurring 19 times, summed to 89.74. Therefore, on the average, $89.74/30 = 3.0$ units of demand were backlogged and did not contribute to average delay, defined by

$$D = \frac{\textit{Total Steady-State Inventory Level}}{\textit{Steady-State Demand}}.$$

- Terminal inventory on hand, $X_2 > 0$ occurring 11 times out of 30, summed to 41.93. Thus, $\overline{X}_2 = 41.93/30 = 1.40$, averaged 9.33 lower $(10.73 - 1.40)$ than did the standard.

- Replenishment, $X_3 = 8$, occurred 21 times during the 30 months, and at an average of $\overline{X}_3 = (21)(8)/30 = 5.60$, was 0.53 lower $(16.0/30)$.

- Demand, X_4, was fully met except that 8.46 SKUs were backlogged at the end of month 30. Thus, $\overline{X}_4 = (184.0 - 8.46)/30 = 5.85$.

- An order quantity of $X_5 = 14$ was placed 13 times out of 30 and averaged 0.07 lower $(1.99/30)$.

- Average terminal on order, $\overline{X}_7 = 440.8/30 = 14.69$, was 8.56 higher (256.8/30). Thus, delay introduced by palletizing indicates a potential storage problem for the receiving department.

- Terminal delayed demand, X_9, averaged 3.77 higher (113.0/30).

- For the order delay subsystem, $\beta_1 = \overline{X}_3/\overline{X}_7 = 5.60/14.69 = 0.38$.

- For the demand delay subsystem, $\beta_2 = 0.61$ is identical to the order policy model.

- For the storage subsystem, $\beta_3 = \overline{X}_4/\overline{X}_2 = 5.85/1.40 = 4.18$, thus exceeding $\beta^* = 1.0$ of the ideal inventory system because demand (or partial demand) was met from replenishment without delay when the storage subsystem was at or near stockout in 18 of the 30 months. One alternative for reducing the potential for stockout is to reduce the inventory review period to $\Delta t \leq 1/\beta_3 = 0.24$.

- Average total delay for Scenario 1 is $D_t = 1/\beta_1 + 1/\beta_2 + 1/\beta_3 = 2.63 + 1.64 + 0.24 = 4.51$, compared to 3.75 for the random demand model and 3.79 for the order policy model. The turnover rate for an equivalent $L_{1,1}$ model is $\beta_e = 1/D_t = 0.22$.

- Relative to the order policy model, the pallet policy increased order delay from $D_1 = 1.0$ to 2.63 and reduced storage delay from $D_3 = 1.15$ to 0.24, a net increase of 0.72. Materials handling should be redesigned to reduce the net increase in delay.

Scenario 2 Backlogging of demand was prevented with the possibility that demand may not always be met. Inversion and automatic reinversion occurred in 8 of the 30 months. An abbreviated 30-month comparison of the two data bases is presented in Table 6.5 of Appendix D. Data are given in SKUs. Analysis of results follows.

- Terminal inventory on hand, $X_2 > 0$ occurring 18 times out of 30, summed to 73.08. Thus, $\overline{X}_2 = 73.08/30 = 2.43$, averaged 8.30 lower (248.92/30) than standard.

- Replenishment, $X_3 = 8$, occurred 17 times during the 30 months and at an average of $\overline{X}_3 = (17)(8)/30 = 4.53$, was 0.16 lower (48.0/30).

- Demand, X_4, that was not satisfied amounted to 41.46, with stockout developing in 12 of the 30 months. Therefore, $\overline{X}_4 = (184 - 41.46)/30 = 4.75$.

- An order quantity of $X_5 = 14$ was placed only 10 times. Thus, $\overline{X}_5 = (14)(10)/30 = 4.67$, and averaged 1.47 lower (43.99/30).

- Average terminal on order, $\overline{X}_7 = 360.8/30 = 12.03$, was 5.89 higher (176.8/30). Again, delay introduced by palletizing indicates a potential storage problem for the receiving department.

- Terminal delayed demand, X_9, averaged 2.40 higher (72.12/30), with $\overline{X}_9 = 253.0/30 = 8.43$.

- For the order delay subsystem, $\beta_1 = \overline{X}_3/\overline{X}_7 = 4.53/12.03 = 0.38$.

- For the demand delay subsystem, $\beta_2 = \overline{X}_5/\overline{X}_9 = 4.67/8.43 = 0.55$.

- For the storage subsystem, $\beta_3 = \overline{X}_4/\overline{X}_2 = 4.75/2.43 = 1.96$, thus exceeding $\beta^* = 1.0$ of the ideal inventory system because demand (or partial demand) was met from replenishment without delay when the storage subsystem was at or near stockout in 12 of the 30 months. Again, the alternative exists for reducing the potential for stockout: reduce the inventory review period to $\Delta t \leq 1/\beta_3 = 0.51$.

- Average total delay for Scenario 2 is $D_t = 1/\beta_1 + 1/\beta_2 + 1/\beta_3 = 2.63 + 1.82 + 0.51 = 4.96$, compared to 4.51 for Scenario 1. The turnover rate for an equivalent $L_{1,1}$ model is $\beta_e = 1/D_t = 0.20$.

EXERCISES

1. Calibrate a production (inventory accumulation) model that calls for manufacturing 10,000 units the first year and then accumulating to 1,000,000 units during the next nine years.

a. Assume accumulation follows a Malthusian (declining) growth curve. Plot the resulting inventory accumulation.

b. Assume accumulation follows an exponential growth curve. Add a plot of the resulting inventory accumulation to the plot in (a).

c. What conditions do you think might favor each model?

2. The Malthusian form, $X_2 = X_{\max} - (X_{\max} - X_1)e^{-DRt}$, is the solution of the differential equation $dX/dt + DR \cdot X = DR \cdot X_{\max}$. Compare values of X_2 from the inventory accumulation model with X_2 values obtained from the differential equation solution.

3. Discuss differences in X_2 response obtained from the inventory accumulation model by modifying the values of $X_1^0 = 1000$, $DR = 0.20$, and $X_{\max} = 6000$ in the following manner:

   a. $X_1^0 = 0$ and 6000 with DR and $X_{\max}$ fixed.

   b. $DR = 0.10$ and 0.30 with X_1^0 and $X_{\max}$ fixed.

   c. $X_{\max} = 12,000$ and $24,000$ with X_1^0 and DR fixed.

4. Compare values of X_2 of the inventory reduction model with X_2 values from the differential equation solution, $X_2 = X_1^0 e^{-DRt}$.

5. Explore and discuss the effect in the inventory reduction model of changing the initial value of $X_1^0 = 6,000$ to $X_1^0 = 3,000$.

6. With X_1^0 fixed at 6,000, find the value of DR that reduces inventory level, X_2, to 0.000 in 20 months.

7. Describe the effect that X_1^0 has on transient response of the ideal inventory model when starting from $X_1^0 = 1000$, 3000, and 5000.

8. Recalibrate the ideal inventory system model for $\beta = 0.20$ to obtain a displaced ideal, and generate a corresponding 20-month data base.

   a. Compare the X_2 response to that of the ideal inventory system model by plotting both on the same graph.

   b. Determine the efficiency for this displaced ideal using Equation 6.5.

   c. Interpret the transient response in terms of the delay parameter D and the shape parameter α.

9. With respect to Figure 6.6, derive and solve the differential equation for dX/dt, starting with $\Delta X/\Delta t = X_3 - X_4$.

10. In general, each subsystem balance equation, written as a difference equation, provides the basis for deriving a differential equation that describes how the subsystem's level changes as an explicit function of time. However, the knowledge gained by writing and solving the system's differential equations becomes too costly for even the relatively simple order delay model. But to illustrate the point, let x and y be defined as *inventory on hand* and *inventory on order*, respectively, for the order delay model.

    a. Starting with expressions for $\Delta x/\Delta t$ and $\Delta y/\Delta t$, perform a macro-to-micro transformation of the model to obtain its differential equation form: $dx/dt + \beta x - (1/D)y = 0$, and $dy/dt + \beta x + (1/D)y = c_0$

    b. Can you solve the system in (a) for x and y?

11. With respect to the order delay model:

    a. Use Equation 6.5 to find the overall efficiency of the order delay model.

    b. Compute the expected delay from the time a unit is ordered (SS_1) until being released to meet demand (SS_2). Use this value to check your answer in (a).

12. Investigate the dynamic response of the order delay model for 25 months that result from changes in its parameters. Recalibrate to accommodate the changes required. Compare runs by plotting the X_3 response on the same graph.

    a. Demonstrate the effect that D has on the response for fixed values of α and β. (For example, with $\alpha = 1$ and $\beta = 0.2$, make runs with $D = 1, 2,$ and 3.)

    b. Fixing D at 1 and β at 0.2, run the model with $\alpha = 0, 0.5,$ and 1.0.

13. Use Equation 6.5 to calculate *total actual average system inventory* for the random demand model. *Hint*: Both inventory on hand and inventory on order are included in the inventory system. Check your answer by computing actual averages from the appendix data.

14. Conduct further analysis of the random demand model to verify that any steady-state properties that the system might exhibit depend entirely on exogenous demand. These properties can be investigated by extending the 30-period Poisson demand sequence by adding 20 additional months of constant demand: $X_4 = 6$ SKUs. Then process these 50 exogenous data with the random demand model and record your observations.

15. Add inventory cost $X_{10} = c_0 + c_1 X_5 + c_3(X_1 + X_2/2)$ as Equation 6 to the random demand model defined in the text, with $c_0 = \$50$, $c_1 = \$2$, and $c_3 = \$8.33$. Generate a complete data base that includes X_{10}. Display X_{10}.

16. Add inventory cost $X_{10} = 5.57X_5 + 8.33(X_1 + X_2/2)$ as Equation 5 to the order policy model.

    a. Verify that 5.57, the coefficient of the ordering cost term in Equation 5, correctly adapts $c_0 = \$50$ and $c_1 = \$2$ to the order policy model.

    b. Generate a complete data base that includes X_{10}, and compare with X_{10} of the random demand model. (See Exercise 6.15.)

17. Using library resources, write a two-page paper to expand on the *push* versus *pull* philosophies.

18. Select some measure (or measures) of performance and modify the three-policy model to obtain improved performance. Provide a rationale for your measure(s).

APPENDIX D

Table 6.2 Data Base for Random Demand Inventory System

Month	X_1	X_2	X_3	X_4	X_5
	X_6	X_7	X_8	X_9	
1	7.539913	11.557484	7.017571	3.000000	6.044503
	7.893333	6.920265	7.566755	4.522252	
2	11.557484	14.943597	5.386113	2.000000	3.681501
	6.920265	5.215652	4.522252	2.840751	
3	14.943597	12.691020	4.747422	7.000000	4.227167
	5.215652	4.695397	2.840751	5.613584	
4	12.691020	10.198150	5.507130	8.000000	6.409056
	4.695397	5.597323	5.613584	7.204528	
5	10.198150	8.682381	6.484231	8.000000	7.469685
	5.597323	6.582777	7.204528	7.734843	
6	8.682381	9.536950	6.854570	6.000000	7.156562
	6.582777	6.884769	7.734843	6.578281	
7	9.536950	11.869444	6.332493	4.000000	5.718854
	6.884769	6.271130	6.578281	4.859427	
8	11.869444	14.178278	5.308835	3.000000	4.239618
	6.271130	5.201913	4.859427	3.619809	
9	14.178278	13.848698	4.670420	5.000000	4.079873
	5.201913	4.611365	3.619809	4.539936	
10	13.848698	10.972555	5.123856	8.000000	5.693291
	4.611365	5.180800	4.539936	6.846645	
11	10.972555	4.071916	7.099362	14.000000	9.231097
	5.180800	7.312535	6.846645	11.615548	
12	4.071916	6.536055	8.464139	6.000000	9.743699
	7.312535	8.592095	1.615548	7.871849	
13	6.536055	6.807216	8.271160	8.000000	7.914566
	8.592095	8.235501	7.871849	7.957283	
14	6.807216	6.075569	8.268353	9.000000	8.304855
	8.235501	8.272004	7.957283	8.652428	
15	6.075569	10.635285	7.559716	3.000000	6.768285
	8.272004	7.480573	8.652428	4.884143	

continued on next page

continued

16	10.635285	12.746368	6.111083	4.000000	4.589428
	7.480573	5.958918	4.884143	4.294714	
17	12.746368	13.028340	5.281972	5.000000	4.529809
	5.958918	5.206756	4.294714	4.764905	
18	13.028340	12.220812	5.192473	6.000000	5.176603
	5.206756	5.190886	4.764905	5.588302	
19	12.220812	14.191269	4.970456	3.000000	4.725534
	5.190886	4.945964	5.588302	3.862767	
20	14.191269	15.487913	4.296644	3.000000	3.575178
	4.945964	4.224498	3.862767	3.287589	
21	15.487913	14.538993	4.051080	5.000000	3.858393
	4.224498	4.031811	3.287589	4.429196	
22	14.538993	15.533376	3.994384	3.000000	3.952798
	4.031811	3.990225	4.429196	3.476399	
23	15.533376	11.994463	4.461086	8.000000	4.984266
	3.990225	4.513404	3.476399	6.492133	
24	11.994463	8.841138	5.846676	9.000000	7.328089
	4.513404	5.994817	6.492133	8.164044	
25	8.841138	6.995477	7.154339	9.000000	8.442696
	5.994817	7.283175	8.164044	8.721348	
26	6.995477	11.056521	7.061044	3.000000	6.814232
	7.283175	7.036363	8.721348	4.907116	
27	11.056521	10.414749	6.358228	7.000000	5.604744
	7.036363	6.282879	4.907116	6.302372	
28	10.414749	8.132803	6.718054	9.000000	7.201581
	6.282879	6.766407	6.302372	8.100791	
29	8.132803	6.673266	7.540464	9.000000	8.400527
	6.766407	7.626470	8.100791	8.700264	
30	6.673266	7.539913	7.866646	7.000000	8.133509
	7.626470	7.893333	8.700264	7.566755	

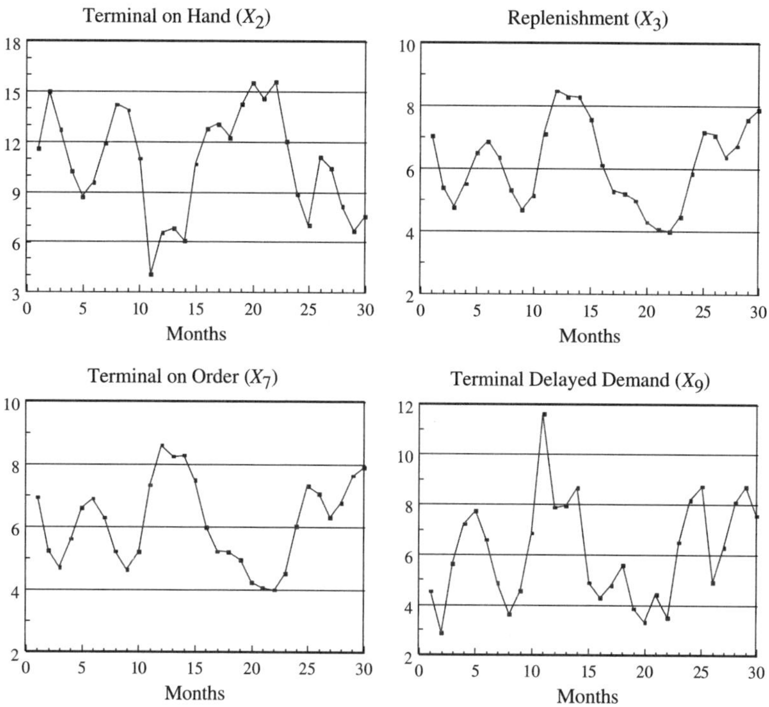

Figure 6.16 Continuation: Secondary Times Series for Random Demand Model

Table 6.3 Performance Test: Order Policy Model

Key: X_2 Terminal on Hand
X_3 Replenishment
X_5 Order Quantity
X_7 Terminal on Order
X_9 Terminal Delayed Demand

SOLUTION FOR MONTH 1

VARIABLE	VALUE	CHANGE	SUM*
X_2	8.694299	−2.863185	−2.863185
X_3	4.154386	−2.863185	−2.863185
X_5	0.000000	−6.044503	−6.044503
X_7	3.738947	−3.181318	−3.181318
X_9	10.566755	6.044503	6.044503

SOLUTION FOR MONTH 11

VARIABLE	VALUE	CHANGE	SUM*
$X2$	−0.532059	−4.603976	−39.320000
$X3$	4.405895	−2.693467	−4.603976
$X5$	0.000000	−9.231097	−7.951207
$X7$	3.965305	−3.347230	−4.938700
$X9$	19.566755	7.951207	44.258700

SOLUTION FOR MONTH 20

VARIABLE	VALUE	CHANGE	SUM*
$X2$	10.541360	−4.946554	−66.015700
$X3$	2.102095	−2.194549	−4.946552
$X5$	0.000000	−3.575178	−7.279164
$X7$	1.891886	−2.332613	−5.179816
$X9$	10.566755	7.279166	71.195510

SOLUTION FOR MONTH 30

VARIABLE	VALUE	CHANGE	SUM*
$X2$	3.937573	−3.602341	−109.560600
$X3$	8.995192	1.128546	−3.602341
$X5$	14.000000	5.866491	−1.999999
$X7$	9.495673	1.602340	−3.442108
$X9$	9.566755	2.000000	113.002600

* SUM = Cumulative algebraic value of change

Table 6.4 Performance Test: Three-Policy Model with Backlogging

Key: X_2 Terminal on Hand
X_3 Replenishment
X_5 Order Quantity
X_7 Terminal on Order
X_9 Terminal Delayed Demand

SOLUTION FOR MONTH 1

VARIABLE	VALUE	CHANGE	SUM*
X_2	4.539913	−7.017571	−7.017571
X_3	0.000000	−7.017571	−7.017571
X_5	0.000000	−6.044503	−6.044503
X_7	7.893333	−0.973068	−0.973068
X_9	11.566755	6.044503	6.044503

SOLUTION FOR MONTH 11

VARIABLE	VALUE	CHANGE	SUM*
X_2	−10.000000	−8.532004	−116.611400
X_3	8.000000	0.900638	−8.532004
X_5	0.000000	−9.231097	−7.951207
X_7	7.893333	0.580798	72.352740
X_9	19.566755	7.951207	44.258700

SOLUTION FOR MONTH 20

VARIABLE	VALUE	CHANGE	SUM*
X_2	4.539913	−10.948000	−240.481000
X_3	0.000000	−4.296644	−10.948000
X_5	0.000000	−3.575178	−7.279164
X_7	7.893333	3.668835	169.285600
X_9	12.566755	7.279166	71.195510

SOLUTION FOR MONTH 30

VARIABLE	VALUE	CHANGE	SUM*
X_2	−8.460087	−16.000000	−369.802600
X_3	0.000000	−7.866646	−16.000000
X_5	14.000000	5.866491	−1.999999
X_7	21.893333	14.000000	256.800000
X_9	9.566755	2.000000	113.002600

*SUM = Cumulative algebraic value of change

Table 6.5 Performance Test: Three-Policy Model without Backlogging

Key: X_2 Terminal on Hand
X_3 Replenishment
x_4 Demand
X_5 Order Quantity
X_7 Terminal on Order
X_9 Terminal Delayed Demand

SOLUTION FOR MONTH 1

VARIABLE	VALUE	CHANGE	SUM*
X_2	4.539913	−7.017571	−7.017571
X_3	0.000000	−7.017571	−7.071571
X_4	3.000000	0.000000	0.000000
X_5	0.000000	−6.044503	−6.044503
X_7	7.893333	0.973068	0.973068
X_9	11.566755	6.044503	6.044503

SOLUTION FOR MONTH 11

VARIABLE	VALUE	CHANGE	SUM*
X_2	0.000000	−4.071917	−98.470650
X_3	8.000000	0.900638	−24.532000
X_4	8.000000	−6.000000	−20.460090
X_5	0.000000	−9.231097	−21.951210
X_7	9.893333	2.580798	60.352740
X_9	13.106668	1.491120	38.117920

SOLUTION FOR MONTH 20

VARIABLE	VALUE	CHANGE	SUM*
X_2	10.000000	−5.487914	−159.199500
X_3	8.000000	3.703356	−34.948000
X_4	3.000000	0.000000	−29.460090
X_5	0.000000	−3.575178	−35.279160
X_7	3.893333	−0.331165	109.285500
X_9	9.106668	5.819079	49.913950

SOLUTION FOR MONTH 20

VARIABLE	VALUE	CHANGE	SUM*
X_2	1.000000	−6.539914	−248.920200
X_3	8.000000	0.133354	−48.000000
X_4	7.000000	0.000000	−41.460090
X_5	0.000000	−8.133509	−43.999990
X_7	11.893333	4.000000	176.800000
X_9	10.106668	2.539913	72.120210

*SUM = Cumulative algebraic value of change

CHAPTER 7

Modeling Corporate Assets

7.1 INTRODUCTION

For the most part, corporations owe their survival or failure to decision making, the primary role of corporate management. The "language" of management is *money*. Mtm systems analysis and modeling of a company's assets and cash flow yield useful information for corporate policy definition and decision making.

Utilization of assets is related to a myriad of company operating decisions that include budgeting, manpower loading, dividend posting, capital plowback, equipment acquisition, and levels of basic and applied research. Each operating decision impacts the health of a company. Decisions are guided by policy, and, unfortunately, selecting a superior policy from among alternatives is often difficult. Furthermore, the effect of decision making under inferior policies may result in restructuring or bankruptcy for the company. Systems analysis and modeling provide a holistic approach for improving asset utilization. Either implicitly or explicitly, critical decision parameters can be included in a systems model, and system performance can be analyzed and adjusted to improve asset utilization. Improved utilization leads to higher earnings and company growth.

Those outside of the corporate structure do not have access to corporate data such as liquid assets and operational (labor and material) costs since these data are highly confidential. Nonetheless, the examples of this chapter present two levels of synthesis of an adaptable, generic, corporate, systems model for assets planning without reference to any particular, corporate organization or structure. Although the context is drawn from manufacturing, the concepts are applicable to assets generated by other means. Initial emphasis focuses on an $L_{1,1}$ level asset flow model, followed by extension to an $L_{2,4}$ level corporate assets planning model.

7.2 ASSET GROWTH MODEL

A manufacturing company is analyzed as a systemic whole. Raw materials are processed into marketable products for sale to its customers. Analysis at the $L_{1,1}$ level is useful for stimulating conceptualization of the system. For a starting scenario, management wishes to transform a 10-year aggregate sales forecast into an asset growth plan for the 10 years. However, additional detail is needed to actually plan and develop business strategies. Knowledge items that support synthesis of an $L_{1,1}$ model follow.

7.2.1 Knowledge Base

- $L_{1,1}$ circumscription of a manufacturing company
- Fundamental units of analysis:

  Space: Territory containing company facilities (plant, employees, etc.), vendors, and customer market.

  Mass: Aggregate product expressed as *asset flow* (from raw materials to work-in-process to production output) in million-dollar units.

  Time: Δt = one year

- Number of subsystems: one
- Schematic: See Figure 7.1.
- System variables:

  X_1 = initial WIP (work in process) inventory level

  X_2 = terminal WIP inventory level

  X_3 = total raw material cost (production input)

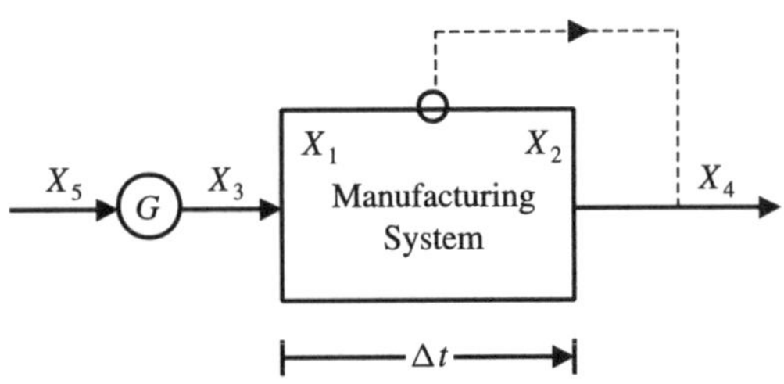

Figure 7.1 Manufacturing System Block Diagram

X_4 = delayed X_3 (production output)

X_5 = sales forecast

- Vector table for $L_{1,1}$ model:

X_j	*IT* Pair	Exo	Endo	Equation Type	Equation No.
X_1	*I*	*		—	—
X_2	*T*		*	balance	1
X_3			*	modification	2
X_4			*	regression	3
X_5		*		—	—

- Market condition: No close competition, the company is producing for a *seller's market* that readily absorbs increasing levels of production.

- Pricing policy—based on the concept of *value added*: The cost of raw materials, X_3, provides the basis for sales pricing. Their total value (including scrap) is to be fully recovered. Furthermore, survival requires a return on investment (ROI), as a percentage of X_3, for value added by those facility assets committed to production. Similarly, additional costs (value added by labor, inventories, and marketing) are to be recovered by markups:

  Raw materials: 100%

  ROI in facilities: 24%

  Labor: 40%

  WIP inventory: 5%

  Marketing: 5%

- Dynamic form for production delay for X_4: Under $L_{1,1}$ simplification, production output is subject to a Δt production delay. Each year's input constitutes a production lot that generates revenue from sales during the following year. Thus, delay is of the transportation type with $D = \alpha = 1.0$.

- Data: Table 7.1 contains primary data, a 10-year sales forecast for exogenous variable, X_5, which (because of the seller's market) also represents *gross revenue*. The manufacturing facilities have the capability to produce yearly forecasts by the end of each year.

- Specifications:

  $X_1^0 = 0$

  Planning horizon: 10 years

Table 7.1 Sales Forecast in Millions of Dollars

Year	Forecast
1	0.879300
2	1.419300
3	1.820000
4	2.220000
5	2.620000
6	3.020000
7	3.420000
8	3.820000
9	4.220000
10	4.620000

7.2.2 Calibration

7.2.2.1 Model Equations

1. $X_1 + X_3 = X_2 + X_4$
2. $X_3 = GX_5$
3. $X_4 = c_1X_1 + c_2X_2$

$$X_1 = \begin{cases} 0 & \text{initially,} \\ X_2 & \text{of previous year thereafter} \end{cases}$$

- Equation 2 relates the sales forecast, X_5, to raw materials, X_3: $X_5 = (1/G)X_3$. From the pricing policy, $1/G = 1.00 + 0.24 + 0.40 + 0.05 + 0.05 = 1.74$, representing a 174% markup above raw materials cost.

- Equation 3 introduces a delay for production output: $X_4 = c_1X_1 + c_2X_2 = 1/D[\alpha X_1 + (1-\alpha)X_2)]$, transportation type delay with $D = \alpha = 1.0$, leading to $c_1 = 1.0$ and $c_2 = 0.0$.

7.2.3 Simulation Tests

For validation, the model was subjected to three simulation tests.

7.2.3.1 Data Base Generation

Sales forecasts were processed through the model to simulate secondary data for the 10 years. If a client corporation is involved, corporate expertise is elicited,

Table 7.2 Data Base for Asset Growth Model

Year	X_1	X_2	X_3	X_4	X_5
1	0.000000	0.505345	0.505345	0.000000	**0.879300**
2	0.505345	0.815690	0.815690	**0.505345**	1.419300
3	0.815690	1.045977	1.045977	0.815690	1.820000
4	1.045977	1.275862	1.275862	1.045977	2.220000
5	1.275862	1.505747	1.505747	1.275862	2.620000
6	1.505747	1.735632	1.735632	1.505747	3.020000
7	1.735632	1.965517	1,965517	1.735632	3.420000
8	1.965517	2.195402	2.195402	1.965517	3.820000
9	2.195402	2.425287	2.425287	2.195402	4.220000
10	2.425287	2.655172	2.655172	2.425287	4.620000

and these secondary data are subjected to a TT test. Failure of the test requires knowledge base revision and restructuring of the model. Table 7.2 contains the complete five-variable data base for this application of the model.

Adjacent years must be used to compare production output with sales forecast. For example, applying the 174% markup to $X_4 = 0.505345$, production in year 2, produces a match of the sales forecast $X_5 = 0.879300$ for year 1:

$$X_5 = (1.74)(0.505345) = 0.879300.$$

Although current year X_5 values "drive" the model, values in the X_5 column also represent revenue realized the following year. Consequently, if this system were to meet the present year's forecast, production must begin one year in advance and then maintain that lead time throughout the 10 years.

Examination of asset growth is assisted by making some simple calculations. After recovery of all direct cost components, only the indirect component based on the 24% ROI in facilities generates asset growth. Asset growth will be computed for the first two years as a baseline for the $L_{2,4}$ model and its 24-month planning horizon. The sum of raw materials for years one and two are:

$$\sum_{t=1}^{2} X_3^t = 0.505345 + 0.815690 = 1.321035.$$

Multiply this sum by 0.24 to obtain asset growth over the two years:

$$\text{Growth} = (0.24)(1.321035) = 0.317048, \quad \text{or } \$317{,}048.$$

7.2.3.2 History Matching

The role of endogenous X_3 was interchanged with exogenous X_5 to check computed values of X_5 against primary data values for X_5. The model was again run, drawing values for X_3 from the data base. Cumulative algebraic values of change (relative to the data base) were computed for X_2, X_4, and X_5. Cumulative error (distortion) over the 10 years was minimal: −$2.00 for X_5, and $0.00 for both X_2 and X_4.

7.2.3.3 Perturbation of Raw Materials Cost

Total raw material cost, X_3, consists of "discounted" values of X_5 (see model Equation 2). Again, the roles of X_5 and X_3 were inverted but now with knowledge from the previous test that negligible distortion will be introduced by inversion. However, rather than inputting *standard* values for X_3 from the data base, an entirely new series was posited. Raw materials cost, X_3, was perturbed to represent changing production volumes, increasing from 100,000 units to 500,000 units over the 10-year planning horizon. Unit cost of material decreased with increasing volume, so that $X_3 = (production\ volume) \times (unit\ cost) \times 10^{-6}$ in millions of dollars. Table 7.3 lists new production volumes and unit costs used to calculate values for X_3.

The new X_3 series was processed by the model, and the new data base is displayed in Table 7.4. X_5 represents *gross revenue*, generated in a *seller's market*.

Unit price for the finished product is obtained by calculating $(X_5 \times 10^6)$ divided by (production volume of previous year) for each year. Table 7.5 shows the one-year offset and unit price in dollars.

7.2.3.4 Analysis of Results

Two applications were demonstrated for the asset growth model and its pricing policy in a seller's market: (1) keying asset growth to sales forecast data, by sim-

Table 7.3 X_3 Cost Data in Millions of Dollars

Year	No. Units	Unit Cost	X_3 Cost
1	100,000	8.00	0.800000
2	100,000	8.20	0.820000
3	200,000	6.00	1.200000
4	200,000	6.15	1.230000
5	300,000	5.00	1.500000
6	300,000	5.22	1.566000
7	400,000	4.00	1.600000
8	400,000	4.18	1.672000
9	500,000	3.00	1.500000
10	500,000	3.20	1.600000

Table 7.4 Data Base Resulting from X_3 Perturbation

Year	X_1	X_2	X_3	X_4	X_5
1	0.000000	0.800000	0.800000	0.000000	1.392000
2	0.800000	0.820000	0.820000	0.800000	1.426800
3	0.820000	1.200000	1.200000	0.820000	2.088000
4	1.200000	1.230000	1.230000	1.200000	2.140200
5	1.230000	1.500000	1.500000	1.230000	2.610000
6	1.500000	1.566000	1.566000	1.500000	2.724840
7	1.566000	1.600000	1.600000	1.566000	2.784000
8	1.600000	1.672000	1.672000	1.600000	2.909280
9	1.672000	1.500000	1.500000	1.672000	2.610000
10	1.500000	1.600000	1.600000	1.500000	2.784000

Table 7.5 Unit Price

Year	No. Units	Gross Revenue	Unit Price
1	100,000	0	0
2	100,000	1.392000	13.9200
3	200,000	1.426800	14.2680
4	200,000	2.088000	10.4400
5	300,000	2.140200	10.7010
6	300,000	2.610000	8.7000
7	400,000	2.724840	9.0828
8	400,000	2.784000	6.9600
9	500,000	2.909280	7.2732
10	500,000	2.610000	5.2200
11	0	2.784000	5.5680

ulating raw material costs over time, and (2) simulating gross revenue (and unit price) over time, based on projected unit costs of raw material. Thus, even this $L_{1,1}$ model has potential to produce valuable information for an actual company in that data required for these two applications are generally available. However, the $L_{1,1}$ model does not provide sufficient detail to support a business plan. Moreover, *horizontal expansion* of this model to internalize calculations for the above perturbation application requires little additional effort (see Exercise 5).

However, greater benefit can be achieved through *vertical expansion* of the $L_{1,1}$ model. The pricing policy will be replaced by additional structure, inventories will be modeled explicitly, and an asset pool will be added.

7.3 CORPORATE ASSETS PLANNING MODEL

Vertical expansion of the previous asset growth model raises several questions:

- Input to the manufacturing system was (value of) raw materials. By what structure is raw material passed to manufacturing?
- Output from the manufacturing system was delayed product. By what structure is product transferred to customers?
- If $L_{1,1}$ throughput represents asset flow, then how are assets related to system structures?

Answers to these questions are obtained by conducting a more detailed level of systems analysis, thereby expanding the knowledge base.

7.3.1 Expanded Knowledge Base

In general, companies have a well-defined material flow. Raw materials are purchased from vendors and stored as inventory. When the product is manufactured, raw materials are drawn from inventory and used for manufacturing. Manufacturing output goes to a finished product inventory. Then the marketing/sales department draws from this inventory for sales to individual customers. Revenue from sales of the product goes to pay for fixed and variable costs that are incurred during the manufacturing process. Thus, asset flow, if properly managed, will produce asset growth in terms of net revenue.

The asset growth model consisted of only a single manufacturing system and a pricing policy. However, a corporate assets planning model will require structural expansion in order to achieve finer detail. More structure can be provided by defining additional subsystems. However, merely adding subsystems in a horizontal expansion of the $L_{1,1}$ model does not accommodate finer detail. But rather, finer detail is enabled by vertical expansion in which one or more units of analysis (space, mass, time) is subdivided. To better utilize additional detail made possible by structural enlargement, the time unit is refined from years to months. The expanded knowledge base follows:

- $L_{2,4}$ circumscription of a manufacturing company.
- Fundamental units of analysis:

  Space: Territory company facilities (plant, employees, etc.), vendors, and customers.

  Mass: Aggregate product expressed as *asset flow* (from raw materials to work-in-process to finished products) in million-dollar units.

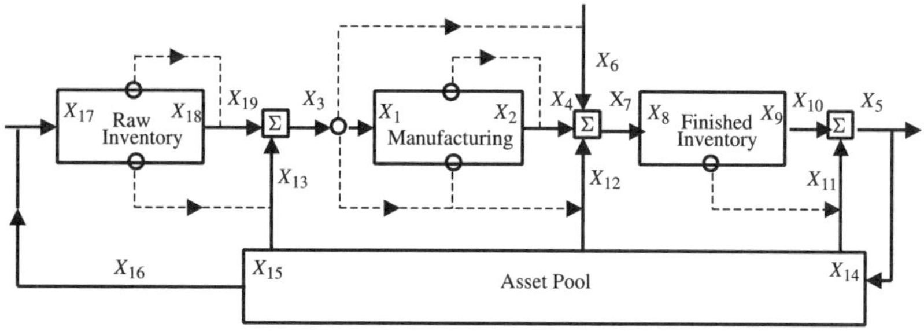

Figure 7.2 Corporate Assets System Block Diagram

Time: Δt = one month

- Schematic: See Figure 7.2.
- System variables: (commensurate unit = cost = one-million dollars)

X_1 = initial WIP (work-in-process) inventory level

X_2 = terminal WIP inventory level

X_3 = total raw material cost (production input)

X_4 = delayed X_3 (production output)

X_5 = sales forecast (revenue assets)

X_6 = ROI in facilities

X_7 = total finished product cost

X_8 = initial finished product inventory level

X_9 = terminal finished product inventory level

X_{10} = finished product output

X_{11} = finished product inventory storage cost

X_{12} = manufacturing cost (WIP storage, labor)

X_{13} = raw material storage cost

X_{14} = initial asset level

X_{15} = terminal asset level

X_{16} = raw material order cost

X_{17} = initial raw material inventory level

X_{18} = terminal raw material inventory level

X_{19} = delayed X_{16} (order cost)

- Four subsystems:

  SS_1: Raw inventory

  SS_2: Manufacturing work-in-process (WIP)

  SS_3: Finished inventory

  SS_4: Asset pool

- Vector table for $L_{2,4}$ model:

X_j	*IT* Pair	Exo	Endo	Equation Type	Equation No.
X_1	*I*	*		—	—
X_2	*T*		*	balance	1
X_3			*	equivalence	2
X_4			*	regression	3
X_5		*		—	—
X_6			*	regression	4
X_7			*	equivalence	5
X_8	*I*	*		—	—
X_9	*T*		*	balance	6
X_{10}			*	equivalence	7
X_{11}			*	regression	8
X_{12}			*	regression	9
X_{13}			*	regression	10
X_{14}	*I*	*		—	—
X_{15}	*T*		*	balance	11
X_{16}			*	regression	12
X_{17}	*I*	*		—	—
X_{18}	*T*		*	balance	13
X_{19}			*	regression	14

- Pricing policy—based on the concept of *value added*. The company's pricing policy can also be applied on a monthly basis. Pricing policy in the $L_{1,1}$ model was implemented through a single regression equation, $X_5 = 1.74X_3$, associated with the single module manufacturing system. For the $L_{2,4}$ model, this single equation will be divided into four regression equations, one associated with each of the four subsystems.

Total raw material costs (including scrap) are to be fully recovered, plus markups to cover value added and other costs:

Raw materials: 100%

ROI in facilities: 24%

Labor: 40%

Raw material storage: 2.5%

WIP inventory: 2.5%

Finished inventory storage and marketing: 5%

- Order policy: Order only enough raw material, based on the current month's sales forecast, to maintain positive inventory and WIP levels.
- Dynamic form for X_4, delayed output from manufacturing subsystem: For *constant* month-to-month, steady-state input of raw material to the manufacturing subsystem, input for the *current* month emerges as production output during the *following* month. Therefore, $D_1 = 1.0$. Furthermore, when raw material input changes from month to month, the subsystem responds with relative rapidity (low inertia) to these changes, modeled by $\alpha_1 = 0.9$.
- Dynamic form X_{19}, delayed output from raw inventory subsystem: Each raw material order is a regular monthly input to the raw material inventory subsystem and is fully output from the subsystem with a delay of one month. Thus, $D_2 = 1.0$. Zero inertia is exhibited by this nonfabricating subsystem, so that $\alpha_2 = 1.0$.
- Data: The sales forecast (primary data), X_5, covers the first 24-month *time fence* of the 10-year planning horizon. Forecasts were obtained by the decomposition method [43]: $Forecast = MT$, where linear trend T is given by Equation 7.1.

$$T = 0.050000 + 0.003750t. \tag{7.1}$$

Monthly indices M are 0.92, 1.00, 1.11, 1.22, 1.25, 1.12, 1.02, 0.95, 0.90, 0.87, 0.84, 0.80, for January through December, respectively. The resulting time series data are presented in Table 7.6. For example, the forecast for month 4 is obtained from $1.22T = 1.22[0.050000 + (0.003750)(4)] = 0.079300$.

- Initial assets total 1.0 million dollars, distributed as follows:

$X_1^0 = 0.040000$

Table 7.6 Sales Forecast in Millions of Dollars

Month	Forecast	Month	Forecast
1	0.049450	13	0.090850
2	0.057500	14	0.102500
3	0.067988	15	0.117937
4	0.079300	16	0.134200
5	0.085937	17	0.142188
6	0.081200	18	0.131600
7	0.077775	19	0.123675
8	0.076000	20	0.118750
9	0.075375	21	0.115875
10	0.076125	22	0.115275
11	0.076650	23	0.114450
12	0.076000	24	0.112000

$$X_8^0 = 0.060000$$

$$X_{14}^0 = 0.8500004$$

$$X_{17}^0 = 0.050000$$

7.3.2 Calibration

7.3.2.1 Model Equations

1. $X_1 + X_3 = X_2 + X_4$
2. $X_3 = X_{13} + X_{19}$
3. $X_4 = c_1 X_1 + c_2 X_2$
4. $X_6 = c_3 X_3$
5. $X_7 = X_4 + X_6 + X_{12}$
6. $X_7 + X_8 = X_9 + X_{10}$
7. $X_5 = X_{10} + X_{11}$
8. $X_{11} = c_4 X_8 + c_5 X_9$
9. $X_{12} = c_6 X_1 + c_7 X_2 + c_8 X_3$
10. $X_{13} = c_9 X_{17} + c_{10} X_{18}$
11. $X_5 + X_{14} = X_{11} + X_{12} + X_{13} + X_{15} + X_{16}$

12. $X_{16} = c_{11} X_5$

13. $X_{16} + X_{17} = X_{18} + X_{19}$

14. $X_{19} = c_{12} X_{17} + c_{13} X_{18}$

$$X_1 = \begin{cases} 0.040000 & \text{initially,} \\ X_2 & \text{of previous month thereafter} \end{cases}$$

$$X_8 = \begin{cases} 0.060000 & \text{initially,} \\ X_9 & \text{of previous month therafter} \end{cases}$$

$$X_{14} = \begin{cases} 0.850000 & \text{initially,} \\ X_{15} & \text{of previous month thereafter} \end{cases}$$

$$X_{17} = \begin{cases} 0.050000 & \text{initially,} \\ X_{18} & \text{of previous month thereafter} \end{cases}$$

- Equation 3 introduces continuous delay in X_3 for production output:

$$X_4 = c_1 X_1 + c_2 X_2 = \frac{1}{D_1}[\alpha_1 X_1 + (1 - \alpha_1) X_2].$$

  With $D_1 = 1.0$ and $\alpha_1 = 0.9$, it follows that $c_1 = 0.9$ and $c_2 = 0.1$.

- Equation 4, a modification of X_3, produces ROI in manufacturing facilities:

$$X_6 = G_1 X_3.$$

  From the company's pricing policy, $G_1 = 0.24 = c_3$.

- Equation 8, a scaled average of X_8 and X_9, gives finished product inventory storage cost:

$$X_{11} = G_2 \left(\frac{X_8 + X_9}{2} \right).$$

  From the pricing policy, $G_2 = 0.05$, yielding $c_4 = c_5 = 0.025$.

- Equation 9, a composite of two dynamic forms, provides manufacturing cost:

$$X_{12} = \underbrace{c_6 X_1 + c_7 X_2}_{T_1} + \underbrace{c_8 X_3}_{T_2}.$$

  $T_1 = G_3 (X_1 + X_2)/2$ is a scaled average of WIP. From the pricing policy, $G_3 = 0.025$, so that $c_6 = c_7 = 0.0125$.

  $T_2 = G_4 X_3$ is labor cost. From the pricing policy, the markup for labor is $G_4 = 0.40 = c_8$.

- Equation 10, a scaled average of X_{17} and X_{18}, defines raw material storage cost:

$$X_{13} = G_5 \left(\frac{X_{17} + X_{18}}{2} \right).$$

From the pricing policy, $G_5 = 0.025$, resulting in $c_9 = c_{10} = 0.0125$.

- Equation 12 implements the company's order policy. X_5 is modified to define raw inventory order cost:

$$X_{16} = G_6 X_5$$

is withdrawn from the asset pool. The rate of withdrawal for a given month is a fraction of the revenue assets generated by sales, X_5, for that month. The fraction, $G_6 = c_{11}$, is to be inferred from the first simulation test.

- Equation 14 produces a transportation delay in X_{16} for raw material (order cost):

$$X_{19} = c_{12} X_{17} + c_{13} X_{18} = \frac{1}{D_2} [\alpha_2 X_{17} + (1 - \alpha_2) X_{18}].$$

Given that $D_2 = \alpha_2 = 1.0$. Thus, $c_{12} = 1.0$ and $c_{13} = 0.0$.

7.3.3 Simulation Tests

Each model (at whatever level of articulation) must be verified by the analyst and his or her peers, and validated to the satisfaction of the domain expert(s) before the model is used to solve problems. The more scenarios for which a model is validated, the greater the confidence that can be placed in the model. The urgency of validation is related to the value of the problem being solved. For example, it is nonsense to spend $10,000 validating a model that will solve a problem worth $1,000. On the other hand, it may well be worth spending $100,000 to validate a model that solves a problem worth $1,000,000. The ability to specify which and how many validation tests comes only from modeling experience. Three illustrative simulation tests are presented for the $L_{2,4}$ model.

7.3.3.1 Data Base Generation

Initially, the value for raw material procurement coefficient, c_{11}, in Equation 12, $X_{16} = c_{11} X_5$, was set equal to 0.6 (approximately, $1/1.74$). A "better" value for c_{11} was imputed through iterative solutions of the model, guided by the following system-based heuristic:

> *All four trends established by IT solution pairs must be positive to accommodate growth but must be kept relatively small.*

Table 7.7 Trends and Betas for the Corporate Assets Planning Model

Subsystem	Percent Trend	Beta
Raw Inventory	1.17	0.99
Manufacturing	2.15	0.98
Finished Inventory	2.15	1.06
Asset Pool	1.06	0.14

Justification:

- Sales forecast data exhibit a positive trend.
- Inventory and WIP levels must be minimized and no monthly level can be negative.
- Backlogging must not occur.

The model was iteratively solved for secondary data using this criterion, yielding $c_{11} = 0.585$ and a *standard* data base for all 19 variables. Data base values for selected months are listed in Table 7.14 of Appendix E. Selective plots of secondary data are included in Figure 7.3 of Appendix E, where small, positive trends are apparent. See Exercise 8 for the effect of trend on β. IT trends and empirical β's (turnover rates) for each subsystem are presented in Table 7.7; they were computed by Equations 8.1 and 6.5, respectively. For example, for the asset pool

$$\beta = \frac{\overline{X}_{11} + \overline{X}_{12} + \overline{X}_{13} + \overline{X}_{16}}{\overline{X}_{15}} = 0.143890.$$

Observe the following:

- Small positive trends were realized.
- Turnover rates for raw inventory, manufacturing, and finished inventory subsystems approximate to $\beta = 1.0$, as for the ideal inventory system.
- Delay for X_{10}, finished product output, is inferred from $D = 1/\beta$

$$\frac{1}{1.06} = 0.943 month.$$

- The economists' *velocity of money* for the asset pool is 1/0.14, or approximately one turnover of assets each seven months. Thus, velocity should be increased by moving some of the assets to an external investment portfolio (see Exercise 10).

Table 7.8 Asset Redistribution After 24 Months

Subsystem	Initial Percent	Final Percent	Asset Level
Manufacturing	4	5.2	0.068660
Finished Inventory	6	7.8	0.103572
Asset Pool	85	82.1	1.089268
Raw Inventory	5	4.9	0.065520
Total	100	100.0	1.327020

Total asset growth for the two years is obtained from the sum of the terminal asset levels, $X_2 + X_9 + X_{15} + X_{18} = 1.327020$. After 24 months, the assets were distributed as shown in Table 7.8. Subtracting the 1.0 million of initial assets:

$$\text{Growth} = 1.327020 - 1.0 = 0.327020, \text{ or } \$327,020, \text{ or } 32.70\%.$$

Asset growth is directly proportionate to X_3. Compared to \$317,048 of the $L_{1,1}$ model, growth is 3.1% higher, due to the difference in the way production input, X_3, was calculated: Markup for raw inventory storage cost was included in the $L_{2,4}$ model but was absent from the $L_{1,1}$ model because the raw inventory subsystem was nonexistent.

7.3.3.2 Perturbation of Forecast

For the first scenario of Phase Two, the sales forecast was perturbed by doubling the slope of the linear trend. New forecast data, displayed in Table 7.9, were obtained by changing the slope from 0.00375 to 0.00750 in Equation 7.1. Seasonal indices remained the same. These data replaced the X_5 data in the *standard* data base of Table 7.14 in Appendix E. Again, the $L_{2,4}$ model was iteratively solved using the previous criterion for selecting a "best" value for c_{11}. A new value for c_{11} was found to be 0.620.

Table 7.10 presents the solution for the final period (month 24), where "Sum" is *cumulative algebraic value of change*, accumulated over the 24 periods. Average monthly change due to perturbation in X_5 is readily obtained by dividing "Sum" for each solution variable by 24 months. For example, terminal finished inventory, X_9, averaged $-0.143409/24 = -0.005975$, or \$5,975, lower.

The asset trend, level, and distribution for each subsystem for the 24 months are shown in Table 7.11.

$$\text{Growth} = 1.502569 - 1.0 = 0.502569, \text{ or } \$502,569, \text{ or } 50.26\%.$$

Compared to the first trend slope, the new trend slope produces 53.68% greater growth in total assets. If the two trends represent upper and lower bounds, asset growth of between 32.70% and 50.26% can be projected for the immediate time fence of 24 months.

Table 7.9 New Forecast Data in Millions of Dollars

Month	Forecast	Month	Forecast
1	0.052900	13	0.135700
2	0.065000	14	0.155000
3	0.080475	15	0.180375
4	0.097600	16	0.207400
5	0.109375	17	0.221875
6	0.106400	18	0.207200
7	0.104550	19	0.196350
8	0.104500	20	0.190000
9	0.105750	21	0.186750
10	0.108750	22	0.187050
11	0.111300	23	0.186900
12	0.112000	24	0.184000

Table 7.10 Perturbation Test: Month 24

Variable	Value	Change	Sum
$X2$	0.118763	0.050103	0.726444
$X3$	0.118752	0.050143	0.731456
$X4$	0.118862	0.049747	0.681349
$X6$	0.028501	0.012035	0.175548
$X7$	0.197834	0.083087	1.167019
$X9$	0.163326	0.059754	−0.143409
$X10$	0.176370	0.069355	1.107262
$X11$	0.007630	0.002645	−0.008662
$X12$	0.050471	0.021305	0.310117
$X13$	0.002874	0.001218	0.018432
$X15$	1.106400	0.017132	0.186470
$X16$	0.114080	0.048560	0.761586
$X18$	0.114080	0.048560	0.761586
$X19$	0.115878	0.048925	0.713026

7.3.3.3 *MRP Offset for Lead Time*

In the previous two tests, the current month's sales forecast was used in Equation 12 of the model to determine the order rate for the current month. However, this practice ignores the average total delay (lead time) in transforming raw material into finished product. To better utilize sales forecast data, this next scenario employs an MRP (material requirements planning) feature [43]: that of offsetting the forecast by an assumed lead time of two months. (See Exercise 8 for computation

Table 7.11 New Asset Distribution after 24 Months

Subsystem	Percent Trend	Percent Total	Asset Level
Manufacturing	3.93	7.9	0.118763
Finished Inventory	5.65	10.9	0.163326
Asset Pool	1.13	73.6	1.106400
Raw Inventory	3.14	7.6	0.114080
Total		100.0	1.502569

Table 7.12 Sales Forecast and Two-Month Offset

Month	X_5	X_{20}
1	0.000000	0.049450
2	0.000000	0.057500
3	0.049450	0.067988
4	0.057500	0.079300
⋮		
22	0.118750	0.115275
23	0.115875	0.114450
24	0.115275	0.112000
25	0.114450	0.000000
26	0.112000	0.000000

of average total delay.) Equation 12 (order policy) was replaced by $X_{16} = c_{11}X_{20}$, where X_{20} is a two-month offset of the X_5 series, an additional exogenous variable extending the number of months to 26 and the number of variables to 20. The exogenous data series is shown in Table 7.12.

For comparison to the standard data base, c_{11} was reset to 0.585. No assets were distributed as initial inventories, so that $X_1^0 = X_8^0 = X_{17}^0 = 0$. A new data base reflecting the *MRP offset* was generated. Data for selected months are presented in Table 7.15 of Appendix E. Because demand $X_5 = 0$ zero during the first two-month start-up period, no revenue was generated. However, transient response to X_{16} (see Equation 12) produced a small trickle of finished product inventory (X_8, X_9), incurring small storage costs by Equation 8: $X_{11} =$ \$7 and \$587. As a result, transient, negative values were obtained for X_{10} for months one and two, as readily seen by examining Equation 7: $X_{10} = X_5 - X_{11}$. Moreover, these values, $X_{10} = -\$7$ and -\$587, added \$594 to finished product inventory by Equation 6. Thus, by X_{11} and Equation 11, \$594 was subtracted from the asset pool but was subsequently returned to the pool as revenue during month three via Equations 6, 7, and 11.

Table 7.13 Asset Distribution After 26 Months

Subsystem	Initial Percent	Final Percent	Asset Level
Manufacturing	0	0.45	0.006050
Finished Inventory	0	5.46	0.072603
Asset Pool	100	94.09	1.252139
Raw Inventory	0	0.00	0.000000
Total	100	100.0	1.3330792

Table 7.13 shows the distribution of assets after 26 months. Subtracting the 1.0 million of initial assets:

$$\text{Growth} = 1.330792 - 1.0 = 0.3330792, \text{ or } \$330,792, \text{ or } 33.08\%.$$

Compared to $327,020 of the standard data base of Table 7.14, growth is 1.15% higher even though c_{11} was the same. Thus, an additional $3,772 of assets were generated by incorporating an MRP offset.

7.3.4 Conclusions

The corporate asset model illustrated in this chapter, although generic, has a wide range of applicability to small and medium manufacturers. Large manufacturers, in most cases, need to extend the systems analysis to permit explicit modeling of major product groups. In addition to product group subsystems, most likely to be included are subsystems for warehouse and distribution centers, transportation, labor, management, stock, and utilities. Expanded circumscription can allow elaborate modeling of the company and the potential for testing many planning scenarios. Once the model is developed, calibrated, and validated, various optimization studies can be performed. Benefiting from the fact that all of the model's equations are linear, linear or quadratic programming can be used for optimization.

Synthesis of a corporate assets planning model through two levels has shown the ease by which the macro-to-micro approach can be applied in analyzing and solving a common industrial problem. The assets distribution problem was readily described by linear rate and level equations. The most demanding (but not difficult) task for the model and analyst was the experimental determination of "best" value for raw material procurement coefficient, c_{11}.

The use of a two-month offset in the sales forecast to drive the model illustrated a noteworthy principle:

> *Obtaining additional knowledge is a worthwhile venture if it produces a net positive return.*

Knowledge items used to calibrate these equations are readily available in any company from the accounting department, the business plan, and from management. Although only two policies (pricing and ordering) were included in the model, a myriad of policies can be simulated by this procedure. Verification and validation of the model is also easily performed (as compared to other types of models). Furthermore, this model is easily communicated and understood by management, due to the simplicity of its linear equations.

EXERCISES

1. Using library resources (e.g., *Harvard Business Review*), write a one- or two-page discussion of corporate assets to include
    a. types of assets
    b. debt-to-asset ratios.
2. Consult several operations management texts such as [36] and [43] to discover various ways to define *aggregate* product.
3. Identify two or three methods by which sales may be forecasted and the features that distinguish each method.
4. The manager for production planning is asking that the asset growth model ($L_{1,1}$) be modified to accommodate one production lot every six months for the 10-year planning horizon. Which of the two following modification schemes would you recommend, and why?
    a. Change the dynamic form for production delay from $D = \alpha = 1.0$ to $D = 0.5$ and $\alpha = 0.9$.
    b. Vertical expansion to $L_{2,1}$ level with $\Delta t = 0.5$ year, $D = 1$, and $\alpha = 0.9$.
5. For the perturbation application, expand the asset growth model ($L_{1,1}$) horizontally to internalize costs and revenues, and hence profit.
    a. Generate an 11-year data base for the model.
    b. When does break even occur?
    c. How much profit accumulates from the asset growth plan?
6. Sketch a triangle to represent Mtm analysis as in Figure 1.20 of Chapter 1, and locate the $L_{1,1}$ and $L_{2,4}$ models. Explicitly represent the expansion from one subsystem to four subsystems.

7. From a dynamics viewpoint, sales forecast, X_5, drives both the $L_{1,1}$ and $L_{2,4}$ models but with comparable difference. Explain.

8. Explore the effect of trend on the empirical β's presented in Table 7.7, using the manufacturing subsystem as an example of a four-variable inventory module. If the magnitude of the trend is negligibly small, then β can be estimated by applying Equation 6.5 to steady-state averages:

$$\beta = \frac{\overline{X}_4}{\overline{X}_2}.$$

On the other hand, if an IT pair exhibits stationary trend, a better estimate of turnover rate is given by:

$$\beta = \frac{2\,\overline{X}_4}{(\overline{X}_1 + \overline{X}_2)}.$$

   a. Run the standard 19-variable, corporate assets planning model ($L_{2,4}$ with $c_{11} = 0.585$) to obtain a complete data base and average values. Verify that your data base matches the partial listing in Table 7.14.
   b. Based on D for the manufacturing subsystem, what is the expected value of β? Calculate an estimate for β if trend is neglected. Repeat if trend is not neglected.
   c. Let $Tr \times 100\% = \big((\overline{X}_2/\overline{X}_1) - 1\big) \times 100\%$ define the percent trend for the IT pair of the example inventory module. Show by derivation that $2(1 + Tr)/(2 + Tr)$ is a trend correction factor in $\beta = 2(1 + Tr)/(2 + Tr)\overline{X}_4/\overline{X}_2$.
   d. Suggest a heuristic that states when to apply the correction factor.
   e. Use the averages in part (a) to estimate the combined production flow time (delay) for the three subsystems that involve materials handling.

9. Employ the standard 19-variable, corporate assets planning model ($L_{2,4}$ with $c_{11} = 0.585$) to recommend a "safe" minimum value for X_{14} (modifying only X_{14}^0) so that the asset pool subsystem can still function to provide *operating capital.*

10. Modify use of the corporate assets planning model ($L_{2,4}$) in the following manner: (1) Remove excess initial assets from SS_4 (see Exercise 7.9). (2) Relocate these excess assets to an independent system representing an external investment portfolio that yields a guaranteed return of 12% per annum, compounded monthly. (3) *Hint*: Use a linear accumulator. Operate the two models and determine the following:

    a. improved asset growth that will result if the initial \$1,000,000 is split into internal and external investments
    b. increased velocity of money for the asset pool subsystem.

APPENDIX E

Table 7.14 Data Base for Corporate Assets Planning Model

Month	X_1	X_2	X_3	X_4	X_5
	X_6	X_7	X_8	X_9	X_{10}
	X_{11}	X_{12}	X_{13}	X_{14}	X_{15}
	X_{16}	X_{17}	X_{18}	X_{19}	
1	0.040000	0.049988	0.050987	0.040999	0.049450
	0.012237	0.074755	0.060000	0.089031	0.045724
	0.003726	0.021519	0.000987	0.850000	0.844290
	0.028928	0.050000	0.028928	0.050000	
2	0.049988	0.031544	0.029710	0.048144	0.057500
	0.007130	0.068178	0.089031	0.104549	0.052661
	0.004839	0.012903	0.000782	0.844290	0.849627
	0.033637	0.028928	0.033637	0.028928	
3	0.031554	0.024282	0.034555	0.031827	0.067988
	0.008293	0.054765	0.104549	0.096348	0.062966
	0.005022	0.014645	0.000918	0.849627	0.857257
	0.039773	0.033637	0.039773	0.033637	
12	0.045614	0.045925	0.045957	0.045645	0.076000
	0.011030	0.076201	0.111966	0.117914	0.070253
	0.005747	0.019527	0.001116	0.914437	0.919587
	0.044460	0.044840	0.044460	0.044840	
13	0.045925	0.045702	0.045680	0.045903	0.090850
	0.010963	0.076284	0.117914	0.109021	0.085177
	0.005673	0.019417	0.001220	0.919587	0.930979
	0.053147	0.044460	0.053147	0.044460	
22	0.071495	0.069661	0.069477	0.071312	0.115275
	0.016675	0.117541	0.083457	0.090062	0.110937
	0.004338	0.029555	0.001690	1.054552	1.066808
	0.067436	0.067787	0.067436	0.067787	
23	0.069661	0.069165	0.069116	0.069611	0.114450
	0.016588	0.115589	0.090062	0.095840	0.109802
	0.004648	0.029382	0.001680	1.066808	1.078596
	0.066953	0.067436	0.066953	0.067436	
24	0.069165	0.068660	0.068609	0.069115	0.112000
	0.016466	0.114747	0.095840	0.103572	0.107015
	0.004985	0.029166	0.001656	1.078596	1.089268
	0.065520	0.066953	0.065520	0.066953	

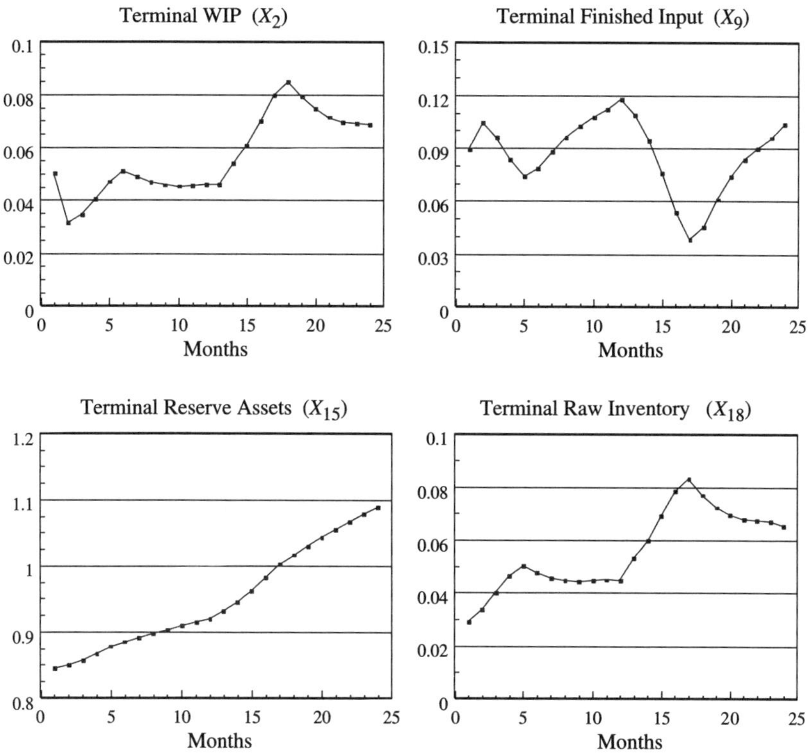

Figure 7.3 Selected Plots of Secondary Data (Millions of Dollars)

Table 7.15 MRP Offset Data Base for Corporate Assets Planning Model

Month	X_1	X_2	X_3	X_4	X_5
	X_6	X_7	X_8	X_9	X_{10}
	X_{11}	X_{12}	X_{13}	X_{14}	X_{15}
	X_{16}	X_{17}	X_{18}	X_{19}	X_{20}
1	0.000000	0.000329	0.000362	0.000033	0.000000
	0.000087	0.000268	0.000000	0.000275	−0.000007
	0.000007	0.000149	0.000362	1.000000	0.970555
	0.028928	0.000000	0.028928	0.000000	0.049450
2	0.000329	0.027039	0.029710	0.003000	0.000000
	0.007130	0.022356	0.000275	0.023219	−0.000587
	0.000587	0.012226	0.000782	0.970555	0.923321
	0.033637	0.028928	0.033637	0.028928	0.057500
3	0.027039	0.033872	0.034555	0.027723	0.049450
	0.008293	0.050599	0.023219	0.025589	0.048230
	0.001220	0.014583	0.000918	0.923321	0.916277
	0.039773	0.033637	0.039773	0.033637	0.067988
⋮					
13	0.045925	0.045702	0.045680	0.045903	0.076650
	0.010963	0.076284	0.049676	0.051848	0.074112
	0.002538	0.019417	0.001220	0.975675	0.976002
	0.053147	0.044460	0.053147	0.044460	0.090850
14	0.045702	0.053756	0.054561	0.046508	0.076000
	0.013095	0.082670	0.051848	0.061348	0.073170
	0.002830	0.023068	0.001414	0.976002	0.964728
	0.059962	0.053147	0.059962	0.053147	0.102500
⋮					
24	0.069165	0.068660	0.068609	0.069115	0.115275
	0.016466	0.114747	0.109384	0.114452	0.109679
	0.005596	0.029166	0.001656	1.052902	1.066238
	0.065520	0.066953	0.065520	0.066953	0.112000
25	0.068660	0.066550	0.066339	0.068449	0.114450
	0.015921	0.112596	0.114452	0.118420	0.108628
	0.005822	0.028226	0.000819	1.066238	1.145822
	0.000000	0.065520	0.000000	0.065520	0.000000
26	0.066550	0.006050	0.000000	0.060500	0.112000
	0.000000	0.061407	0.118420	0.072603	0.107224
	0.004776	0.000907	0.000000	1.145822	1.252139
	0.000000	0.000000	0.000000	0.000000	0.000000

Part III

Stochastic Models

CHAPTER 8

Stochastic Analysis

8.1 INTRODUCTION

Systems exhibit variation over time and are characterized by *states*: "level" variables that are subject to change during each time increment. System variables (levels and their associated rates) are quantified at uniformly spaced increments of time. Variables are classified as *discrete* or *continuous* by *count* or by *measurement*, respectively. Natural systems exhibit *stochastic* properties, uncertainties created by factors that normally cannot be ignored. Model structures that accommodate stochastic properties exhibit a spectrum running from completely deterministic to completely nondeterministic. On the one hand, stochastic data can be input to a deterministic model that produces *identical* output for repeated runs using the same input data. On the other hand, stochastic properties can be built into a model that produces *typical* output for repeated runs using the same input cues.

Data that characterize systems are drawn from infinite sequences of state values. Each sequence is commonly referred to as a *time series*. If the sequence exhibits randomness, its state values are generated by a stochastic variable. Thus, the sequence constitutes a *stochastic time series* if its variation is characterized by a probability distribution function (pdf). Without randomness, the sequence is known as a *deterministic time series*. Each pdf allocates probability to events (outcomes or states within a *population*) as a dimensionless *weight*, measured on the interval (0,1). A pdf is characterized by *moments of inertia*, first-order, second-order, and higher.

Moments of a pdf are defined by mathematical *expectation*, symbolized by the expectation operator E:

$$E\{\varphi(x)\} = \begin{cases} \sum_{\text{all } x} \varphi(x) f(x) & \text{for } x \text{ discrete,} \\ \int_{\text{all } x} \varphi(x) f(x)\, dx & \text{for } x \text{ continuous,} \end{cases}$$

where $\varphi(x)$ is any function of x, and $f(x)$ is the pdf of x.

The *first* moment of inertia, μ, locates a point that represents the center of gravity, or "mass," of the pdf and is called the *mean* or *expected value* of x:

$$\mu = E(x).$$

Similarly, the *second* moment of inertia, σ^2, provides a measure of the spread of weight, or "mass," relative to the mean, and is called the *variance* of x:

$$\sigma^2 = E[(x - \mu)^2].$$

Each stochastic time series is classified as *stationary* or *nonstationary*. Stationarity requires that the first two moments of the pdf (μ and σ^2) be time invariant. A series is nonstationary if either moment varies with time. If for a given duration of time, a time series exhibits the same kind of stationarity or nonstationarity, it is further classified as *homogeneous*.

8.2 STOCHASTIC REPRESENTATION

In Mtm systems analysis, the term *population* may refer to a *single entity* or a *related, multiple-entity set*, used interchangeably with *state space*. Entities exhibit discrete or continuous variation and their respective realizations as observed outcomes are limited only by their frequency or density within the population. Values associated with an entity set (population or state space) may derive from a *stochastic process*. A stochastic process contains one or more stochastic variables. Let X_j^t be defined as a stochastic variable. In unbounded time, it will generate an infinite sequence, the time series $\{X_j^t\}$ with pdf, $f(x_j)$. Moreover, any finite sequence generated by the stochastic process is also a time series: $\ldots, X_j^1, X_j^2, \ldots, X_j^N, \ldots$. Thus, N successive observations are regarded as a *sample realization* from the *state space* of the infinite time series. Because any value from the state space of the process can appear as any one of the N observations, state space is spanned by an N-coordinate axis system, and the sample is said to exhibit N *degrees of freedom*.

8.2.1 Mean, Variance, and Standard Deviation

Moments of a pdf can be estimated from the sample realization. An estimator provides an *unbiased* point estimate of the moment if the expected value of the estimator is mathematically equal to the moment. Only the mean and variance are used to represent stochastic processes.

8.2.1.1 Sample Mean

The arithmetic average of the sample, $\overline{X}_j = \widehat{\mu}_j$, is called the *sample mean* and provides an estimate of the first moment of inertia:

$$\overline{X}_j = \frac{1}{N}\sum_{t=1}^{N} X_j^t.$$

Because $E(\overline{X}_j) = \mu_j$, $\overline{X}_j = \widehat{\mu}_j$ is an unbiased estimator of μ_j.

8.2.1.2 Sample Variance

The mean squared deviation of observations from the sample mean, $s_j^2 = \widehat{\sigma}_j^2$, is called the *sample variance* and provides an estimate of the second moment of inertia

$$s_j^2 = \frac{1}{N}\sum_{t=1}^{N}(X_j^t - \overline{X}_j)^2.$$

One of the N degrees of freedom is "used up" in that all N observations are required to calculate $\overline{X}_j$. If N is replaced by $N-1$, then s_j^2 becomes an unbiased estimator for σ_j^2: $E(s_j^2) = \sigma_j^2$.

8.2.1.3 Sample Standard Deviation

The root-mean square of the differences between observations and the sample mean, $s_j = \widehat{\sigma}_j$, is called the *sample standard deviation*:

$$s_j = \sqrt{\frac{1}{N}\sum_{t=1}^{N}(X_j^t - \overline{X}_j)^2}.$$

Two or more standard deviations from the mean are considered to be a significant departure. Even if N is replaced by $N-1$, s_j is a *biased* estimator for σ_j, since $E(s_j) \neq \sigma_j$. Nonetheless, its use as a point estimator is justified for large N.

8.2.2 Autocorrelation and Cross-Correlation

As a theoretical concept, correlation, or dependence, may exist *within* observations of a single time series or *between* two or more time series. *Autocorrelation*, ρ_k, provides a measure of correlation within a single time series. Similarly, *cross-correlation*, $\rho_{i,j}$, expresses a measure of correlation between two time series. Both autocorrelation and cross-correlation are dimensionless, ranging in value from -1.0 to $+1.0$. Values are obtained from a ratio of variances, measured in the same units, and hence are dimensionless. Because *primary* time series exist as *sample* realizations, only estimates can be obtained for autocorrelation and cross-correlation of the infinite series. Any standard text on time series can be consulted for greater detail, as, for example, [40].

8.2.2.1 Sample Autocorrelation

Autocorrelation between pairs, X_j^t and X_j^{t-k}, separated by k Δt's is defined by the ratio of *autocovariance* to variance, $\rho_k = \gamma_k/\sigma^2$. If, for a given population, individual values are dependent on past levels (states) of the process, then autocorrelation is present, and the probability distribution is *conditional*, indicated by $P\{X_j^t/X_j^1, X_j^2, \ldots, X_j^{t-1}\}$.

On the other hand, if individual values of X_j^t are not dependent on past levels, then *autocorrelation* is not present. The probability distribution of the process is *absolute* and is denoted by $P\{X_j^t\}$. Some "populations" must be subdivided to obtain homogeneous groupings. Such populations can be *stratified* into subpopulations to which absolute probabilities then apply.

Sample autocovariance is given by

$$\widehat{\gamma}_j^k = \frac{1}{N} \sum_{t=k+1}^{N} [(X_j^t - \overline{X}_j)(X_j^{t-k} - \overline{X}_j)].$$

Noting that degrees of freedom cancel, sample autocorrelation is provided by the ratio

$$\widehat{\rho}_j^k = \frac{\sum_{t=k+1}^{N}[(X_j^t - \overline{X}_j)(X_j^{t-k} - \overline{X}_j)]}{\sum_{t=1}^{N}(X_j - \overline{X}_j)^2}.$$

If ρ_k is not significantly different from zero, then autocorrelation does not exist, and the probability distribution of the stochastic process is absolute. Process probability can be represented by an empirical or analytical distribution such as the Poisson, exponential, or normal distributions. To determine if ρ_k is significantly different from zero, its point estimator, $\widehat{\rho}_j^k$, is plotted versus k. If $\widehat{\rho}_j^k$ falls outside of plus or minus two standard deviations from zero, autocorrelation is found to exist, and $\widehat{\rho}_j^k$ is its estimate.

8.2.2.2 Sample Cross-Correlation

Cross-correlation between $\{X_i\}$ and $\{X_j\}$ is defined by the ratio of covariance to root-mean variance,

$$\rho_{i,j} = \frac{\gamma_{i,j}}{\sqrt{\sigma_i^2 \sigma_j^2}}.$$

Sample covariance is found from

$$\widehat{\gamma}_{i,j} = \frac{1}{N} \sum_{t=1}^{N}[(X_i^t - \overline{X}_i)(X_j^t - \overline{X}_j)].$$

Similarly, sample cross-correlation is defined by the ratio

$$\widehat{\rho}_{i,j} = \frac{\sum_{t=1}^{N}[(X_i^t - \overline{X}_i)(X_j^t - \overline{X}_j)]}{\sqrt{\sum_{t=1}^{N}(X_i - \overline{X}_i)^2 \sum_{t=1}^{N}(X_j - \overline{X}_j)^2}}.$$

8.2.3 Markov Process

Given a stationary stochastic process and its present and past level, if the conditional probability of any future level of the process depends only on the present level (i.e., $k = 1$), the process is classified as a *first-order Markov process* [24]. The conditional probabilities form a probability transition matrix that gives the probability of achieving the next level of the process, given the current level.

8.2.4 Monte Carlo Simulated Sampling

Methods for introducing stochastic (or random) variation into a model that involve an element of chance are referred to as *Monte Carlo simulated sampling* techniques [27]. Chance is introduced via the principle of *simple random sampling*:

> *A sequence of observations, $X_j^1, X_j^2, \ldots, X_j^N$, constitutes a simple random sample of size N if each observation has an equal chance of being selected, and these N random variables are independent.*

Crude Monte Carlo simulated sampling involves drawing observations from the *cumulative probability distribution*, $C(X_j) = P(x \leq X_j)$:

$$C(X_j) = \begin{cases} \sum_{x=-\infty}^{X_j} P(x) & \text{for } x \text{ discrete} \\ \int_{-\infty}^{X_j} f(x)\,dx & \text{for } x \text{ continuous,} \end{cases}$$

Stratified Monte Carlo simulated sampling consists of drawing observations from the $C(X_j)$ of each stratum represented by an independent subpopulation, or from the $C(X_j)$ of each row of a Markov transition matrix.

Simple random sampling is implemented by using the *uniform distribution* of random decimals in the selection of N observations from the (sub) population represented by its *cumulative probability distribution*, $C(X_j)$. Simulated sampling is schematically represented in Figure 8.1. The rectangle at the left represents the uniform distribution, rotated to conform its decimal fraction range (0,1) to cumulative probability of the same range. $C(X_j)$ is *monotonic increasing* and is typical of the cumulative probability distribution of a continuous random variable. For a discrete random variable, $C(X_j)$ exhibits step increases.

8.3 MODELING STOCHASTIC PROCESSES

Many systems are impacted by stochastic components. Systems modelers have sought innovative ways to introduce stochastic time series into their models. Monte Carlo simulated sampling techniques are easy to implement and are widely accepted. Models are capable of realistic response whether empirical and analytical probability distributions are used to simulate primary time series.

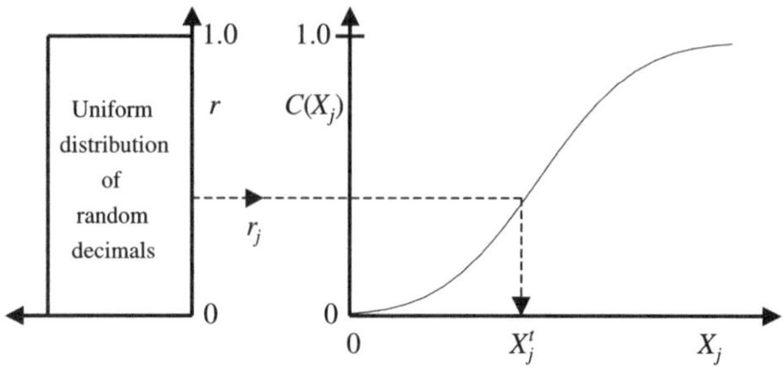

Figure 8.1 Monte Carlo Simulated Sampling

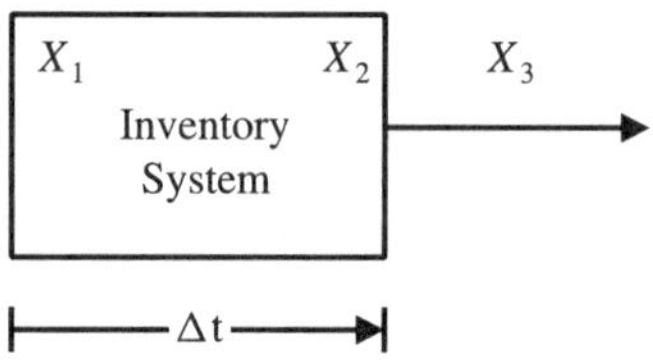

Figure 8.2 Simple Inventory System

Two distinct approaches exist for modeling stochastic processes and will be illustrated for a simple inventory reduction system. Simplest possible structure was selected for the system so that distinguishing features can only be attributed to one approach or the other. The system, which experiences independent, stochastic demand for X_3 items during each inventory period, $\Delta t = 1$, is presented in Figure 8.2, where

X_1 = initial inventory level,

X_2 = terminal inventory level, and

X_3 = inventory depletion rate (stochastic demand).

Let $\mathbf{X}$ = vector of system variables, and partition into vectors $\mathbf{Y}$ and $\mathbf{Z}$, where

$$\mathbf{Y} = (X_1, X_2)^T = \textit{endo} \text{ partition, and } \mathbf{Z} = (X_3) = \textit{exo} \text{ partition.}$$

The model consists of one balance equation and its initialization relationship:

$$X_2 = X_1 - X_3$$

$$X_1 = \begin{cases} X_1^0 & \text{at } t = 0, \\ X_2 & \text{of previous } \Delta t \text{ thereafter} \end{cases}$$

8.3.1 Conventional Approach

Stochastic variables are built into the model prior to run time as *endogenous stochastic forms*, cumulative probability distributions presumed to possess the

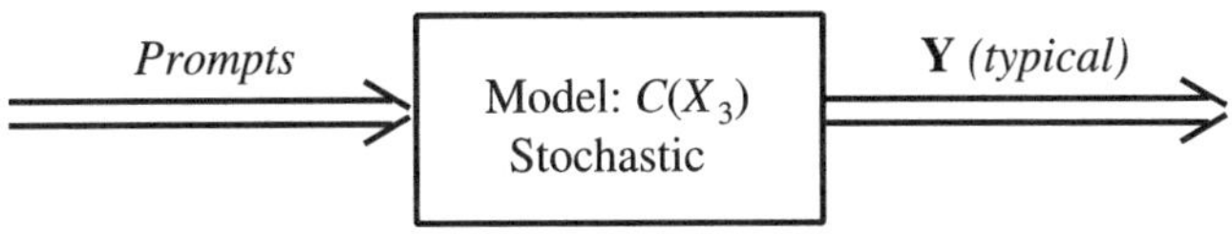

Figure 8.3 Stochastic Model

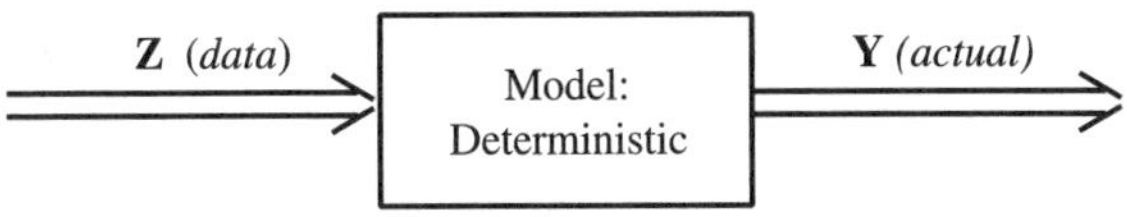

Figure 8.4 Stochastic Input to Model

desired statistical properties. Consequently, modeling effort is concentrated on producing the desired *effect*, with the result that *cause*, which forms the core of the physical system, is disregarded. Hypothetically, the model produces typical results, and modelers are obligated to demonstrate that model response is not significantly different from that of the physical system. Most simulation texts present statistical validation tests, as for example, reference [29].

For this simple inventory system, demand is approximated by an analytical pdf, the Poisson probability distribution:

$$P(X_3 = n) = \frac{e^{-\lambda}\lambda^n}{n!}, \quad n = 0, 1, 2, \ldots.$$

Sequence $\{X_3\}$ is a stochastic time series. It is stationary because $\mu_3 = \sigma_3^2 = \lambda$, where λ is the mean, constant, demand rate. The *cumulative* probability of demand is expressed as

$$C(X_3) = P(n \leq X_3) = \sum_{n=0}^{X_3} \frac{e^{-\lambda}\,\lambda^n}{n!}.$$

Figure 8.3 displays the distinguishing feature of the conventional approach—the model is *stochastic*, the result of building the cumulative Poisson distribution into the model as an endogenous stochastic form to provide Monte Carlo simulated sampling. The model is prompted by an input file containing all parameters pertinent to the simulation, such as estimated mean and variance $\widehat{\lambda}$, time horizon T, or number of observations N, and output format. Model output **Y** is stochastic, but its realistic values will not match the actual values. They are at best typical.

8.3.2 Systems Approach

Stochastic variables are input to the model during run time as *exogenous stochastic forms*, primary data occurring as time series. Effort is concentrated on modeling both *cause* and *effect* relationships.

Figure 8.4 depicts the unique feature of the systems approach—the model is *deterministic*, the result of assigning X_3 to the exogenous partition. Demand, even though *stochastic*, is given exactly, by inputting the vector $\mathbf{Z}$, which contains the historical time series. Although model output $\mathbf{Y}$ is stochastic, its values match actual values, excepting for distortion introduced in their calculation. In the absence of historical data, the time series can be generated externally via Monte Carlo simulated sampling from an assumed distribution, as in the current illustration.

In general, use of exogenous stochastic forms is attended by both advantages and disadvantages.

8.3.2.1 Advantages

1. Inputting stochastic variation in the form of a time series bears direct correspondence with input to or output from the physical system.

2. The model, consisting of a linear system of equations, transforms input into an output that accurately represents the nonlinear, stochastic response of the physical system, that is, it reproduces *actual* results rather than producing typical results.

3. The model consists of relationships that represent *cause* and *effect*.

4. During the debugging (validation) phase, deterministic model structure with stochastic input has a persuasive advantage. Output from the deterministic model must match the actual, historical record. Any deviation beyond the precision capability of the computer indicates structural deficiency in the model and is easily pinpointed. However, output from the stochastic model only approximates certain properties of the historical record—for example, mean and standard deviation. Structural deficiency tends to become confounded with sampling error so that precise debugging is not possible.

8.3.2.2 Disadvantages

1. More structure is required to accommodate both *cause* and *effect*.

2. All but the simplest models include equations that must be calibrated.

Whether stochastic forms are endogenous or exogenous depends upon availability of primary data and desired use of the model. Some commercial simulation languages do, indeed, permit use of both forms. For example, Simscript II.5 (CACI Products Company) features five levels of programming languages and accepts stochastic forms as both *internal* and *external* processes.

8.3.3 Confidence Intervals

Although the structure of Mtm systems models is linear and deterministic, stochastic input produces stochastic output. When population means and standard deviations are to be estimated from model output, confidence intervals may be used. Confidence intervals for means and standard deviations of the output time series are determined in the following manner. Make 10 or more runs of the model, using independent stochastic sequences. Then use standard statistical procedures for computing the confidence intervals [41].

To clear the systems model at the end of a run, add N_0 additional periods of zero input to the *exo* file, where N_0 is greater than or equal to the time duration for transient *tail-off* to zero. To start the systems model in steady state, solve for the steady-state initial conditions, using trend analysis.

8.4 TREND ANALYSIS

A stochastic time series is nonstationary if either mean or variance changes with time. All series exhibit a trend: either *zero* or *nonzero*. A nonzero trend may be *linear* or *nonlinear* and is established when the mean does not remain constant. Analysis is confined to the duration of time for which the same *type* of trend prevails. In a systems model, trends may influence both rates and states (levels). Trends are established by dynamic forms. For example, the impact of various dynamic forms (as rate variables) on the initial state of a system is registered by the resulting displacement of the terminal state. Moreover, system states interact with endogenous dynamic forms. Thus, it is possible for systems models to include complex cause-effect relationships, that is, for rates to impact levels and vice versa.

A trend is classified as *homogeneous* within the region of analysis if deviation above and below the trend line does not exceed two standard deviations (or less, if greater homogeneity is required). When trend is present in a system, all cross-correlated variables are influenced by that trend. However, trend or its absence is always displayed by the IT pair(s) of the system.

8.4.1 Percent Trend

Trend is inherent to each stochastic time series of the system. Trend can be discovered through statistical analysis of the series or appear as a knowledge base item. Given the IT pair X_I, X_T. It follows that for the average period

$$\overline{X}_T = k\,\overline{X}_I$$

where

$k < 1 \Rightarrow$ decreasing trend,

$k = 1 \Rightarrow$ zero trend, and

$k > 1 \Rightarrow$ increasing trend.

Trend may be expressed in percent

$$\text{Percent trend} = (k-1) \times 100\% = \left(\frac{\sum_{t=1}^{N} X_T^t}{\sum_{t=0}^{N-1} X_I^t} - 1 \right) \times 100\%. \tag{8.1}$$

8.4.2 Zero Trend

Let IT pair X_I and X_T represent a stochastic time series. Two cases exist by which the pair can exhibit zero trend.

Case 1 *Primary data exist for X_I and X_T.*
IT pair X_I and X_T exhibits *zero* trend over the region of analysis if

- $k = (\sum_{t=1}^{N} X_T^t)/(\sum_{t=1}^{N} X_I^t) = 1.0$ to acceptable approximation,
- deviation above and below $\overline{X}_T$ or $\overline{X}_I$ does not exceed two standard deviations.

Case 2 *Primary data do not exist for X_I and X_T.*
Secondary data sequences can be generated for the IT pair X_I and X_T such that zero trend is exhibited if

- X_I is designated *exo* and X_T is designated *endo*,
- X_I, X_T satisfy the *zero-trend theorem*

$$\overline{X}_I = \overline{X}_T \Longleftrightarrow X_I^0 = X_T^N. \tag{8.2}$$

PROOF Let $\{X_I^t\}$ and $\{X_T^t\}$, $t = 1, 2, \ldots, N$ designate sample realizations for the IT pair.

Then

$$\overline{X}_I = \frac{1}{N} \sum_{t=0}^{N-1} X_I^t = \frac{1}{N}(X_I^0 + X_I^1 + X_I^2 + \cdots + X_I^{N-1})$$

and

$$\overline{X}_T = \frac{1}{N} \sum_{t=1}^{N} X_T^t = \frac{1}{N}(X_T^1 + X_T^2 + \cdots + X_T^{N-1} + X_T^N)$$

Substitute into left-hand side of Equation 8.2 to establish forward implication:

$$X_I^0 + X_I^1 + X_I^2 + \cdots + X_I^{N-1} = X_T^1 + X_T^2 + \cdots + X_T^{N-1} + X_T^N$$

By definition,

$$X_I^t = \begin{cases} X_I^0 & \text{at } t = 0, \\ X_T^t & \text{of previous } \Delta t \text{ thereafter.} \end{cases}$$

Consequently,; $X_I^t = X_T^t$ for $t = 1$ to $N-1$, and identical terms cancel. Therefore, $X_I^0 = X_T^N$. Proof of the reverse implication is similar.

8.4.2.1 Five-step procedure for establishing zero trend

1. Select trial value for X_I^0.
2. Run model to obtain X_T^N and $k = \overline{X}_T/\overline{X}_I$.
3. Check k for convergence to 1: When $k = 1$ to specified accuracy, stop.
4. Reset X_1^0 to X_T^N of step 2.
5. Return to step 2.

8.4.3 Nonzero Trend

Let IT pair X_I and X_T represent stochastic time series. Again, two cases exist by which the pair can exhibit nonzero trend.

Case 1 *Primary data exist for X_I and X_T.*
IT pair X_I and X_T exhibits *nonzero* trend (linear or nonlinear) over the region of analysis if

- k is significantly different than 1.0,
- deviation above and below the trend line does not exceed two standard deviations (or less, if greater homogeneity is required).

Case 2 *Primary data do not exist for X_I and X_T.*
Nonzero trends can be built into secondary data sequences for IT pair X_I and X_T by superimposing their characteristic variations upon trend lines established by select modules, as for example:

- linear accumulator for a homogeneous, linear trend,
- polynomial accumulator for a homogeneous, nonlinear trend,
- various exponential dynamic forms.

See Exercise 18 for establishing convergence to a specified linear trend.

8.5 KNOWLEDGE ACQUISITION VIA INFERENCE MATRIX

Mtm systems analysis and modeling are carried out by modelers who must alternate between two roles or find capable assistance. As a *knowledge engineer*, systems analysis must be performed. As a *domain expert*, knowledge items must be acquired. And again as a *knowledge engineer*, items must be assembled into a knowledge base. However, the modeler, with limited knowledge of the domain, may not be able to construct a valid model without assistance from a domain expert. Identification of appropriate rate or regression equations, standard or nonstandard dynamic forms, is critical to model structure. In the absence of primary data, knowledge acquisition and reverse regression apply, "given system knowledge, regression coefficients can be found that generate secondary data consistent with that knowledge."

From years of experience, domain experts possess knowledge about relationships between system variables that ranks next in value to knowledge of primary data. An *inference matrix* is a mechanism for transferring domain expertise into a simple, weighted set of relationships known as regression equations. With respect to the model, these relationships become dynamic forms. Thus, the inference matrix serves as a knowledge acquisition tool to infer possible *physical* and or *statistical correlations* between dependent and independent variables. Circumscription limits the number of variables in a model, and *independent* variables selected for each regression equation are restricted to these variables.

The inference matrix allows the modeler or knowledge engineer, under the guidance of a domain expert or experts, to designate the relative "magnitude" and "direction" of relationships between variables of a model. It provides a tool for extracting knowledge in communicable form. The matrix graphically "asks" the domain expert to reason. The expert visualizes potential relationships and is asked to evaluate; is this relationship strong, moderate, weak, or nonexistent?; in what direction is the correlation, positive or negative?; and is delay involved?

Table 8.1 displays the format of an inference matrix. *Dependent* variables, Y_i, for which dynamic forms have not been identified, are listed by row and model variables by column. To infer a regression relationship for Y_i, a code symbol is placed in the column of each potential independent variable, X_j, for which correlation, either positive or negative, is believed to exist. Strength of the correlation between each Y_i and selected X_j is coded "S," "M," or "W," designating *strong*, *moderate*, or *weak* correlation, respectively. Furthermore, the code contains valuable information about partial derivatives: their algebraic sign and relative importance. Delayed or lagged outputs are explicitly coded by including "L^D," where "D" is the delay parameter—for example,"$+SL^1$." An asterisk and footnote can be added to distinguish statistical from physical correlations.

Table 8.1 Inference Matrix

Dep. Vars.	X_1	X_2	...	X_n	K
Y_1					
Y_2					
⋮					
Y_k					

The *net effect* of independent variables omitted from a regression equation by circumscription can also be included. Any such effect judged to be significant can be indicated by placing a "Y" for *yes* in the K column, thus including the *constant form* K as a vertical axis intercept in the equation.

Because each possible correlation is considered, the inference matrix will yield more than enough information from which to infer model structure. Five rules help to sort out the key correlations.

1. *Parsimony* (sparing selection) is based on a 14th-century Latin heuristic *Occam's razor* and translated, "Things should not be multiplied without good reason" [48]. Parsimony is achieved by including only those correlations that most directly represent relationships between variables.

2. *Physical* correlation is preferred to *statistical* correlation.

3. IT pairs combine (regress) as weighted averages.

4. *Multicolinearity* is to be avoided. It is caused by including too many independent variables in a regression equation [37].

5. *Singularity* prevents solution and must be avoided. It is caused by including (1) too few variables, resulting in a zero row or column in the endogenous coefficient matrix, or by (2) including a row (or column) that is a linear combination of other rows (or columns).

An *interaction matrix*, having the same format as the inference matrix, provides a useful tool for organizing knowledge acquisition for large modeling projects. Cell entries can be used to identify

- personnel assignments to make up interdisciplinary teams to research possible correlations between variables,

- data sources, literature sources, expertise sources, and so on.

8.6 REGRESSION ANALYSIS

Chapter 2 introduced the principles of *forward* and *reverse* regression. Reverse regression applies if any primary data are unavailable and leads to the calibration of appropriate dynamic forms. Illustration of reverse regression techniques is a major focus of this text, and is covered elsewhere in the text.

Forward (or standard) regression, on the other hand, involves fitting equations to primary data. However, standard regression is rarely desirable in systems modeling. First of all, standard regression applies only if primary data are available to cover at least those dependent and independent variable(s) that are correlated. If applicable, standard regression can be used to supplement the number of equations in an under-specified systems model. Two possibilities exist. In Phase One, to complete the data base, one equation is required for each *secondary* variable. Consequently, if by small chance all variables are primary, the model will consist only of balance equations. Addition of equations to the model can then be accomplished by forward regression. Second, Phase Two scenarios may require more equations than were established by Phase One. Again, forward regression can be used.

Whenever primary data series are available, it is preferable to designate them as *exogenous stochastic forms*, thereby eliminating the otherwise unavoidable regression error. Nonetheless, standard regression provides a way to extend the number of equations, so a brief introduction follows. For more extensive treatment, consult any applied statistics text, as, for example, reference [27]. Regression divides into linear and nonlinear analyses. By their special design, Mtm models are multivariable and linear. Therefore, only linear regression in one or more variables is of interest.

8.6.1 Linear Regression

Linear regression has the general multivariate form

$$X_k = c_0 + \sum_{j=1}^{n} c_j X_j.$$

Let $\{X_k^p\}$ designate the primary, stochastic time series for X_k that is to be regressed upon independent variables X_j $(j = 1, 2, \ldots, n)$ with $j \neq k$ and $c_j = 0$ wherever cross-correlation does not exist. At each point in time t, the difference between observation $X_k^{p,t}$ and calculated X_k^t is defined as *regression error*. In fitting a curve to $\{X_k^p\}$, the objective is to minimize the *error* sum of squares (SSE)

$$\text{SSE} = \sum_{t=1}^{N}[X_k^{p,t} - (c_0 + \sum_{j=1}^{n} c_j X_j^t)]^2.$$

Differentiation is used to find the set of coefficients that minimize SSE

$$\frac{\partial\,\text{SSE}}{\partial c_j} = 0, \quad \text{for} \quad j = 0 \text{ to } n.$$

The resulting set of normal equations is solved for the $n + 1$ coefficients.

Fitting of data to linear regression equations is easily performed using a computer and commercial statistical software such as SPSS (SPSS, Inc.). In addition to coefficient output, several measures indicate goodness of fit [4]:

- *Standard error of estimate*, regression error expressed in standard deviation units.
- *Coefficient of multiple determination*, the proportion of variation in the dependent variable explained by the set of independent variables.
- *Coefficient of partial determination*, the proportion of variation in the dependent variable explained by each independent variable.
- *Coefficient of partial correlation*, the strength of association between the dependent and an independent variable while controlling for, or holding constant, the other independent variables.
- Probabilities for testing significance of regression and regression coefficients.

EXERCISES

1. What characterizes a time series as stochastic?
2. How is a probability distribution function characterized?
3. What is a stationary time series?
4. Define a stochastic process.
5. Relate degrees of freedom to sample realization.
6. What connotation does the term *population* carry in Mtm analysis?
7. Distinguish between a population mean and a sample mean.
8. Suggest a reason for using a biased estimator for variance in the analysis of stochastic time series.
9. What is a first-order Markov process?

10. Many telephone polls violate the principle of simple random sampling. Can you explain why?

11. Generate an $N = 30$ realization from a discrete Poisson process with $\lambda = 6$, using Monte Carlo simulated sampling. Plot the sequence. Is it stochastic? If yes, why?

12. What is the unique difference between the conventional and systems approach to modeling stochastic processes?

13. When are confidence intervals used in systems modeling?

14. Use the data provided below to test the statement, "Difference between averages is equal to average difference."

X_1	X_2
3.12	4.56
4.56	5.24
5.24	6.18
6.18	7.00

15. What is a homogeneous trend?

16. Verify Equation 8.1 for percent trend.

17. Distinguish between the two cases that arise in analysis of zero and nonzero trends.

18. Show that for case 2,

$$X_I^0 = \frac{X_T^N + (1-k)(N\overline{X}_T - X_T^N)}{k}, \qquad k \neq 1.$$

Suggest a procedure for convergence for this nonzero trend case.

19. What purpose does an inference matrix serve?

20. Define parsimony, multicolinearity, and singularity.

21. How might a modeler determine whether to use *forward* or *reverse* regression?

22. The following table presents primary data for three stochastic variables of a systems model, X_2, X_5, and X_7.

    a. Use a statistical computer package to evaluate the inference that

$$X_2 = \beta_0 + \beta_1 X_5 + \beta_3 X_7.$$

b. From the output in (a), discuss the measures that indicate goodness of fit.

c. Do you recommend use of this regression equation? Explain your reasoning.

t	X_2	X_5	X_7
1	.538	.597	.944
2	.522	.509	.841
3	.513	.663	.911
4	.529	.751	.768
5	.565	.739	.718
6	.594	.598	.634
7	.600	.556	.735
8	.584	.594	.858
9	.554	.667	.673
10	.553	.776	.609
11	.595	.754	.604
12	.637	.689	.457
13	.641	.498	.612
14	.647	.643	.642
15	.634	.681	.849
16	.629	.778	.733
17	.638	.829	.672
18	.662	.751	.594
19	.675	.704	.559
20	.658	.633	.604
21	.629	.663	.678
22	.625	.709	.571
23	.648	.763	.490
24	.682	.681	.747
25	.676	.627	.651
26	.655	.667	.645
27	.640	.804	.609
28	.668	.782	.705
29	.678	.707	.453
30	.692	.653	.382

CHAPTER 9

Work Physiology Model

9.1 INTRODUCTION

Ergonomics and human factors engineering have contributed to remarkable increases in productivity that improve the health, safety, and well-being of workers and end-users. Driven by management's concern for improvement in terms of work productivity, absenteeism, workman's compensation costs, and worker morale, human resources will continue to be a critical component of industrial economics. Improvement involves eliminating hazards, reducing risk of injury, and making tasks more comfortable. Better human resource management requires tools to assist managers and engineers in making better matches between workers and tasks. This chapter draws from a master's thesis by Robert R. Wehrman [55] in which he applied systems engineering to human physiology to synthesize a work physiology model that describes a worker's physiological response to a given task. Dr. Robert J. Marley, Wehrman's major professor, served as domain expert.

Physiological models are primarily applied to repetitive lifting where it is presumed that the lifting load is well within the physical strength of the worker. The respiratory (oxygen transport) system is the key limiting factor impacting these lifting tasks. Knowledge of physiology, understanding the human body, is critical to modern work design. At best, it has been difficult to quantify physiological demands on the human body. Although it is possible to measure some physical parameters, such as oxygen uptake of a worker while performing a task, costs restrict such practice to research laboratories. Given a known workload and physiological capacity of a worker, an effective method for worker screening is sought by researchers. Most commonly, the approach has been to relate motion analysis to oxygen uptake by the worker.

Many researchers have modeled oxygen uptake as a function of energy expenditure, but the majority of these models have been *static* regression models developed from empirical observations. Hagan et al. [22] conducted a comparative study of 12 regression models for steady-state oxygen uptake and energy expenditure during horizontal treadmill tests and found their predictions to be similar. Disadvantageously, these models were applicable only to specific segments of the population and for horizontal running.

Garg et al. [20] have derived a series of static regression models that expand applicability to numerous, common, manual materials handling jobs. Although robust in applying to jobs without previous experimental measure, their industrial application was limited. Furthermore, these models were static, applying only to steady-state workloads.

Not all industrial work achieves steady state. The body's response to workload increase is transient. Stored energy supplies are consumed to meet short-term demand, whereas oxygen uptake increases to a level required to meet the new demand. A similar transient response occurs when the workload is decreased. Researchers have defined models, such as that by Morton [35], that account for rate of change of oxygen uptake with respect to time. However, these models are still static, applying only to engaging or disengaging constant workloads.

Ergonomists are continually looking for better ways to determine the best combination of job parameters for reducing physical stress. Specific individuals or population segments are tested against computer simulated tasks to quantify the risk of injury. These data from task simulation coupled with systems analysis provide adequate knowledge-based support for modeling the human at work. A work physiology model provides a mechanism for quantifying actual workload during simulation of a specific task by a worker. That is, given the workload in energy units, the model predicts oxygen uptake; or, given oxygen uptake, the model derives the energy-equivalent workload. For cost effectiveness and accuracy, a macro systems model can be used in a variety of applications.

Wehrman, a mechanical engineer, explored the use of systems analysis for modeling physiological response to changing workloads for tasks involving large, rhythmic muscle movements for both transient and steady-state responses. He observed that mathematical formulation and iterative solution of Mtm models are in some ways analogous to *finite element analysis* (FEA), commonly employed by mechanical engineers. In FEA, each finite element is similar to a subsystem. FEA is typically applied to mechanical systems having geometric properties, heat distributions, stress distributions, and so on, that are too complicated to describe and solve by traditional differential equation techniques. Finite elements are established as linear forms for parameters such as heat and stress. Thus, the system is described by a balanced set of equations and unknowns. However, Mtm and FEA differ in application. In FEA, algorithms automatically generate finite elements to fully circumscribe the system. Computational resources are the limiting factor

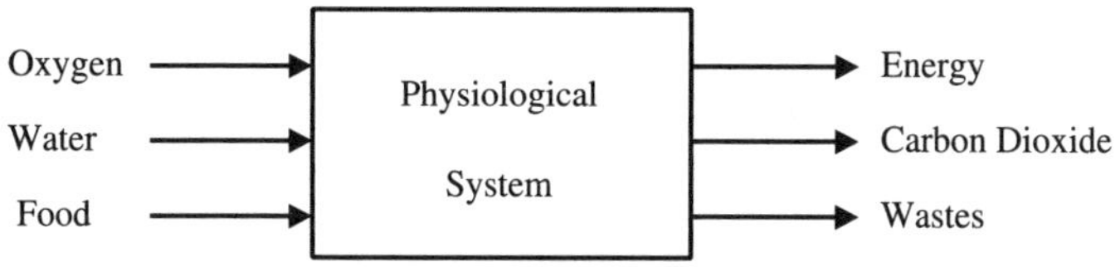

Figure 9.1 Macro Block Diagram of Work Physiology Model

in the number of elements that can be handled. In Mtm, no generic algorithm can automatically divide a system into subsystems. The modeler must explicitly relate each variable component of the system to a specific subsystem, relying on knowledge of the system. The distinguishing feature of an FEA model is its many generic finite elements as opposed to a few unique and specific subsystems in an Mtm model.

9.2 SYSTEMS ANALYSIS

A human being consists of a triad of interacting systems: physiological, psychological, and pneumatological. Each system is composed of many subsystems. For example, the physiological system includes such commonly known subsystems as skeletal, muscular, nervous, digestive, excretory, respiratory, circulatory, metabolic, and many others. Micro analysis requires a detailed breakdown to an elemental level at which chemical reactions and electrical phenomena can be isolated and modeled. Macro analysis, on the other hand, approaches modeling at the system/subsystem level at which system modules and interacting subsystems can be modeled. All of the aforementioned subsystems are involved in the performance of work by a human. A macro block diagram of work physiology is presented in Figure 9.1, showing only those inputs and outputs most directly related to physical work. Furthermore, the metabolic subsystem, if interfaced with the oxygen uptake rate of the respiratory subsystem, provides an adequate basis for an $L_{1,1}$ work physiology model that will accurately represent metabolic response and oxygen uptake of a worker. Horizontal expansion of this $L_{1,1}$ model to include other subsystems is beyond the scope of this chapter but will readily yield to analysis and modeling.

9.3 KNOWLEDGE BASE

- $L_{1,1}$ circumscription of the metabolic system and its interface with the respiratory system.
- Fundamental units of analysis:

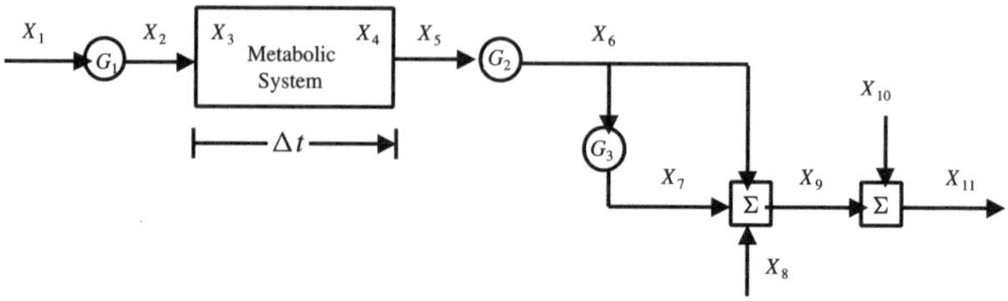

Figure 9.2 Block Diagram of Work Physiology Model

Space: Work station

Mass: One worker. Mass attributes are expressed in units of *watts*, *kilocalories* (kcal) per Δt, and oxygen uptake in *liters* per Δt.

Time: $\Delta t = 0.5$ minute

- Number of subsystems: one
- Schematic: See Figure 9.2
- System variables:

$X_1 =$ workload in watts

$X_2 =$ input workload in kcal per Δt, a rate

$X_3 =$ initial energy debit in kcal, a level

$X_4 =$ terminal energy debit in kcal, a level

$X_5 =$ output workload in kcal per Δt, a rate

$X_6 =$ workload in equivalent liters of oxygen per Δt

$X_7 =$ heat production in equivalent liters of oxygen per Δt

$X_8 =$ resting metabolism (RM) in liters of oxygen per Δt

$X_9 =$ oxygen demand in liters per Δt

$X_{10} =$ adaptive response in liters of oxygen per Δt

$X_{11} =$ oxygen uptake in liters of oxygen per Δt

- Primary data: X_1; X_{11}
- Auxiliary data: RQ = (nonprotein) respiratory quotient, the ratio of volume of carbon dioxide to volume of oxygen consumed during Δt [3]. RQ varies from Δt to Δt.

- Vector table for model structure:

X_j	IT Pair	Exo	Endo	Equation Type	Equation No.
X_1		*		—	—
X_2			*	modification	1
X_3	I	*		—	—
X_4	T		*	balance	2
X_5			*	regression	3
X_6			*	modification	4
X_7			*	modification	5
X_8			*	constant form	6
X_9			*	equivalence	7
X_{10}			*	equivalence	8
X_{11}		*		—	—

- Dynamic form for continuous delay for X_5.
- Constant form for resting metabolism for X_8.
- Definitions and relationships:

  1 watt = 14.33 gram calories per minute, a rate.

  Energy yield of oxygen: $Y = 1.333 \cdot \text{RQ} + 3.766$ in kcal per liter of oxygen [55].

  Resting metabolism (RM), steady-state component of oxygen uptake of a worker sitting upright in a chair, composed only of the volume required to maintain sitting posture. RM is assumed to be constant.

  Mechanical efficiency of external work, [3]

$$E = \frac{\textit{External Work}}{\textit{Gross Output} - \text{RM}}.$$

  E varies in level from Δt to Δt.

9.4 GENERIC MODEL

Based on the preceding knowledge items, the generic form for the $L_{1,1}$ work physiology model is the following

1. $X_2 = G_1 X_1$
2. $X_2 + X_3 = X_4 + X_5$

3. $X_5 = 1/D\,[\alpha\, X_3 + (1 - \alpha)\, X_4]$

4. $X_6 = G_2 X_5$

5. $X_7 = G_3 X_6$

6. $X_8 = \text{RM}$

7. $X_9 = X_6 + X_7 + X_8$

8. $X_{11} = X_9 + X_{10}$

$$X_3 = \begin{cases} X_3^0 & \text{initially,} \\ X_4 & \text{of previous } \Delta t \text{ thereafter} \end{cases}$$

Equations 3, 4, 5, and 6 contain generic parameters. Calibration involves "fitting" the model to a specific worker. All equations requiring calibration are discussed in sequence as follows:

Equation 1 $X_2 = G_1 X_1$ to convert from watts to kcal per Δt.

$$G_1 = 0.01433 \frac{\text{kcal}/\Delta t}{\text{watt}} \cdot \frac{1 \text{ min}}{2\Delta t} = 0.007165 \frac{\text{kcal}/\Delta t}{\text{watt}}$$

Equation 3 $X_5 = 1/D\,[\alpha\, X_3 + (1 - \alpha)\, X_4]$. D is the time (a lag or delay expressed in Δt units) that elapses between the performance of a Δt work element and its translation into an oxygen deficit (demand) by the metabolic system. Convergence to steady state is controlled by α, with rapid convergence when α is close to unity and slow convergence when α is close to zero. Values for D and α are to be selected that best represent the worker's adaptive response to elements of external work.

Equation 4 $X_6 = G_2 X_5$ to convert from kcal per Δt to liters of oxygen per Δt.

$$G_2 = 1/\text{Yield} \frac{\text{liters of oxygen}}{\text{kcal}},$$

and is computed for each Δt.

Equation 5 $X_7 = G_3 X_6$ to relate oxygen consumed in performing external work to internal heat production. G_3 is a function of efficiency and is computed for each Δt. Applying the definition of mechanical efficiency to the model

$$E = \frac{X_6}{X_{11} - \text{RM}}. \tag{9.1}$$

$X_{11} - \text{RM}$ can be shown to be equivalent to $X_6 + X_7$ if X_{10} can be neglected.

From Equations 6, 7, and 8, $X_{11} - \text{RM} = X_6 + X_7 + X_{10}$. X_{10} is the corrective increment of adaptive response required to match the worker's oxygen uptake (X_{11}) with oxygen demand (X_9) generated by work elements during each Δt. At all times, involuntarily, through the process of *homeostasis*, the body seeks to maintain internal consistency. Because of homeostasis, X_{10} will exhibit small, transient excursions from zero in response to an external workload. Consequently, dropping X_{10} from the denominator of Equation 9.1 will introduce but slight approximation error only during Δt's that exhibit transient response. Reduction in magnitude of the X_{10} series is discussed in the next section.

Solving $E = X_6/(X_6 + X_7)$ for X_7 yields $X_7 = (1 - E)/E\ X_6$ so that $G_3 = (1 - E)/E$. Because E is also a variable, changing from Δt to Δt, Equation 5 is of second-order nonlinearity. Wehrman introduced an approximation by providing a constant value for E, calculating an average efficiency based on the last two minutes of each task's (work load's) duration. However, this source of approximation error can be eliminated.

An exogenous value for the numerator of Equation 9.1 at time t, X_6^t, was obtained by substituting the previously computed value of X_6^{t-1}. Fortunately, the calibration process made some compensation for this substitution. Any remaining approximation error was confounded with distortion and experimental error and was distributed innoculously across the endogenous variables. Then, for each Δt, efficiency was computed by

$$E = \frac{X_6^{t-1}}{X_{11}^t - RM}. \tag{9.2}$$

Equation 6 $X_8 = \text{RM}$ RM is a constant whose value was inferred during the calibration process, along with values for D and α that best represent the worker's adaptive response to elements of external work.

9.5 CALIBRATION AND VALIDATION

Wehrman's study included five male subjects ranging in age from 23 to 28 years, each performing the same five tasks. The work physiology model of this chapter was calibrated for subject 1 for tasks 1, 2, and 5. The model was validated for each of the tasks by coupling task elements in watts and the ensuing oxygen demand to oxygen uptake by the worker, while minimizing the adaptive response associated with those elements. Subject 1, whose age, height, and weight were 26 yr, 180.3 cm, and 91.2 kg, respectively, had a physical work capacity of 4.45 liters of oxygen per minute. The five tasks were as follows

1. Cycle ergometer
2. Arm-crank ergometer

3. Squat/arm lift
4. Load carrying
5. Random combination of tasks 3 and 4

A Monark Ergometer model 818E was used for tasks 1 and 2. Tasks 3 and 4 simulated an industrial task of lifting and carrying boxes. A SensorMedics model 2900c breath-by-breath metabolic measure cart was used for oxygen and carbon dioxide measurements. Auxiliary data (RQ) and primary data (X_1, X_{11}) are presented in Appendix F, Tables 9.2 and 9.3, respectively.

9.5.1 Physiological Functions Simulated by the Model

Input to the model consists of work elements in watts, primary data series X_1. Output from the model, primary data series X_{11}, yields total oxygen uptake by the worker (subject 1). So that output can be coupled to input, the model must be able to simulate the following physiological functions:

- Introduce delay between each Δt work element and the ensuing demand for oxygen: Equation 3.
- Generate a rise or fall in the amount of heat released by metabolism to correlate with the worker's efficiency for each Δt work element: Equations 4 and 5.
- Minimize SAD $= \sum_{i=1}^{n} |X_{10}|$, sum of the absolute deviations of X_9 from X_{11}. Minimizing SAD is consistent with homeostasis, in this instance, maintaining metabolic equilibrium. Although the X_{10} series cannot be driven to zero, individual values are relatively small, attributable to experimental error and extraneous factors that also generate oxygen demand.

9.5.2 Calibration Procedure

Calibration parameters were adjusted with the objective of minimizing SAD. Because work elements are not encountered during the initial rest period, $X_3^0 = 0$ to initialize the metabolic response system. At the beginning of each Δt iteration of the model, values for X_1 and X_{11} are supplied from the exogenous data base, and new values are calculated for G_2, and G_3. The following steps comprised the calibration procedure:

1. Assign initial values to the calibration parameters:

   Resting metabolism: RM $= 0.5$ liters of oxygen per Δt

   Delay: $D = 1.0 \equiv 0.5$ minute

Shaping parameter: $\alpha = 0.5$

2. Make repetitive runs of the model, adjusting RM to obtain an initial reduction of SAD.

   RM values must be updated for each Δt in calculating E for G_3 in Equation 5 and in the constant form for X_8 in Equation 6.

3. Make repetitive runs of the model, adjusting D to obtain further reduction of SAD.

4. Make repetitive runs of the model, adjusting α to obtain continued reduction of SAD.

5. Repeat steps 2 through 4 until SAD has been minimized to a reasonable number of stable decimal places (at least 4 decimal places).

9.5.3 Task 1

The cycle ergometer provided the workload for task 1. Figure 9.3 displays the Δt work element, X_1, consisting of the following 19-minute sequence: 2 minutes of rest, 5 minutes of work at 75-watt level, 4 minutes at 100-watt level, 4 minutes at 150-watt level, concluding in 4 minutes of rest.

Calibration established the following values:

- Minimum SAD = 1.563500 liters of oxygen for the 19 minutes.
- $X_8 = \text{RM} = 0.592$ liters of oxygen per Δt.
- $D = 5.070 \equiv 2.535$ minutes.
- $\alpha = 0$.

Table 9.4 of Appendix F lists secondary data for X_2, X_4, X_5, X_6, X_7, X_9, and X_{10}.

Figure 9.4 displays the input work elements (X_2) to be transformed by the metabolic system into a response (X_5) that meshes with the worker's oxygen uptake. The delay of 2.535 minutes accounts for the lag between introduction of a work element and the worker's metabolic response to that work element. Slow convergence to steady state is indicated by $\alpha = 0$ and is apparent in Figure 9.4 in that the worker's oxygen demand did not reach steady state before the workload stepped to the next level. Furthermore, recovery from the debt inflicted by the workload was not completed by the end of the 4-minute rest period, as indicated by X_5 not having returned to zero.

The two closely intertwined series (X_9, X_{11}) of Figure 9.5 indicate that the oxygen demand generated by task 1 very nearly accounted for all of the worker's oxygen uptake. The X_{10} series is the difference between X_{11} and X_9. Each X_{10}

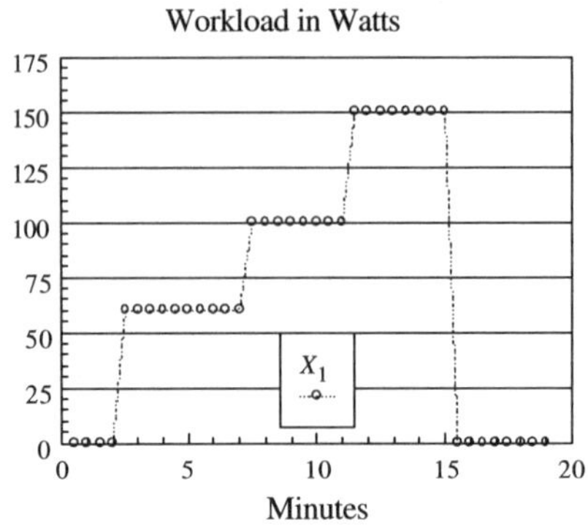

Figure 9.3 Task 1 Workload

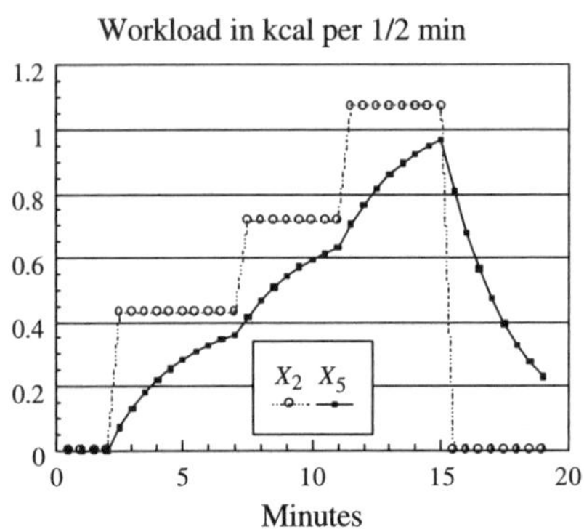

Figure 9.4 Task 1 Metabolic System Delay

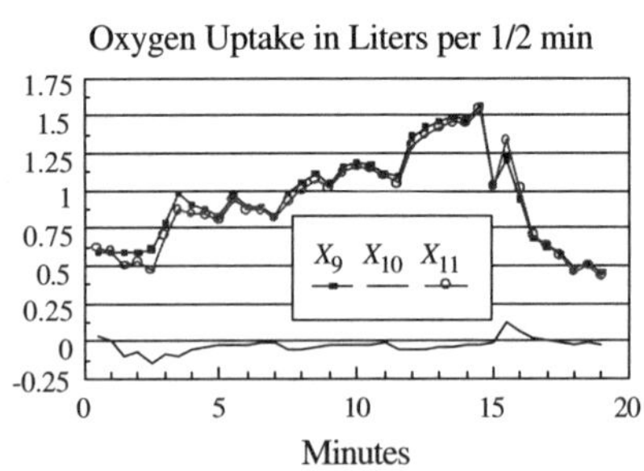

Figure 9.5 Task 1 Oxygen Uptake

value is the corrective increment of adaptive response required during Δt to match the worker's oxygen uptake, X_{11}, with oxygen demand, X_9, generated by each of the work elements. Visual inspection for close match and the small magnitude of X_{10} validates this calibration for subject 1 and task 1.

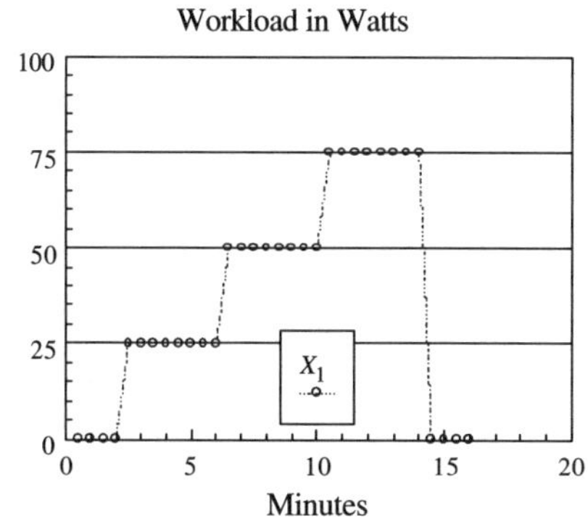

Figure 9.6 Task 2 Workload

9.5.4 Task 2

The arm-crank ergometer was used to establish the workload for task 2. Figure 9.6 shows the Δt work element, X_1, comprising a 16-minute sequence: 2 minutes of rest, 4 minutes of work at 25-watt level, 4 minutes at 50-watt level, 4 minutes at 75-watt level, concluding in 2 minutes of rest.

Applying the task 1 model to task 2 yielded SAD $= 1.583855$. However, an improved value for SAD was obtained by recalibrating the model to task 2.

- Minimum SAD $= 0.896877$ liters of oxygen for the 16 minutes.
- RM $= 0.404$ liters of oxygen per Δt.
- $D = 3.52 \equiv 1.76$ minutes.
- $\alpha = 0$.

Secondary data are listed in Table 9.5 of Appendix F.

Input workload (X_2) and the metabolic response (X_5) to that workload are given in Figure 9.7. The delay of 1.76 minutes was significantly smaller than the task 1 delay. Although the workload of task 2 averaged a little less than half the workload of task 1, most likely the difference can be attributed to the much smaller muscle group of the arm used to turn the crank. Again, α was zero and the worker's oxygen demand did not achieve steady state during any of the 4-minute workload durations. Debt repayment was not complete at 16 minutes but required approximately 2 more minutes.

In Figure 9.8, X_9, the oxygen demand generated by the workload accounts for nearly all of the worker's oxygen uptake, X_{11}. Greatest departures from zero by the X_{10} series ($X_{11} - X_9$) occur prior to and after encountering the work elements. High correlation of work-induced oxygen demand with the worker's oxygen uptake validates this calibration for subject 1 and task 2.

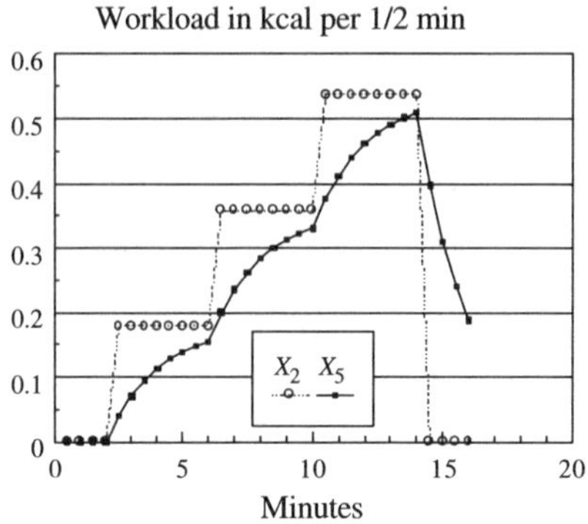

Figure 9.7 Task 2 Metabolic System Delay

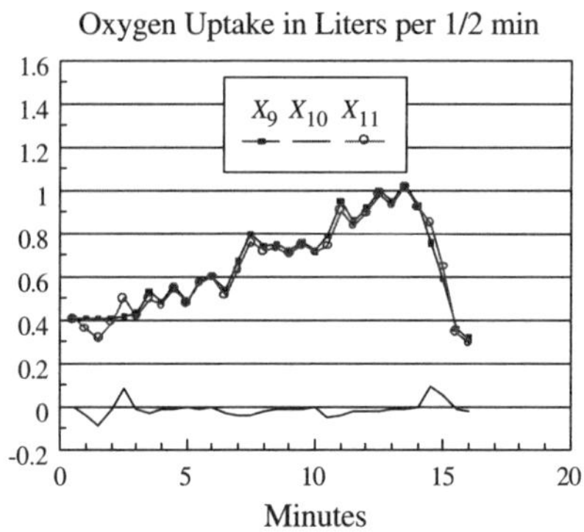

Figure 9.8 Task 2 Oxygen Uptake

9.5.5 Task 5

Simulation of an industrial task, lifting and carrying boxes, made up the task 5 sequence displayed in Figure 9.9. Obviously, the X_1 series lacks homogeneity, which if ignored, will provide a test for robustness. Work elements, X_1, consisted of the 20-minute sequence: 3 minutes of rest, 3 minutes of workload at 7 watts, 2 minutes at 3 watts, 5 minutes at 2 watts, 1 minute of rest, 3 minutes at 10 watts, ending in 3 minutes of rest.

Applying the task 2 model to task 5 yielded SAD $= 1.716540$. Again, recalibration produced an improved model:

- Minimum SAD $= 1.032495$ liters of oxygen for the 20 minutes.
- RM $= 0.2815$ liters of oxygen per Δt.
- $D = 22.46 \equiv 11.23$ minutes.
- $\alpha = 0$.

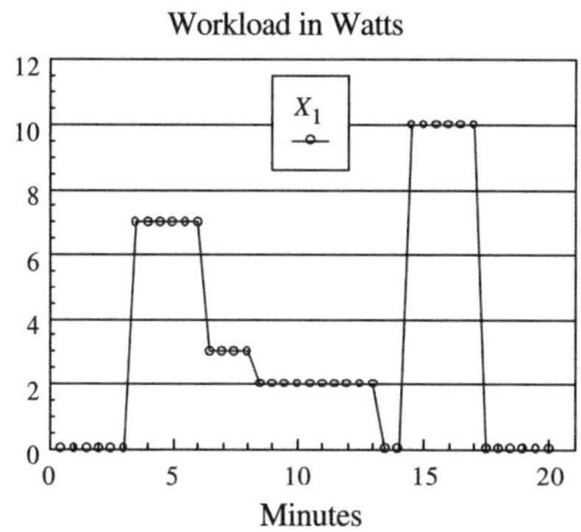

Figure 9.9 Task 5 Workload

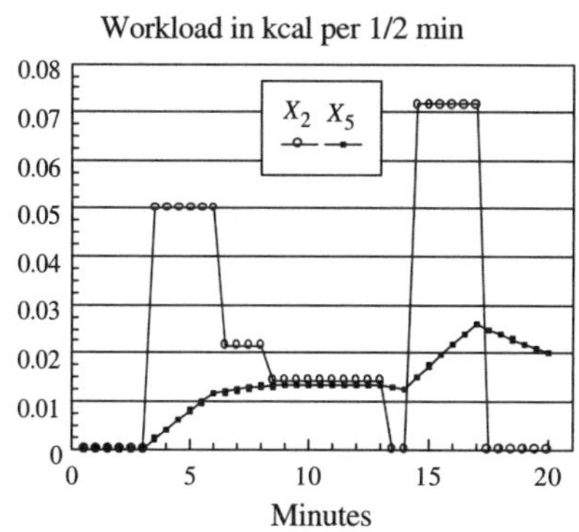

Figure 9.10 Task 5 Metabolic System Delay

See Table 9.6 of Appendix F for secondary data.

Task 5 was an order of magnitude lighter than tasks 1 and 2, and the slow metabolic response (X_5) of the worker reflected the lightness of the workload as evidenced by comparing the two curves of Figure 9.10. The long delay of 11.23 minutes can be attributed to the nature of the task, a light load distributed over multiple muscle groups, that only slowly incur a metabolic debt. Appropriately, α was again zero.

Figure 9.11 also indicates a slow response to the 7-watt work elements comprising the first component of the task. This 3-minute task noticeably caused the worker's oxygen uptake to lock onto the oxygen demand by the metabolic system. The effect is exhibited by the close coincidence of the two curves (X_9 and X_{11}), from 6.5 minutes and beyond. This effect is also observed in the X_{10} series, where homeostatic correction is nil after 6.5 minutes. Debt repayment was complete by the end of the final 3 minutes of rest, another indication of light task.

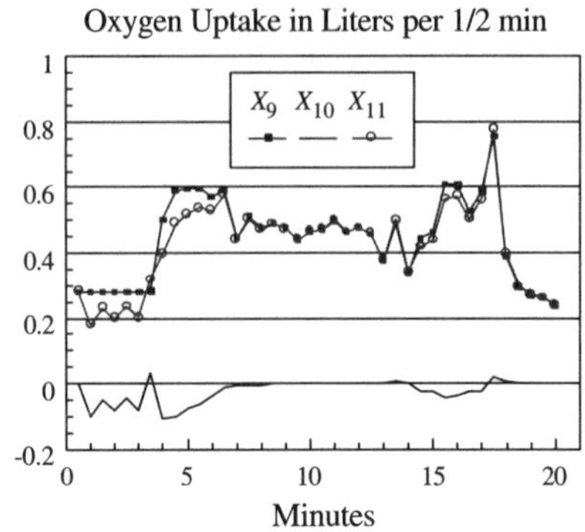

Figure 9.11 Task 5 Oxygen Uptake

Despite nonhomogeneity, a single calibration of the model over the 20-minute sequence produced close coincidence and demonstrated both robustness and validity. For closer coincidence between the two curves, divide the X_1 series (task 5) into two 10-minute segments. Then repeat calibration of the model to subject 1 for each segment.

9.5.6 Efficiency Comparison

Theoretically, efficiency is defined by Equation 9.1. A value for efficiency is needed prior to each iteration to calibrate G_3 so that Equation 5 of the model can be solved. However, a value of X_6 obtained by solving the model for that iteration is required by Equation 9.1 in order to calibrate G_3. Consequently, Equation 9.2 was derived to avoid this dilemma. Thus the question may be raised, how closely do these two equations compare? Table 9.1 provides a comparison of the two equations based on the work elements of task 1 for minutes 3 through 17. Examination of the table shows that efficiency values vary only in the sixth decimal place. Most probably this variation is within the single-precision accuracy encountered in solving two different but equivalent equations via floating-point arithmetic (see Exercise 5).

9.5.7 Workload Quantification

Can the model quantify the workload encountered by a worker given oxygen uptake, resting metabolism, and respiratory quotient? Two tests using role inversion were performed for task 1: history matching for X_1 and perturbation of X_{10}. A *standard* data base, consisting of primary data for X_1 and X_{11} plus secondary data, was obtained from an initial run of the calibrated model for task 1. Because values used to calculate G_3 are not available during these particular test runs, G_3 values were saved to a file.

Table 9.1 Efficiency Comparison

Period	Equation 9.1	Equation 9.2
5	0.000000	0.000000
6	0.177416	0.177416
7	0.124151	0.124151
8	0.187906	0.187906
9	0.237218	0.237217
10	0.320311	0.320312
11	0.218842	0.218843
12	0.295224	0.295224
13	0.312225	0.312225
14	0.407867	0.407866
15	0.288311	0.288311
16	0.267306	0.267307
17	0.260231	0.260232
18	0.316702	0.316703
19	0.269219	0.269219
20	0.266516	0.266515
21	0.283676	0.283676
22	0.322449	0.322449
23	0.366844	0.366844
24	0.264690	0.264690
25	0.261843	0.261843
26	0.262201	0.262201
27	0.264501	0.264501
28	0.277690	0.277690
29	0.258768	0.258768
30	0.578696	0.578695

9.5.7.1 History Matching for X_1

X_1 was inverted from *exo* to *endo*, thus to become a solution variable in the model. To compensate for the loss of an *exo* variable, X_{10} (corrective response) was inverted from *endo* to *exo*. The history matching test was performed in the following manner:

- Set the solution vector to (1, 2, 4, 5, 6, 7, 8, 9).
- Rerun the task 1 model using the previous values for RM, G_2, and G_3, inputting *exo* variables X_{10} and X_{11} from the data base.
- Compare the new X_1 series to the historical (standard) values.

Outcome of the role inversion reproduced X_1 values that exhibited three to five decimal-place accuracy relative to historical values. Table 9.7 of Appendix F shows signed values of change and their cumulative sum. Model values averaged (39) 10^{-6} watts lower than the standard over the 19-minute period of record, supporting the interpretation that distortion introduced by inverting the roles of X_1 and X_{10} was negligible.

9.5.7.2 Perturbation of X_{10}

Adding or subtracting work elements disrupts the metabolic system and causes it to experience transient response. Response is controlled by homeostasis, the process of maintaining constancy of the internal environment while adjusting to the new energy state. After accounting for continuous delay by the metabolic system and efficiency related heat production, X_{10} provides the minimal amount of correction (in liters of oxygen per Δt) required to equate workload to oxygen uptake. The purpose of the perturbation test, then, was to discover the effect on X_1 (the work profile) of incrementally reducing X_{10} to zero in $n = 5$ iterations of the model. Subsequently from Equation 8, when X_{10} is fully removed, $X_9 = X_{11}$, and the model is being driven by oxygen uptake, X_{11}, totally excluding the homeostatic correction modeled by the X_{10} series. The test procedure consisted of the following steps:

1. Set the solution vector to (1, 2, 4, 5, 6, 7, 8, 9) and $n = 1$.
2. Reduce the magnitude of the *standard* X_{10} series by $n \cdot 20\%$.
3. Rerun the task 1 model using previous values for RM, G_2, and G_3, inputting perturbed X_{10} and standard X_{11} from the data base.
4. Compare the perturbed $(X_1 - P)$ series to the standard X_1 series.
5. Set $n = n + 1$ and go to step 2, terminating with $n = 5$.

Comparison of each new X_1 series ($n = 1, \ldots, 5$) showed rapidly growing departures (outliers) from $X_1 = 0$ during the initial rest period. However, X_1 values corresponding to work elements remained within the three- to five-decimal-place accuracy observed for the history matching test for X_1. In Figure 9.12, in the total absence of X_{10} (when $n = 5$), outliers that appear in the initial work profile $(X_1 - p)$ displace the first work element. Thereafter, however, X_1 consistently coincides with the standard profile, although exhibiting a delay of one Δt. Thus, the X_{10} series is clearly seen to moderate response to the work profile. When X_{10} is totally present, history matching is achieved as in the previous test. But when X_{10} is totally removed, work elements are delayed by $\Delta t = 0.5$ minute, the minimum time resolution capability of this $L_{1,1}$ model.

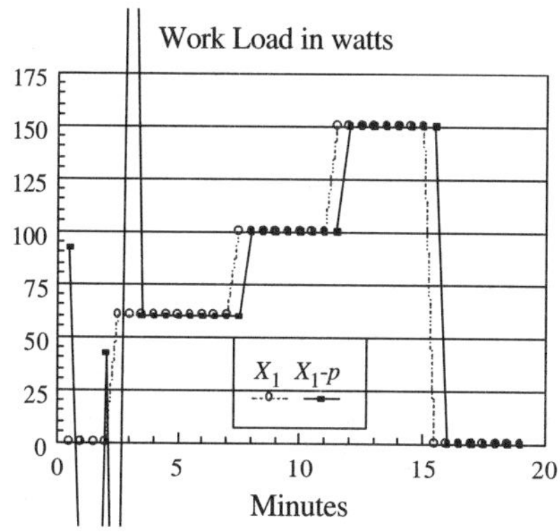

Figure 9.12 Workload Response to $X_{10} = 0$

9.6 CONCLUSIONS AND COMMENTS

The usefulness of a macro-level approach for analyzing and modeling the human at work has been demonstrated. With the work physiology model as an example, researchers are offered a challenge to synthesize models of systems that interact with the metabolic system—digestive, respiratory, and circulatory—to provide a broader-based work physiology model. Ergonomists are challenged to experimentally measure and tabulate energy expenditures for macro elements of industrial tasks.

Perturbation tests demonstrated that the model accurately predicted work profile in terms of energy expenditure when being driven by oxygen uptake. The calibration procedure can be improved by taking measures to mitigate or *filter* outliers in the primary X_{11} and auxiliary RQ data. Filtering will greatly smooth the transient effects caused by omitting the X_{10} series when quantifying workloads. A three-period moving average is likely to produce an adequate filter.

Calculating new values for G_2 and G_3 prior to each iteration preserved nonlinearities introduced by variability in their determinants, RQ and E, respectively. As a principle, existence of variability in any coefficient of an Mtm systems model is symptomatic of a system too narrowly circumscribed. Thus, a time variant coefficient compensates for exclusion of an adjacent system that might otherwise have introduced nonlinearity via its time varying rates and levels. In the current model, the metabolic system was interfaced through Equation 4, containing G_2, with a rate variable from the excluded respiratory system. Efficiency was supplied by Equation 5, containing G_3, as a means of generating heat production in lieu of including respiratory, circulatory, and other interacting systems.

Extension of the model to include other systems will provide a tool for further theoretical inquiry. For example, the value of D of the present model, representing delayed response to work elements by the metabolic system, reveals information about the worker's musculatory, respiratory, and circulatory systems just as much as do basal and resting metabolism rates. Coefficients of an expanded model will similarly provide parameters by which tasks can be assessed for risks to individual workers.

Practical application of a work physiology model in an industrial setting is dependent on measurement of energy requirements for an array of tasks and the subsequent calibration of the model to include 95 percent of the labor pool. A generally accepted physiological workload that can be maintained for an eight-hour workday without experiencing fatigue is approximately $1/3 \cdot$ PWC (physical work capacity). Thus, PWC would be determined for each worker, and the model would be used to rate each task in the task array according to its proportion of the allowable $1/3 \cdot$ PWC.

EXERCISES

1. What purpose does a work physiology model fulfill?

2. What are the similarities between FEA and Mtm analysis?

3. Minimum squared error (MSE) is often used in curve fitting but was bypassed in favor of SAD for the work physiology model. Can you explain why?

4. Explain the meaning of X_{10} and its function in calibrating the work physiology model.

5. Using model Equations 6, 7, and 8, verify that $X_6 + X_7$ is a valid substitution for the denominator of Equation 9.1.

6. Although numerators and denominators differ in Equations 9.1 and 9.2, show that they do, indeed, provide the same assessment of efficiency by performing calculations for period 20 in Table 9.1.

7. Given that subject 1 of Wehrman's study had a PWC of 4.45 liters of oxygen per minute:

   a. Should X_9 or X_{11} be used for comparing a task to PWC?

   b. How did each of the three tasks compare to $1/3$ PWC?

8. Heart patient Tom Smith has a PWC of 3.95 liters of oxygen per minute. However, Mr. Smith's doctor has restricted physical labor to no more than $1/4 \cdot$ PWC for extended periods of time. Mr. Smith has an average efficiency of 0.25, an average yield of 4.5 kcal per liter, and a resting metabolism of 0.85 liters of oxygen per minute. Use these parameters in the work physiology

model and run the model to estimate what workload in watts Mr. Smith can handle for extended periods. (*Hint*: Values for X_{10} and X_{11} are not available.)

9. Experiment with improving the calibration procedure by prefiltering the primary X_{11} and auxiliary RQ data for task 1.

10. Using library resources, write a one- or two-page paper to contrast *epidemiological, biomechanical, psychophysical*, and *physiological* approaches to ergonomic studies.

APPENDIX F

Table 9.2 Respiratory Quotient Data for Work Physiology Model

Period	Task 1	Task 2	Task 5
1	0.75	0.87	0.83
2	0.79	0.89	0.86
3	0.77	0.87	0.85
4	0.76	0.95	0.87
5	0.76	0.87	0.88
6	0.73	0.88	0.86
7	0.72	0.91	0.82
8	0.74	0.89	0.90
9	0.75	0.91	0.87
10	0.74	0.90	0.83
11	0.77	0.84	0.79
12	0.79	0.91	0.83
13	0.82	0.90	0.81
14	0.80	0.92	0.84
15	0.81	0.90	0.85
16	0.82	0.93	0.85
17	0.82	0.94	0.87
18	0.84	0.93	0.87
19	0.86	0.95	0.88
20	0.86	0.96	0.87
21	0.88	0.93	0.87
22	0.86	0.94	0.85
23	0.87	0.98	0.84
24	0.85	0.98	0.85
25	0.87	0.98	0.86
26	0.90	0.97	0.85
27	0.92	0.98	0.86
28	0.92	1.00	0.85
29	0.91	0.93	0.89
30	0.93	0.97	0.92
31	0.93	1.08	0.87
32	1.02	0.99	0.85
33	1.16		0.83
34	1.11		0.84

continued on next page

continued

35	1.01	0.84
36	0.98	0.88
37	0.93	0.86
38	0.93	0.87
39		0.88
40		0.90

Table 9.3 Primary Data for Work Physiology Model

	Task 1		Task 2		Task 5	
Period	X_1	X_{11}	X_1	X_{11}	X_1	X_{11}
1	0	0.6210	0	0.4040	0	0.2815
2	0	0.5920	0	0.3585	0	0.1795
3	0	0.4990	0	0.3145	0	0.2305
4	0	0.5275	0	0.3880	0	0.1995
5	60	0.4710	25	0.4980	0	0.2345
6	60	0.6980	25	0.4180	0	0.2005
7	60	0.8700	25	0.4950	7	0.3170
8	60	0.8455	25	0.4710	7	0.3965
9	60	0.8390	25	0.5430	7	0.4915
10	60	0.8035	25	0.4780	7	0.5195
11	60	0.9365	25	0.5670	7	0.5375
12	60	0.8690	25	0.5950	7	0.5295
13	60	0.8710	50	0.5095	3	0.5795
14	60	0.8165	50	0.6305	3	0.4415
15	100	0.9225	50	0.7535	3	0.5045
16	100	1.0070	50	0.7170	3	0.4715
17	100	1.0685	50	0.7290	2	0.4895
18	100	1.0180	50	0.7030	2	0.4735
19	100	1.1270	50	0.7455	2	0.4410
20	100	1.1610	50	0.7100	2	0.4655
21	100	1.1490	75	0.7400	2	0.4725
22	100	1.0985	75	0.9035	2	0.4975
23	150	1.0493	75	0.8340	2	0.4640
24	150	1.2990	75	0.8910	2	0.4765
25	150	1.3685	75	0.9760	2	0.4590
26	150	1.4190	75	0.9335	2	0.3765
27	150	1.4545	75	1.0145	0	0.5005
28	150	1.4475	75	0.9200	0	0.3405

continued on next page

continued

29	150	1.5405	0 0.8450	10	0.4185	
30	150	1.0275	0 0.6445	10	0.4380	
31	0	1.3340	0	0.3410	10	0.5655
32	0	1.0125	0	0.2910	10	0.5725
33	0	0.7015			10	0.5040
34	0	0.6345			10	0.5640
35	0	0.5655			0	0.7760
36	0	0.4565			0	0.3975
37	0	0.4995			0	0.2960
38	0	0.4285			0	0.2700
39					0	0.2610
40					0	0.2365

Table 9.4 Secondary Data for Task 1

Period	X_2	X_4	X_5	X_6	X_7	X_9	X_{10}
1	.000000	.000000	.000000	.000000	.000000	.592000	.029000
2	.000000	.000000	.000000	.000000	.000000	.592000	.000000
3	.000000	.000000	.000000	.000000	.000000	.592000	−.093000
4	.000000	.000000	.000000	.000000	.000000	.592000	−.064500
5	.429900	.359076	.070824	.018806	.000000	.610806	−.139806
6	.429900	.658997	.129980	.034514	.160023	.786537	−.088537
7	.429900	.909507	.179390	.047634	.336044	.975679	−.105679
8	.429900	1.118747	.220660	.058593	.253227	.903820	−.058320
9	.429900	1.293515	.255131	.067746	.217840	.877586	−.038586
10	.429900	1.439492	.283923	.075391	.159977	.827368	−.023868
11	.429900	1.561420	.307972	.081777	.291903	.965680	−.029180
12	.429900	1.663260	.328059	.087111	.207956	.887067	−.018067
13	.429900	1.748323	.344837	.091566	.201703	.885269	−.014269
14	.429900	1.819373	.358851	.095287	.138336	.825623	−.009123
15	.716500	2.118101	.417771	.110932	.273833	.976766	−.054266
16	.716500	2.367616	.466985	.124000	.339887	1.055887	−.048887
17	.716500	2.576024	.508092	.134915	.383528	1.110444	−.041944
18	.716500	2.750099	.542426	.144032	.310755	1.046787	−.028787
19	.716500	2.895495	.571104	.151647	.411638	1.155285	−.028285
20	.716500	3.016938	.595057	.158008	.434857	1.184865	−.023865
21	.716500	3.118374	.615064	.163320	.412407	1.167728	−.018728
22	.716500	3.203099	.631775	.167758	.352504	1.112261	−.013761
23	.074750	3.573096	.704753	.187136	.322988	1.102124	−.052824
24	1.074750	3.882139	.765708	.203321	.564828	1.360149	−.061149
25	1.074750	4.140268	.816621	.216840	.611290	1.420131	−.051631
26	1.074750	4.355872	.859146	.228132	.641934	1.462066	−.043066
27	1.074750	4.535956	.894666	.237564	.660594	1.490158	−.035658
28	1.074750	4.686372	.924334	.245442	.638427	1.475869	−.028369
29	1.074750	4.812008	.949114	.252022	.721906	1.565928	−.025428
30	1.074750	4.916946	.969812	.257518	.187479	1.036997	−.009497
31	.000000	4.106906	.810041	.215093	.404666	1.211760	.122240
32	.000000	3.430315	.676591	.179658	.171567	.943225	.069275

continued on next page

continued

33	.000000	2.865189	.565126	.150060	−.058600	.683460	.018040
34	.000000	2.393164	.472025	.125338	−.089840	.627498	.007002
35	.000000	1.998903	.394261	.104690	−.126824	.569866	−.004366
36	.000000	1.669595	.329309	.087443	−.200620	.478823	−.022323
37	.000000	1.394538	.275057	.073037	−.150298	.514739	−.015239
38	.000000	1.164795	.229743	.061004	−.197569	.455436	−.026936

Note: $X_8 = 0.592000$, to complete the data set of Table 9.4.

Table 9.5 Secondary Data for Task 2

Period	X_2	X_4	X_5	X_6	X_7	X_9	X_{10}
1	.000000	.000000	.000000	.000000	.000000	.404000	.000000
2	.000000	.000000	.000000	.000000	.000000	.404000	−.045500
3	.000000	.000000	.000000	.000000	.000000	.404000	−.089500
4	.000000	.000000	.000000	.000000	.000000	.404000	−.016000
5	.179125	.139496	.039629	.008045	.000000	.412045	.085955
6	.179125	.248129	.070491	.014272	.010563	.428835	−.010835
7	.179125	.332729	.094525	.018985	.102062	.525046	−.030046
8	.179125	.398612	.113242	.022866	.057832	.484699	−.013699
9	.179125	.449919	.127818	.025671	.130380	.560051	−.017051
10	.179125	.489875	.139169	.028026	.052762	.484788	−.006788
11	.179125	.520991	.148009	.030294	.145897	.580191	−.013191
12	.179125	.545223	.154893	.031109	.165029	.600138	−.005138
13	.358250	.703590	.199883	.040253	.096256	.540509	−.031009
14	.358250	.826919	.234920	.047056	.217725	.668781	−.038281
15	.358250	.922964	.262206	.052803	.339384	.796188	−.042688
16	.358250	.997759	.283454	.056626	.279035	.739662	−.022662
17	.358250	1.056007	.300002	.059773	.283286	.747060	−.018060
18	.358250	1.101369	.312889	.062507	.250167	.716674	−.013674
19	.358250	1.136694	.322924	.064170	.286416	.754586	−.009086
20	.358250	1.164204	.330740	.065549	.247029	.716578	−.006578
21	.537375	1.325124	.376456	.075206	.310293	.789498	−.049498
22	.537375	1.450441	.412057	.082099	.463187	.949286	−.045786
23	.537375	1.548034	.439782	.086702	.367406	.858108	−.024108
24	.537375	1.624035	.461374	.090959	.419951	.914909	−.023909
25	.537375	1.683222	.478188	.094274	.498572	.996846	−.020846
26	.537375	1.729315	.491283	.097110	.448323	.949433	−.015933
27	.537375	1.765210	.501480	.098866	.522669	1.025534	−.011034
28	.537375	1.793163	.509421	.099906	.421525	.925431	−.005431
29	.000000	1.396446	.396718	.079253	.270582	.753836	.091164
30	.000000	1.087498	.308948	.061069	.124249	.589318	.055182
31	.000000	.846901	.240597	.046218	−.093898	.356320	−.015320
32	.000000	.659533	.187367	.036842	−.026918	.313924	−.022924

Note: $X_8 = 0.404000$

Table 9.6 Secondary Data for Task 5

Period	X_2	X_4	X_5	X_6	X_7	X_9	X_{10}
1	.000000	.000000	.000000	.000000	.000000	.281500	.000000
2	.000000	.000000	.000000	.000000	.000000	.281500	−.102000
3	.000000	.000000	.000000	.000000	.000000	.281500	−.051000
4	.000000	.000000	.000000	.000000	.000000	.281500	−.082000
5	.000000	.000000	.000000	.000000	.000000	.281500	−.047000
6	.000000	.000000	.000000	.000000	.000000	.281500	−.081000
7	.050155	.048016	.002139	.000440	.000000	.281940	.035060
8	.050155	.093985	.004186	.000843	.219419	.501762	−.105262
9	.050155	.137993	.006147	.001248	.309588	.592336	−.100836
10	.050155	.180125	.008023	.001647	.312418	.595565	−.076065
11	.050155	.220460	.009820	.002038	.314755	.598292	−.060792
12	.050155	.259075	.011540	.002368	.285881	.569749	−.040249
13	.021495	.268605	.011965	.002469	.308193	.592162	−.012662
14	.021495	.277729	.012371	.002532	.161549	.445581	−.004081
15	.021495	.286464	.012760	.002605	.226783	.510888	−.006388
16	.021495	.294826	.013133	.002681	.192866	.477046	−.005546
17	.014330	.295973	.013184	.002676	.205002	.489179	.000321
18	.014330	.297070	.013233	.002686	.190026	.474212	−.000712
19	.014330	.298121	.013279	.002689	.156943	.441132	−.000132
20	.014330	.299127	.013324	.002705	.182415	.466620	−.001120
21	.014330	.300090	.013367	.002714	.188901	.473115	−.000615
22	.014330	.301012	.013408	.002737	.215106	.499343	−.001843
23	.014330	.301894	.013447	.002752	.180782	.465035	−.001035
24	.014330	.302739	.013485	.002753	.192261	.476514	−.000014
25	.014330	.303548	.013521	.002752	.174739	.458991	.000009
26	.014330	.304323	.013556	.002767	.092735	.377002	−.000502
27	.000000	.291345	.012978	.002642	.206450	.490592	.009908
28	.000000	.278921	.012424	.002536	.054102	.338138	.002362
29	.071650	.335621	.014950	.003019	.160056	.444575	−.026075
30	.071650	.389904	.017368	.003479	.176877	.461855	−.023855
31	.071650	.441871	.019682	.003996	.322212	.607707	−.042207
32	.071650	.491623	.021899	.004470	.321056	.607026	−.034526
33	.071650	.539252	.024020	.004930	.240462	.526892	−.022892
34	.071650	.584851	.026051	.005332	.300220	.587052	−.023052
35	.000000	.559911	.024940	.005105	.468308	.754913	.021087
36	.000000	.536034	.023877	.004834	.105020	.391354	.006146
37	.000000	.513175	.022859	.004653	.009304	.295457	.000543
38	.000000	.491292	.021884	.004443	−.015423	.270520	−.000520
39	.000000	.470341	.020951	.004242	−.023815	.261927	−.000927
40	.000000	.450284	.020057	.004039	−.046889	.238650	−.002150

Note: $X_8 = 0.281500$

Table 9.7 History Matching for X_1

Period	VALUE	CHANGE	SUM*
1	−0.000024	−0.000024	−0.000024
2	−0.000004	−0.000004	−0.000028
3	−0.000004	−0.000004	−0.000032
4	−0.000004	−0.000004	−0.000036
5	59.000723	−0.000277	−0.000313
6	60.000209	0.000209	−0.000104
7	60.000222	0.000222	0.000118
8	59.999905	−0.000095	0.000023
9	59.999981	−0.000019	0.000004
10	59.999642	−0.000358	−0.000354
11	60.000405	0.000405	0.000051
12	60.000071	0.000071	0.000122
13	60.000135	0.000135	0.000257
14	59.999262	−0.000738	−0.000481
15	100.000563	0.000563	0.000081
16	99.999469	−0.000531	−0.000449
17	100.000542	0.000542	0.000093
18	100.000029	0.000029	0.000121
19	99.999671	−0.000329	−0.000207
20	100.000051	0.000051	−0.000156
21	100.000436	0.000436	0.000280
22	99.999366	−0.000634	−0.000354
23	150.000480	0.000480	0.000126
24	149.999443	−0.000557	−0.000431
25	150.000542	0.000542	0.000111
26	150.000020	0.000020	0.000131
27	149.999673	−0.000327	−0.000196
28	149.999987	−0.000013	−0.000208
29	150.000107	0.000107	−0.000101
30	149.999820	−0.000180	−0.000282
31	0.000697	0.000697	0.000415
32	−0.000353	−0.000353	0.000062
33	0.002383	0.002383	0.002444
34	−0.006061	−0.006061	−0.003617
35	−0.001304	−0.001304	−0.004921
36	0.003764	0.003764	−0.001157
37	−0.000344	−0.000344	−0.001501
38	0.000021	0.000021	−0.001480

*SUM = Cumulative algebraic value of change

CHAPTER 10

Macro Model of Blue Glacier

10.1 INTRODUCTION

Glaciers are one of the most intriguing elements in our natural environment. To the geologist, glaciers are a never-ending source of fascinating information and discovery. Glaciers are sensitive and dynamic. They move and respond in many different and subtle ways to changes in the environment. They have been larger and more numerous in the past and could possibly return to assume a more extensive role in the future.

Glaciers inherently present great potential for systems modeling as evidenced in the following:

- A glacier is bounded in space by the geographical features of its domain.
- A glacier is a body of ice subject to mass balance.
- A glacier, by its very nature, exhibits change over time.

Accurate systems models are possible for any bounded, conservative system such as a glacier. The Mtm approach provides an effective basis for modeling, since systems analysis and modeling proceed hand in hand. The most macro level of analysis yields a model of simplest detail, providing mass balance for only the system domain (a bounded region of space) over an extended period of time. Progressively more micro levels of analysis yield models that articulate increasingly greater detail. Knowledge about the system domain is obtained by inspection of the system from books and articles describing the system and from knowledge acquisition sessions with domain experts.

Blue Glacier is typical of those studied by glaciologists. Some descriptive information about Blue Glacier, one complete time series, and several isolated data were found in articles by LaChapelle [28], Armstrong [2], and in a document by Sharp [45]. This information and data, though sparse, provided adequate domain knowledge for synthesizing a macro model for Blue Glacier. From the available knowledge items, it was possible to identify a dynamic form for a macro model, calibrate the model, and generate a five-variable data base and extend it to seven variables. Thus, the first use of the model was to produce a complete data base commensurate with the detail articulated by the systems analysis. These data provided a base line for simulation testing of the model.

10.2 SYSTEMS ANALYSIS OF BLUE GLACIER

Glacial mass, although quantified in terms of cubic meters, is also represented by average thickness in meters. A glacier operates on a budget, that is, its ice/water mass satisfies a mass balance equation. The budget year, or equivalently the glacier year, does not start and end on the same calendar date each year. The dividing point between budget years is defined as that time when the accumulation of snow exceeds the ablation (wastage) of older material. Thus, the budget year begins when accumulation becomes established. During a year in which accumulation exceeds ablation, the glacier expands; when ablation predominates, the glacier contracts.

Blue Glacier is a small valley glacier draining from the northeast slope of Mount Olympus, the highest peak in the Olympic Mountains of northwestern Washington State. It is the largest of more than 250 glaciers in this region. In 1986, Blue Glacier was 4.3 km in length, had a volume of 0.5 cubic km, and covered a total area of 5.3 square km from an elevation of 1275 m to 2350 m. It features a high material turnover related to heavy winter snows and strong melting in the summer and exhibits a slightly positive average mass budget (0.3 m thickness) for the 31 years of record.

10.3 KNOWLEDGE BASE

- $L_{1,1}$ circumscription of the Blue Glacier system.
- Fundamental units of analysis:

  Space: Bed (domain) of Blue Glaciers in square meters.

  Mass: One unit = 1 meter (average *thickness* of glacial ice and snow expressed in equivalent *depth* of water)

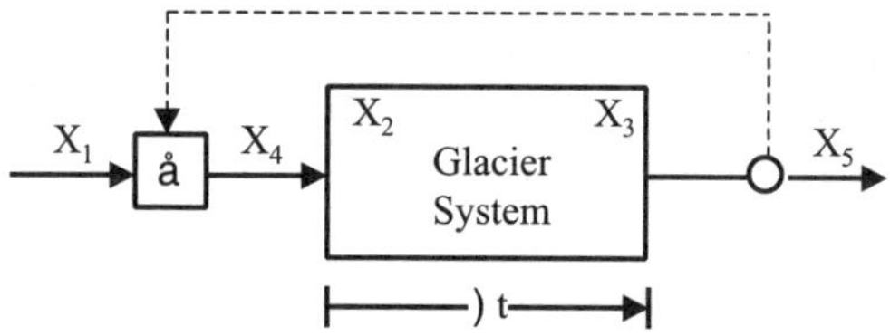

Figure 10.1 Blue Glacier System

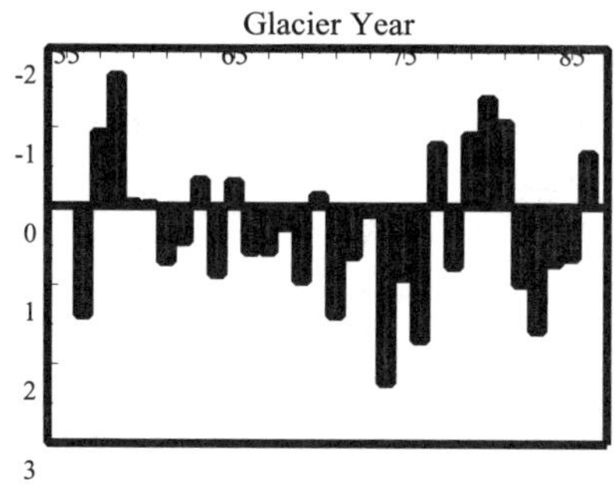

Figure 10.2 Primary Data: Blue Glacier System

Time: $\Delta t = 1$ budget (glacier) year

- Number of subsystems: one
- Schematic: See Figure 10.1.

The $\sum$ block makes X_1 equivalent to the net of X_4 and X_5, and exists solely to accommodate primary data.

- System variables:

X_1 = change in average depth (net rate, $\pm$)

X_2 = initial average depth (level)

X_3 = terminal average depth (level)

X_4 = average depth accumulation (annual input rate)

X_5 = average depth ablation (annual output rate)

- Primary data series, an exogenous stochastic form for X_1:

1. A 31-year realization of $\{X_1\}$, spanning the 31-year time base from 1956 through 1986, was reported by Armstrong [2], page 186, where it was noted that data for 1976, 1977, and 1978 were estimated with an expected accuracy of 0.2 to 0.3 m water equivalent. The X_1 series appears in Figure 10.2 and values are provided in Table 10.1 of Appendix G.
2. One value for X_3 was available from volume (mass) and area data, 0.5 km^3 and 5.3 km^2, respectively, for the end of the 1986 budget year, Armstrong, page 183: $X_3^{1986} = 0.5\ km^3/5.3\ km^2 = 94.34$ m.
3. A slightly positive average mass budget of 0.3 m for the 31 years of record, corresponding to $\overline{X}_1$ of the model was also reported by Armstrong, page 190.
4. One value for average depth ablation for the 1958 season was reported by LaChapelle [28], page 445: $X_5 = 5.2$ m.
5. Area at the end of the 1958 budget year was calculated from total mass ablation and thickness ablation data:

$$\frac{(22.1)\,10^6\ m^3}{5.2\ m} = 4.25\ km^2,$$

   LaChapelle, page 445.

- Vector table for model structure:

X_j	IT Pair	Exo	Endo	Equation Type	Equation No.
X_1		*		—	—
X_2	I	*		—	—
X_3	T		*	balance	1
X_4			*	equivalence	2
X_5			*	regression	3

- Dynamic form for continuous delay for X_5

10.4 CALIBRATION

Based on the above knowledge items, the model consists of three equations:

1. $X_2 + X_4 = X_3 + X_5$
2. $X_4 = X_1 + X_5$

3. $X_5 = c_1 X_2 + c_2 X_3$

$$X_2 = \begin{cases} X_2^{1956} & \text{initially,} \\ X_3 & \text{of previous year thereafter} \end{cases}$$

Calibration involves deriving an initial value for the $(X_2,\ X_3)$ IT pair and values for two regression coefficients.

A value for X_2^{1956} can be inferred from a submodel comprised of Equations 1 and 2:

$$X_2 = X_3 - X_1$$
$$X_3 = \begin{cases} X_3^{1986} = 94.34 & \text{initially,} \\ X_2 & \text{of following year therafter} \end{cases}$$

Values for the X_1 series and X_3^{1986} were available from primary data items (a) and (b), and the submodel was solved iteratively from 1986 to 1956, as indicated by the following sample calculations:

$$X_2^{1986} = 94.34 - (-0.68) = 95.02 = X_3^{1985}$$
$$X_2^{1985} = 95.02 - (0.66) = 94.36 = X_3^{1984}$$
$$\vdots$$
$$X_2^{1957} = 85.61 - (-0.94) = 86.55 = X_3^{1956}$$
$$X_2^{1956} = 86.55 - (1.39) = 85.16.$$

Consequently, the "terminal" value of X_2 for this reverse series yields the starting value, $X_2^{1956} = 85.16$. Furthermore, secondary data for the time series for the $(X_2,\ X_3)$ IT pair were inferred from the X_1 series and X_3^{1986} as the time base was traversed in the reverse direction. A plot of the X_3 series appears in Figure 10.3.

Figure 10.2 strongly infers stationarity for the X_1 time series. Dynamics of the glacier are embedded in this time series. X_1 provides the effect for which the model is to establish cause. Equation 3 generates output from the model. Fluctuations in X_5 (annual, average depth ablation) do not derive directly from X_4 (annual, average depth accumulation) but rather as a delayed loss from the weighted-average mass of the glacier:

$$X_5 = \frac{1}{D}[\alpha\, X_2 + (1 - \alpha)\, X_3] = c_1\, X_2 + c_2\, X_3.$$

Despite annual fluctuations in its average depth (Figure 10.2), Blue Glacier responds with medium inertia, as exhibited by the relative smoothness of its average depth series (see Figure 10.3). Consequently, $\alpha = 0.5$ was selected, which also specified that $c_1 = c_2$. Therefore, set $c_1 = c_2 = c$ in Equation 3. Although the X_5 series was unknown, primary data item (d) provided $X_5 = 5.2$ m for 1958.

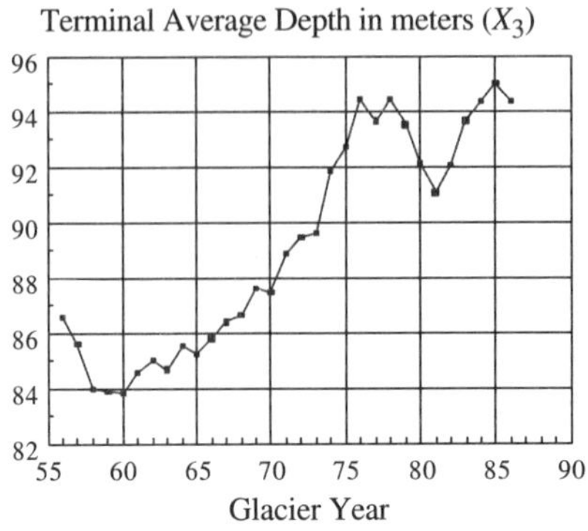

Figure 10.3 Terminal Average Depth: Blue Glacier System

Values for the X_2, X_3 series obtained by the submodel, were 85.61 and 83.95, respectively. Thus, c was calculated from Equation 3:

$$c = \frac{5.2}{85.61 + 83.95} = 0.030668 = c_1 = c_2.$$

A statistic that might otherwise be very difficult to obtain, is now easily inferred: $D = 1/(c_1 + c_2) = 16.30$ years, the average life span (or delay) for one unit of mass of Blue Glacier. Note that if $D \to \infty$, delay becomes infinite and $X_5 \to 0$, there is no ablation and the glacier experiences maximum growth. But as $D \to 0$, ablation soars, and the glacier hastily disappears.

10.4.1 The Tentative Model

1. $X_2 + X_4 = X_3 + X_5$
2. $X_4 = X_1 + X_5$
3. $X_5 = 0.030668X_2 + 0.030668X_3$

$$X_2 = \begin{cases} 85.16 & \text{initially,} \\ X_3 & \text{of previous year thereafter.} \end{cases}$$

10.5 MODEL VALIDATION

An $L_{1,1}$ macro model has been synthesized for Blue Glacier. Its major purpose is to serve as a touchstone or benchmark for models exhibiting greater detail. However, validity must first be established by subjecting the model to simulation testing. Three simulation tests are offered for illustration. These tests provide synthetic data, purportedly representing Blue Glacier dynamics, that invite critique from domain experts. If the model's response to a test receives positive concurrence from the experts, the model is said to have passed a Turing Type test.

10.5.1 Secondary Data Generation

Given the primary X_1 series, Blue Glacier dynamics were reconstructed by the model as secondary data: IT pair (X_2, X_3), X_4, and X_5 (see Figures 10.3 and 10.4 for X_3, X_4, and X_5. The complete data base is presented in Table 10.1 of Appendix G.

Averages for the X_2 and X_3 depth series were calculated: $\overline{X}_2 = 88.86$ m and $\overline{X}_3 = 89.15$ m. Their difference, $89.15 - 88.86 = 0.29$, compares favorably with the 0.3 m (slightly positive) average mass budget listed in primary data item (c). Average increase in average depth is (0.3)/(89.15) = 0.0034, or 0.34 percent per year. The relatively smooth X_3 series (Figure 10.3) displays the net effect of integration of X_4 (accumulation) and X_5 (ablation) over each of the 31 years of record.

The X_4 series (Figure 10.4), representing the annual rate of accumulation, shows sharp differentiation from one year to the next. The X_5 series (Figure 10.4), on the other hand, represents the annual rates of ablation and exhibits great smoothness. Smoothing is realistic in that the glacial body mass is characterized by time persistent caloric inertia. That is, once the snow from a given accumulation season is absorbed into the body of the glacier, thermal energy beyond that of the immediate ablation season may be required to remove an equivalent amount of mass.

LaChapelle [28], page 445, reported a specific annual deficit for 1958 of 1.7 m of water and that this deficit is about 1 to 2 percent of the total glacier mass. The model infers a deficit for 1958 of $X_4 - X_5 = 3.54 - 5.20 = -1.66$ m, or a deficit relative to $X_3 =$ 83.95 m of $(1.66\,\text{m}/83.95\,\text{m})\times 100\% = 1.98$ percent.

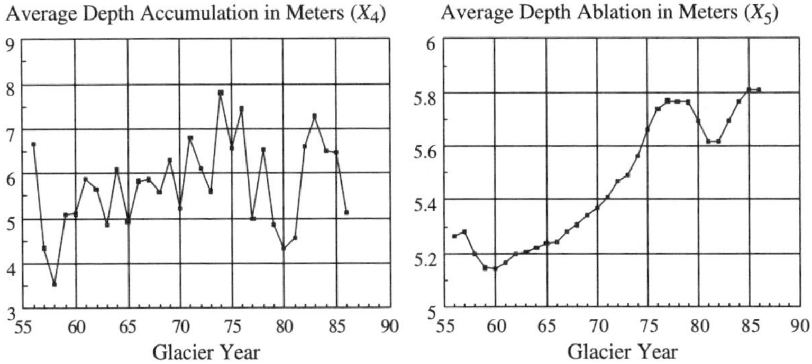

Figure 10.4 Secondary Data: Blue Glacier System

10.5.2 History Matching

Given a complete data base, the accuracy of the computer model and computer combination was tested through history matching. The exogenous role of primary X_1 was interchanged with the endogenous role of X_4. Thus, X_4 was selected to be exogenous (read from the data base), so that values of X_1 were computed by the model. These computed values of X_1 match historical data for X_1 to at least six decimal places. Cumulative algebraic error for the 31 years was -1.0×10^{-6} m, thus confirming the accuracy of the model and computer combination in reproducing the X_1 series. The maximum cumulative algebraic error introduced by this role exchange occurred for secondary variable X_3, averaging -0.67×10^{-6} m/year.

10.5.3 Perturbation

The model can be subjected to many, different perturbation tests. These tests examine the model's response to "What if" types of questions. The purpose of each test is to determine what will happen if one or more of the independent variables of the completed data base are perturbed. To illustrate such a test, the following question was posed: "What if X_4 (average depth accumulation) were reduced to some percent of its data base value so that Blue Glacier exhibits a balanced budget (net of accumulation and ablation is zero) as registered by X_3 (terminal average depth), over the 31 years of record? What percent reduction might effect this balance?" To answer this question, reduce each value of X_4 in the data base by an equal, small percentage. Then run the model as structured in the previous test. Rerun the model, each time increasing the percentage until budget balance is achieved.

The reduction was found to be 11.25 percent. That is, the model suggests that if actual accumulation (the X_4 series) had been 88.75 percent of the data base series, Blue Glacier would have experienced a zero trend, exhibiting the same average depth (X_3) at the end of the 31 years as that which existed at the beginning of the 31 years ($X_2^{1956} = 85.16$ m). The balanced X_3 series appears in Figure 10.5.

10.6 STATIC RELATIONSHIPS

Response of a system model is *dynamic* because systems experience change with respect to time (or space). Nevertheless, for some fixed time duration, a system domain may possess characteristics that are time invariant. Equations that incorporate such characteristics are said to be *static*. Examples of time invariant characteristics for Blue Glacier include elevation levels, glacial bed contours, confining walls, slopes, and so on, as well as physical constants such as the heat of fusion of ice. Utilizing the systems analysis and modeling principles that comprise the

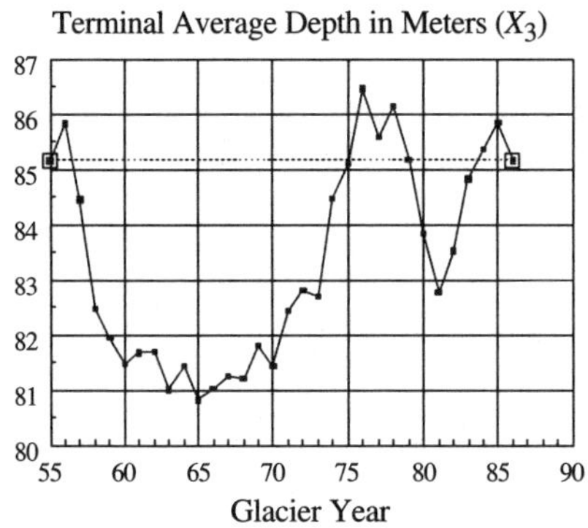

Figure 10.5 X_3 Balanced Budget Series

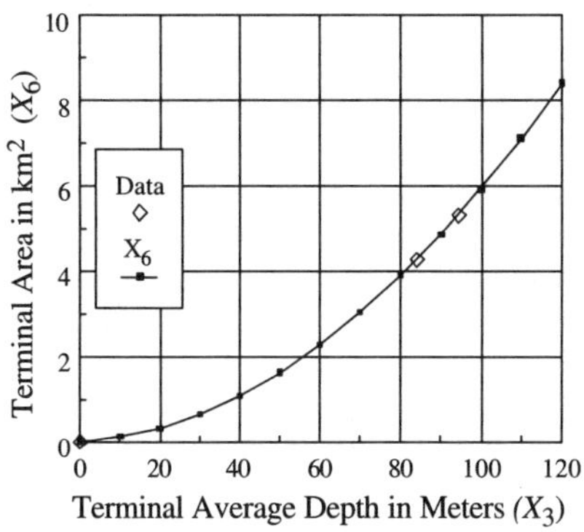

Figure 10.6 Area versus Depth of Blue Glacier

Mtm approach, these characteristics become essential to the circumscription and calibration of more micro models. However, calculations associated with static relationships between variables of this $L_{1,1}$ model provide additional descriptive information and thereby extend its usefulness. Two examples will illustrate.

10.6.1 Area of Blue Glacier

Rate of change of *area* with respect to rate of change in *depth* (water equivalent of *thickness*) obviously is not constant but rather depends on the spatial detail of the glacial bed: the geometry of its cross-sectional contours coupled with elevation change. A minimum of three points are required to illustrate the nonlinear, static relationship between *terminal area* and terminal depth, X_3. Three points are available from primary data items for 1958, 1986, and the coordinate axes origin (zero depth, zero area). See Figure 10.6 where terminal area is designated by X_6. Examination of the smooth curve connecting the three points reveals that a fitted, second-order polynomial would better serve to interpolate between the three points

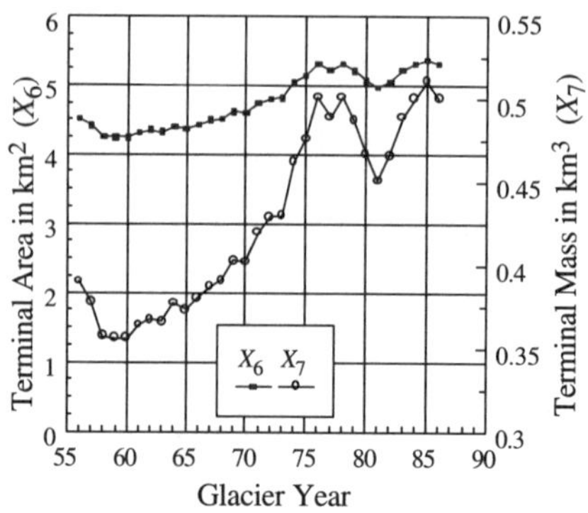

Figure 10.7 Terminal Area and Mass of Blue Glacier

than would a straight line. The *predicting* equation with zero-axis intercept is

$$X_6 = a\,X_3 + b\,X_3^2. \tag{10.1}$$

Two unknown polynomial coefficients, a and b, require two nonzero data points, supplied from 1958 (primary data item e and Table 10.1 of Appendix B), and 1986 (primary data item b):

$$4.25 = 83.95a + (83.95)^2 b$$
$$5.30 = 94.34a + (94.34)^2 b$$

Simultaneous solution yields $a = 0.005715$ and $b = 0.000535$. Figure 10.6 displays the second-order curve for X_6 that passes through the two data points. Equation 10.1 was used to generate a time series for X_6 that estimates terminal areas of Blue Glacier for the 31-year period. See Figure 10.7 and Table 10.2 of Appendix G. In a more micro model, the nonlinear, static relationship supplied by Equation 10.1 would be replaced by dynamic, linear equations that incorporate spatial details of the glacier bed.

10.6.2 Mass of Blue Glacier

Of additional interest is the water equivalent mass of Blue Glacier. The time series, X_7, representing *terminal average mass* of Blue Glacier, is defined by the following equation:

$$X_7 = X_3 \cdot X_6.$$

This series, shown in Figure 10.7, averaged 0.427 km^3 and is listed in Table 10.3 of Appendix G. The mass at the end of the 1986 glacier year is 0.500 km^3 and matches that shown by primary data item (b). To place in perspective, 1 acre-foot of water will cover an acre consisting of 43,560 ft^2 to a depth of 1 foot. In metric equivalence, 1 acre-foot = (1.2334877) 10^3 m^3. Thus, 0.500 km^3 converts to (0.405355) 10^6 acre-feet, about 30 percent of the content of Yellowtail Reservoir lying on the Bighorn tributary of the Yellowstone River in Montana.

10.7 FURTHER MODELING

By design, the macro model of Blue Glacier was not detailed. Consequently, domain experts might readily critique the data generated for its few variables. Upon its validation, the $L_{1,1}$ model provides an acceptable basis for extended modeling of the glacier. Again by design, each additional extension adds limited but significant detail. For example, a second-level model might articulate greater detail by subdividing the $L_{1,1}$ model as follows:

- Space: Divide the domain of Blue Glacier (in square meters) into two subdomains: an accumulation zone and a wastage zone.
- Mass: Retain quantification in cubic meters with representation by average thickness in meters.
- Time: Divide the glacier's budget year into two periods: one for which accumulation exceeds wastage and the other for which wastage exceeds accumulation.

More micro levels of modeling will require that phenomena associated with accumulation and wastage zones be parameterized—for example, rain, snow, hoar frost, refrozen meltwater, radiation, convection, conduction, turbulence, humidity, and condensation—as related to temperature, spatial detail of the glacier bed, and elevation. Just as time is incremented in Δt units, space may be incremented in Δx, Δy, and Δz units. Obviously, much more field data are required, and from a practical standpoint, it becomes paramount that benefits justify costs of further systems analysis and modeling. In the future, glacier systems models may assume a strategic role in the ongoing debate over global warming.

EXERCISES

1. How does the "glacier year" differ from a "water year" that runs from October 1 to October 1?
2. Using library resources, compile a narrative, physical description of one of the Alaskan glaciers.
3. Why is a glacier's equivalent *depth* of water a more convenient measure of mass for a macro model than its average *thickness*?

4. Suppose that in primary data item (b), true terminal average depth varies from $X_3^{1986} = 94.34\,\mathrm{m}$ by as much as $\pm 10\%$. Recalibrate and rerun the model
    a. after decreasing 94.34 by 10%, and then
    b. after increasing 94.34 by 10%.
    c. Graph "error" bands corresponding to 94.34 $\pm 10\%$ for the X_3, X_4, and X_5 series.
    d. Discuss sensitivity of secondary data to possible error in X_3^{1986}.

5. Suppose that in primary data item (d), true average depth ablation varies from $X_5^{1958} = 5.2\,\mathrm{m}$ by as much as $\pm 10\%$. Recalibrate and rerun the model
    a. after decreasing 5.2 by 10%, and then
    b. after increasing 5.2 by 10%.
    c. Graph "error" bands corresponding to 5.2 $\pm 10\%$ for the X_4 and X_5 series.
    d. Discuss sensitivity of secondary data to possible error in X_5^{1958}.

6. Refer to the data points for years 1958 and 1986 used to calibrate Equation 10.1.
    a. Calculate the ratios of area to depth for the two years.
    b. What may be learned from the magnitudes of the ratios in (a) regarding the type of regression relationship between X_6 and X_3 for the 31 years of record?

7. Refer to the data points for years 1958 and 1986 used to calibrate Equation 10.1.
    a. Fit a linear predicting equation of the form $X_6 = a + bX_3$ to the points.
    b. Calculate areas using the linear equation and compare to values from Equation 10.1 presented in Table 10.2 of Appendix B.

8. Synthesize an $L_{1,2}$ model by adding an area subsystem to the $L_{1,1}$ model, assuming linearity between depth and area within the historical range of variation of these variables as in the previous exercise. *Hint*: Use reverse iteration from 1958 to 1956 to discover a starting value for initial area, X_6^{1956}. Generate a complete data base including both subsystems.

9. Suppose that a long-term cooling effect causes Blue Glacier to increase in mass until the nearly vertical canyon walls become the dominant constraint to growth of glacial area. Suggest modifications for Figure 10.6 and Equation 10.1 with potential to better represent the relationship between area and depth.

APPENDIX G

Table 10.1 Data Base for Blue Glacier Model

Year	X_1	X_2	X_3	X_4	X_5
1956	1.390000	85.160000	86.550000	6.656002	5.266002
1957	−.940000	86.550000	85.610000	4.339803	5.279803
1958	−1.660000	85.610000	83.950000	3.540066	5.200066
1959	−.060000	83.950000	83.890000	5.087317	5.147317
1960	−.040000	83.890000	83.850000	5.104250	5.144250
1961	.710000	83.850000	84.560000	5.874798	5.164798
1962	.450000	84.560000	85.010000	5.650373	5.200373
1963	−.340000	85.010000	84.670000	4.863746	5.203746
1964	.870000	84.670000	85.540000	6.090000	5.220000
1965	−.320000	85.540000	85.220000	4.916868	5.236868
1966	.580000	85.220000	85.800000	5.824841	5.244841
1967	.580000	85.800000	86.380000	5.860416	5.280416
1968	.270000	86.380000	86.650000	5.576484	5.306484
1969	.950000	86.650000	87.600000	6.293899	5.343899
1970	−.150000	87.600000	87.450000	5.218433	5.368433
1971	1.390000	87.450000	88.840000	6.796462	5.406462
1972	.640000	88.840000	89.480000	6.108718	5.468718
1973	.100000	89.480000	89.580000	5.591412	5.491412
1974	2.240000	89.580000	91.820000	7.803175	5.563175
1975	.910000	91.820000	92.730000	6.569779	5.659779
1976	1.700000	92.730000	94.430000	7.439823	5.739823
1977	−.790000	94.430000	93.640000	4.977731	5.767731
1978	.770000	93.640000	94.410000	6.537117	5.767117
1979	−.910000	94.410000	93.500000	4.852824	5.762824
1980	−1.370000	93.500000	92.130000	4.322901	5.692901
1981	−1.070000	92.130000	91.060000	4.548071	5.618071
1982	.990000	91.060000	92.050000	6.605617	5.615617
1983	.580000	92.050000	93.630000	7.274434	5.694434
1984	.730000	93.630000	94.360000	6.495277	5.765277
1985	.660000	94.360000	95.020000	6.467906	5.807906
1986	−.680000	95.020000	94.340000	5.127292	5.807292

Table 10.2 Terminal Area in square km of Blue Glacier

Year	Area (X_6)	Year	Area (X_6)
—	—	71	4.73
56	4.50	72	4.79
57	4.41	73	4.81
58	4.25	74	5.04
59	4.24	75	5.13
60	4.24	76	5.31
61	4.31	77	5.23
62	4.35	78	5.31
63	4.32	79	5.21
64	4.40	80	5.07
65	4.37	81	4.96
66	4.43	82	5.06
67	4.49	83	5.23
68	4.51	84	5.30
69	4.61	85	5.37
70	4.59	86	5.30

Table 10.3 Terminal Mass in cubic km of Blue Glacier

Year	Mass (X_7)	Year	Mass (X_7)
—	—	71	0.420
56	0.390	72	0.429
57	0.378	73	0.430
58	0.357	74	0.462
59	0.356	75	0.476
60	0.356	76	0.501
61	0.364	77	0.489
62	0.367	78	0.501
63	0.366	79	0.487
64	0.377	80	0.467
65	0.373	81	0.451
66	0.380	82	0.466
67	0.387	83	0.489
68	0.391	84	0.500
69	0.403	85	0.511
70	0.402	86	0.500

CHAPTER 11

State Water Planning Model

11.1 INTRODUCTION

Proper management of natural resources is essential if economic benefits are to be balanced against environmental costs. However, increasing demands for resource development and utilization have made balanced management increasingly more difficult. Montana, as a Columbia and Missouri headwaters state, depends on water to support its chief industries, agriculture and tourism. From early on, the Montana Legislature has recognized the need for reserving adequate flows to meet increasing demands within the state. Matching funds provided to researchers at Montana State University by the Montana Legislature and the U.S. Department of Interior resulted in the development of a fully operational state water planning model, the first of its kind in all the 50 states [7].

Mtm systems philosophy was used to develop a series of systems models, readily solved by a computer. This philosophy represented a significant departure from the conventional "micro-to-macro" approach utilized by many previous hydrological modeling ventures. The typical practice was to make a differential analysis of flow at a point for an instant of time as the basis for estimating flows for finite elements of space and time and to extend this finite-element approximation to include successively larger geographical regions. Nevertheless, the sheer size of Montana, its complex river basin systems, its unique water-rights structure, and sparsity of data did, indeed, preclude such an approach.

However, the Mtm approach reversed this order and was not beset by these difficulties [10]. Rather than starting from a complex model of a small geographic area and extrapolating to larger areas, development began with some relatively

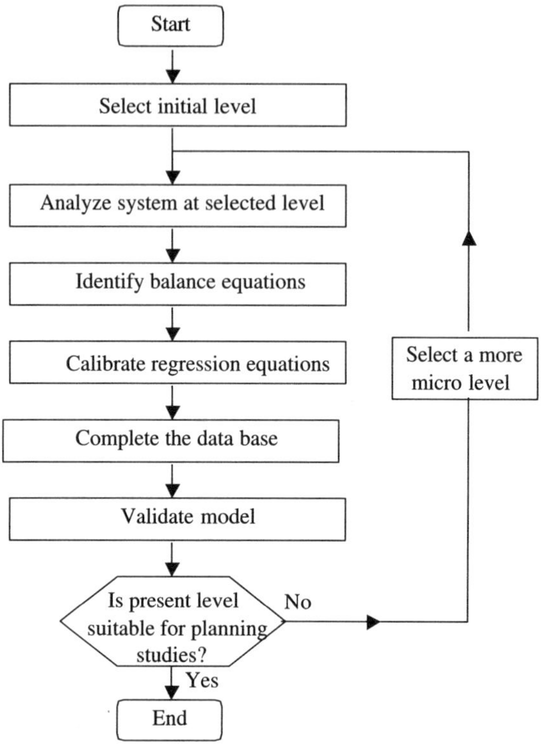

Figure 11.1 Macro-to-Micro Planning Procedure

simple models of large geographical areas. Thus, it was possible to encounter and overcome many problems on a relatively easy level. Progressing from macro to micro levels yielded significant economies in time and effort and was therefore very cost effective. First of all, it encompassed the entire region under consideration, thus ensuring comprehensive systems analysis and modeling. Second, the process of hierarchical analysis was comparatively simple. Third, model development and computer programming were tested and revised at a macro level and then extended confidently to more micro and complex models.

Figure 11.1 presents a block diagram of the Mtm modeling procedure. The "start" block represents the process by which the knowledge engineer becomes familiar with the problem domain (system problem) selected for systems modeling. The "end" block represents transfer of the systems model to the user, or client. Iterating from an initial macro level to a final level of detail makes up the macro-to-micro approach.

Water resources for the entire state were modeled for an annual time unit, using as few variables as consistent with that aggregate level of detail. The model was progressively refined to more micro and complex versions associated with smaller geographical regions and shorter time increments.

Table 11.1 Planning Model Mtm Series

Level	Model Type	Space Unit	Time Unit	Number of Subsystems	Number of Variables	Number of Equations
$L_{1,0}$	Regression	State	1 year	0	5	1
$L_{1,1}$	System	State	1 year	1	6	1
$L_{1,2}$	Systems	State	1 year	2	12	5
$L_{2,2}$	Systems	Basin	1 year	2	13	7
$L_{3,4}$	Systems	Basin	6 months	4	22	12
$L_{4,4}$	Systems	Basin	1 month	4	22	12
$L_{5,2}$	Systems	Subbasin	1 year	2	12	6
$L_{6,5}$	Systems	Subbasin	1 month	5	31	21
$L_{6,6}$	Systems	Subbasin	1 month	6	34	18

Research and development proceeded as if opening a series of black boxes:

- The state of Montana
- Yellowstone River Basin in Montana
- Subbasin 43-Q of the Yellowstone River Basin.

Each geographical unit (the space unit) was viewed as a black box into which water flows, falls, is stored, is consumed, and out of which water flows. Figure 11.2 presents these three geographical units.

A series of nine models was systematically defined and calibrated, each model more complex than the previous model. Compilation of primary data lagged model development, which is not unusual when modeling is based on knowledge derived through systems analysis. An $L_{1,2}$ model, encompassing the entire state, was synthesized six months before data became available for testing the model. Water, the unit of mass, remained constant throughout the series: One unit equals one million acre-feet. The series are summarized in Table 11.1.

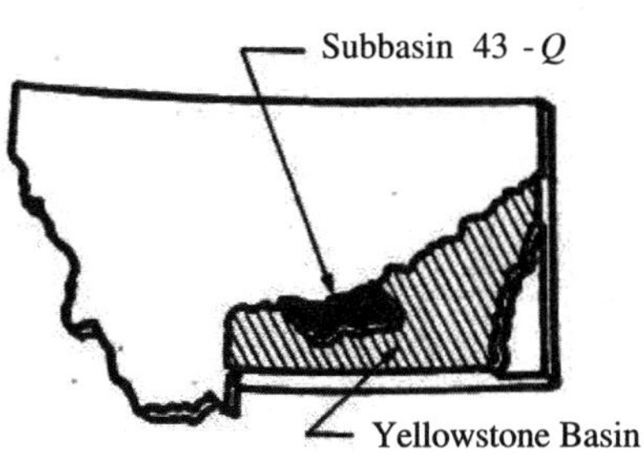

Figure 11.2 State of Montana and Location of Yellowstone River Basin and Subbasin 43-Q

A knowledge base consisting of raw information was compiled for each $L_{i,j}$ level of analysis. Knowledge items consisted of system components, relationships, data even though sparse, and heuristics. Each systems model was derived from a physical, system components basis rather than from the basis of data availability. All essential variables (required to complete mass-balance equations of each model) had a one-to-one correspondence with system components. Although many primary data were unavailable, all essential variables were included in the model. First use of the model was to generate secondary data to substitute for missing primary data, thus providing the advantage of a complete data base. Further advantage of including all essential variables is realized if in the future additional primary data become available. Secondary values may then be replaced by primary values without necessitating major revision of the model.

Water planning models are used to assess effectiveness of potential management policies. A model can circumscribe a single subbasin, multiple basins joined in series, and parallel combinations to cover some or all of a river basin, or even all of a political region such as a state. Montana contains six river basins, and each is segmented into tributary or main-stem subbasins. Subbasin 43-Q is but one of 19 subbasins of the Yellowstone Basin. However, to review development of all nine models is too extensive for a single chapter. Therefore, only synthesis of the $L_{2,2}$ model of the Yellowstone River Basin within Montana is presented.

11.2 SYSTEMS ANALYSIS OF YELLOWSTONE RIVER BASIN

The Yellowstone is a major tributary of the Missouri River, its headwaters arising in Wyoming and entering Montana at its southern border. Its flow leaves Montana at Sidney, just prior to its confluence with the Missouri. Bedrock is overlaid by a floodplain that consists largely of alluvial boulders, gravels, and sand, the product of glacial action and erosion. These alluvium, together with clays and silts, produce a strong correlation between groundwater and surface water flow. High mountains at its upper tributaries receive an abundant snowpack that provides a major source of surface flow and groundwater replenishment. Spring snow melt, accelerated by warm rains, produces peak runoff in May or June. Its water is valued as a major resource for agriculture, industry, cities, recreation, and wildlife. Its upper reaches provide blue ribbon fishing, while its lower reaches are navigable. Yellowtail Reservoir, located on the Big Horn tributary and completed in 1966, has a capacity of 1.375 million acre-feet. The Basin contains numerous other smaller impoundments.

Basic components of the river basin system that belong to the same mass balance were classified into two homogeneous subsystems. Knowledge items, acquired through systems analysis, are listed by numbered item.

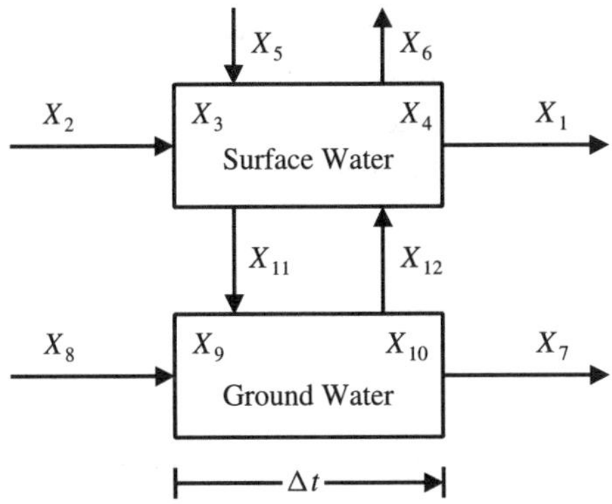

Figure 11.3 Yellowstone River Basin System

11.3 KNOWLEDGE BASE

1. $L_{2,2}$ circumscription of the Yellowstone River Basin system.
2. Fundamental units of analysis:
   - Space: Yellowstone River Basin in Montana
   - Mass: One unit = one million acre-feet of water
   - Time: $\Delta t = 1$ year
3. Two subsystems:

   SS_1: Surface water, all surface flows and impoundments subjected to atmospheric evaporation loss.

   SS_2: Subsurface water, all groundwater flows and aquifers not subjected to atmospheric evaporation loss.
4. Schematic: See Figure 11.3.
5. System variables:

   SS_1: Surface water
   X_1 = surface outflow
   X_2 = surface inflow
   X_3 = initial surface storage
   X_4 = terminal surface storage
   X_5 = precipitation
   X_6 = surface loss
   X_{11} = percolation
   X_{12} = subsurface discharge
   X_{13} = unexplained surface loss

SS_2: Subsurface (or ground) water
X_7 = subsurface outflow
X_8 = subsurface inflow
X_9 = initial available capacity
X_{10} = terminal available capacity
X_{11} = percolation
X_{12} = subsurface discharge

6. Description of variables:

    - X_1 is the flow leaving the Basin at Sidney via the main stem of the Yellowstone.
    - X_2 consists of all surface flows that arise in Wyoming and enter the Montana portion of the Yellowstone River Basin.
    - IT pair (X_3, X_4) is the aggregate of all surface water impoundments such as reservoirs, lakes, and ponds.
    - X_5, precipitation within the Basin, occurs as both rain and snow. However, any difference in snowpack at the beginning of water years (October 1 to October 1) is negligibly small.
    - X_6 lumps together all forms of water vapor loss from the surface water subsystem, such as, evaporative loss from bodies of water, interception loss from nonwater surfaces, and transpiration loss from vegetation.
    - X_7 and X_8 are composite subsurface flows, analogous to X_1 and X_2.
    - IT pair (X_9, X_{10}) represents the potential for storage of groundwater (in millions of acre feet) in the porous strata above the piezometric surface (water table). Thus, their supply/demand roles are reversed, requiring an algebraic sign reversal.
    - X_{11} is composed of all flows entering the groundwater subsystem from the surface water subsystem.
    - X_{12} is the aggregate of all flows leaving the subsurface to the surface, for example, springs and pumping.
    - X_{13}, unexplained surface loss, was included as a component of X_6 during a revision of the model that required explicit definition of two of the several forms of losses.

7. Primary data: A 39-year historical record, years 1929 through 1967, was selected because it contains a classical drought cycle. Raw data from this record were scaled, interpolated, extrapolated, and compiled to obtain exogenous dynamic forms expressed in millions of acre-feet for the five primary variables of the model.

- X_1, X_2 surface flows: U.S. Geological Survey (USGS) stream gage data were used to obtain the composite outflow (X_1) and composite inflow (X_2). In some instances regression techniques and missing data routines were applied.
- X_3, X_4 surface storages: Values for surface reservoir storage were taken from USGS records and summed to reflect aggregate storage in the Yellowstone Basin in Montana. However, some of the smallest reservoirs were excluded because of their negligible size.
- X_5 precipitation: Precipitation depth readings in inches were obtained from U.S. National Weather Service records. Depths were converted into volumes in acre-feet via the Thiessen polygon method for 1000-foot increments of elevation within polygons [44].

Time series plots for four of these five data (X_3 is the same as X_4 of the previous year) are presented in Figure 11.4. All series met the criteria for classification as a homogeneous, stationary time series, with the exception of IT pair X_3, X_4 for the last two years as Yellowtail Reservoir came on line. However, this anomaly is not reflected in the other series. Data for these five primary exogenous forms for the 39-year record are presented in Table 11.3 of Appendix H.

8. Vector table for model structure:

X_j	IT Pair	Exo	Endo	Equation Type	Equation No.
X_1		*		—	—
X_2		*		—	—
X_3	I	*		—	—
X_4	T	*		—	—
X_5		*		—	—
X_6			*	balance	1
X_7			*	regression	2
X_8			*	regression	3
X_9	I	*		—	—
X_{10}	T		*	balance	4
X_{11}			*	regression	5
X_{12}			*	regression	6
X_{13}			*	regression	7

The two balance equations were obtained from Figure 11.3 by assigning *demand* variables as negative components and *supply* variables as positive components, and are presented next.

11.3.1 Surface Water Subsystem

Equation 1 $X_6 = -X_1 + X_2 + X_3 - X_4 + X_5 - X_{11} + X_{12}$ for surface loss. Five of these eight surface variables have primary data. Rate variables X_{11} and X_{12} that link the two subsystems are defined by regression equations. Therefore, X_6 is the only variable left to take definition from Equation 1. As noted in knowledge item 6, X_6 was divided into *explained* and *unexplained* loss components. *Explained* losses required one additional regression equation for X_{13}.

11.3.2 Ground Water Subsystem

Equation 4 $X_{10} = X_7 - X_8 + X_9 - X_{11} + X_{12}$ for terminal available capacity. None of these six subsurface variables have primary data. Rate variables X_7 and X_8 are defined by regression equations. Thus, Equation 4 defines the *IT solution* pair, X_9, X_{10}. Given X_9, solve for X_{10}.

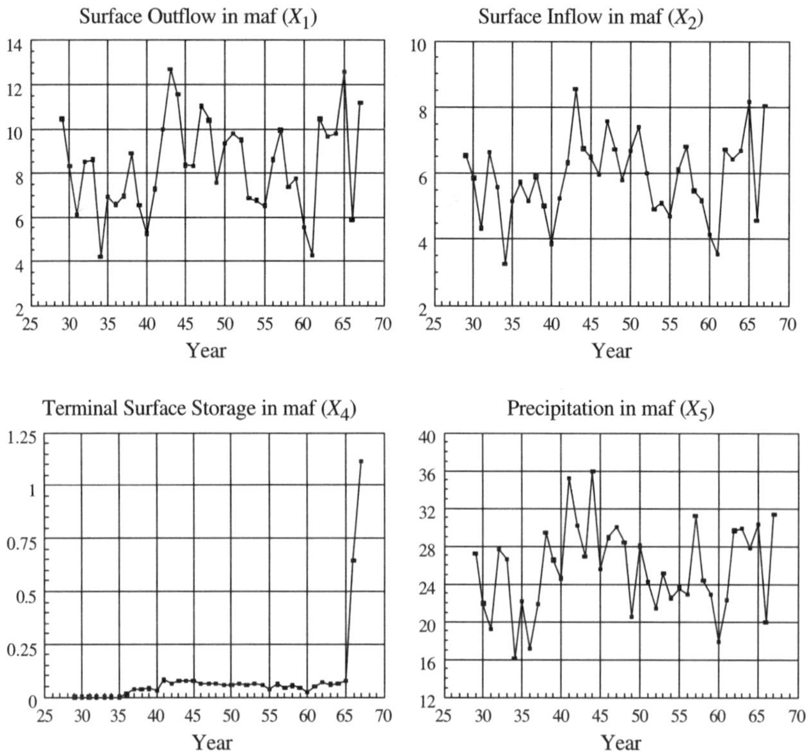

Figure 11.4 Primary Time Series for Yellowstone River Basin

Table 11.2 Yellowstone River Basin Inference Matrix

Endo	X_1	X_2	X_3	X_4	X_5	X_6	X_7	X_8	X_9	X_{10}	X_{11}	X_{12}	X_{13}	K
X_7	**+S**	+W	+W	+W	+W	+W		+W	−M	−M	+S	−S	+M	Y
X_8	+W	**+S**			+M	+W	+M		−W	−W	+M	−M	+W	Y
X_{11}	**+S**	**+S**	**+M**	**+M**	**+S**	+M	+S	+M				+W		
X_{12}	+M	+M			+W	+M	+M	+W	**−S**	**−S**	+W			Y
X_{13}	**−S**	**−S**	**−M**	**−M**	−S	**+1**	−W	−W	−W	−W	−M			

11.4 CALIBRATION

Five regression equations required calibration. For homogeneous stationary time series, the following calibration principle was applicable: "If the systems model is calibrated to the average Δt period of the time series, then it is calibrated to any other Δt period." For calibration purposes only, X_j was replaced by $\overline{X}_j$. Values for the hypothetical average year for the five primary data in millions of acre feet are as follows:

$$\overline{X}_1 = 8.360959$$
$$\overline{X}_2 = 5.874492$$
$$\overline{X}_3 = 0.059273$$
$$\overline{X}_4 = 0.087823$$
$$\overline{X}_5 = 25.581301$$

11.4.1 Knowledge Acquisition via Inference Matrix

Primary data were not available for the groundwater subsystem and may possibly never become available. However, the knowledge and experience of domain experts provided testable substitutes for primary data by exploiting system relationships. An *inference matrix* was used as a knowledge acquisition tool to infer possible *physical* and/or *statistical correlationships* between *dependent* (in this instance, *endogenous*) variables and *independent* variables. However, circumscription confined selection of homogeneous *independent* variables for each of the five regression equations to variables from the X_j column of the vector table. Table 11.2 displays postulated regression relationships based on interviews with hydrologists from the Civil Engineering Department of Montana State University and the Montana Department of Natural Resources and Conservation (DNRC).

Coded elements of the matrix *infer* the existence of correlation, either positive or negative, but make no distinction between physical and statistical correlations. Strength of correlation between each *dependent* variable and each possible *independent* X_j was coded "S," "M," or "W," indicating *strong*, *moderate*, or *weak* correlation, respectively. The impact of independent variables omitted from three of the regression equations by circumscription was acknowledged by including a vertical axis intercept, the *constant form* K, as indicated by "Y" for *yes* in the K column.

Although use of the inference matrix in Table 11.2 provided excellent focus of attention, several sessions were required with the domain experts. Selected correlations are bolded, thus providing explicit regression relationships for Equations 2, 3, 5, 6, and 7. A bolded "1" appears in the X_6 column for X_{13} (Equation 7) because $X_{13} = X_6 - \textit{explained surface losses}$.

11.4.2 Extension of Knowledge Base Using Reverse Regression

All five regression equations pertain to the groundwater subsystem for which there were no primary data to enable curve fitting by forward regression. Therefore, the initial systems analysis was inadequate to support calibration. Consequently, the *reverse regression principle* applies: "Given system knowledge, regression coefficients can be found that generate data consistent with that knowledge." Dynamic forms and heuristics for their calibration were obtained as additional knowledge items. Dynamic forms were identified through systematic, term-by-term analysis of the model's regression equations. Calibration heuristics, as *rules of good judgment*, were obtained by probing domain experts' knowledge of the Yellowstone River Basin.

Each regression equation was visually inspected for standard dynamic forms. Heuristics are enclosed in quotes. Calibration to the average year required use of averages in the following calculations.

Equation 2 $X_7 = \underbrace{c_1 X_1}_{T_1} + \underbrace{c_2}_{T_2}$ for subsurface outflow.

$T_1 = G \cdot X_1$ is a scaled *stochastic form*, where $c_1 = G$ establishes the magnitude of statistical correlation.

$T_2 = c_2$ is a *constant form* that inserts the average effect of those omitted independent variables that account for continued groundwater flow even though surface flow may dwindle to zero. Heuristic knowledge items 9, 10, and 11 apply to both subsurface outflow and inflow:

9. "Although subsurface flow has a *relatively* strong positive correlation with surface flow (compared to other independent variables), its *absolute* value is quite small." Consequently, the selection $G = 0.005$ provided $c_1 = 0.005$.

10. "Subsurface flow is obstructed and quite small compared to free surface flow." Therefore, average subsurface flow activity was estimated to be 2% of average surface flow activity.

$$\overline{X}_7 + \overline{X}_8 = 0.02(\overline{X}_1 + \overline{X}_2) = (0.02)(8.360959 + 5.874492) = 0.284709$$

11. "Subsurface water that flows into the basin through the largely alluvial gravels at the higher elevations are only a little less in volume than subsurface outflow through the largely sedimentary clays at the basin's lower elevations." Thus, average subsurface inflow was taken to be 97% of average subsurface outflow: $\overline{X}_8 = 0.97\,\overline{X}_7$. Applying knowledge items 9, 10, and 11 to Equation 2 yielded:

$$\overline{X}_7 = 0.1445223,\ \overline{X}_8 = 0.1401867,\ \text{and } c_2 = \overline{X}_7 - c_1\,\overline{X}_1 = 0.102718.$$

Equation 3 $X_8 = \underbrace{c_3 X_2}_{T_1} + \underbrace{c_4}_{T_2}$ for subsurface inflow.

Application of knowledge items 9, 10, and 11 yielded similar results for Equation 3. T_1 and T_2 are scaled *stochastic* and *constant* forms, respectively.

For $T_1,\ c_3 = G = 0.005$.

For $T_2,\ c_4 = \overline{X}_8 - c_3\,\overline{X}_2 = 0.110814$.

Equation 5 $X_{11} = \underbrace{c_5 X_1 + c_6 X_2}_{T_1} + \underbrace{c_7 X_3 + c_8 X_4}_{T_2} + \underbrace{c_9 X_5}_{T_3}$ for percolation.

Let PR_j be the percolation rate for $T_j,\ j = 1, 2, 3$.

Then, $T_1 = G \cdot (X_1 + X_2)/2$ is a scaled average of two *stochastic forms*, where $G = PR_1$ provides the magnitude of physical correlation.

$T_2 = 1/D_1\,[\alpha X_3 + (1-\alpha)X_4)]$ is a *continuous delay form* with $\alpha = 0.5$ and $1/D_1 = PR_2$.

$T_3 = G \cdot X_5$ is a scaled *stochastic form*, where $G = PR_3$ provides the magnitude of physical correlation.

12. "One unit of average surface flow produces one *scaled* unit of average percolation. One unit of average surface storage produces only 60% as much percolation as one unit of average surface flow. One unit of average precipitation accounts for as much percolation as one unit of average surface flow."

Thus, $PR_2 = 0.60\,PR_1$ and $PR_3 = PR_1$ yielded

$$\overline{X}_{11} = \left[\left(\frac{\overline{X}_1 + \overline{X}_2}{2}\right) + 0.6\left(\frac{\overline{X}_3 + \overline{X}_4}{2}\right) + \overline{X}_5\right] PR_1 = 32.743156 PR_1$$

and the following new knowledge item.

13. Parameterized coefficients:

$$c_5 = c_6 = 0.5\,PR_1$$

$$c_7 = c_8 = 0.3\,PR_1$$

$$c_9 = PR_1$$

$$\overline{X}_{11} = 32.743156 \ PR_1$$

Equation 6 $X_{12} = \underbrace{c_{10}}_{T_1} - \underbrace{(c_{11}X_9 + c_{12}X_{10})}_{T_2}$ for subsurface discharge.

$T_1 = c_{10}$ is a *constant form* that represents maximum discharge rate.

$T_2 = DR\,(w_1\,X_9 + w_2\,X_{10})$ is an *exponential growth form*, where DR is the exponential discharge rate. Since change in X_{12} is not monotonic, equal weights are appropriate: $w_1 = w_2 = 0.5$. Available capacity grows as groundwater is discharged. In combination, T_1 and T_2 comprise a *Malthusian form*, with X_{12} approaching its maximum discharge rate as available capacity approaches zero.

14. "The maximum subsurface discharge rate is three times its average discharge rate." Therefore, $c_{10} = 3.0\,\overline{X}_{12}$.

15. "Groundwater mining (caused by human depletion activities) was not in evidence during the period of record." Thus, $\overline{X}_9 = \overline{X}_{10}$; that is, a *zero trend* must be generated for the X_9, X_{10} time series by the model.

16. "The groundwater subsystem has a storage capacity such that, on the average, it could accommodate a hypothetical input equal to that volume obtained by adding average surface inflow and average precipitation." Consequently, $\overline{X}_9 = \overline{X}_2 + \overline{X}_5 = 31.4557930$.

Substituting knowledge items 14, 15, and 16 into Equation 6 produced:

$$\overline{X}_{12} = \frac{DR}{2}\,\overline{X}_9 = 15.7278965 \ DR.$$

17. "On the average, 40% of surface outflow from the basin is derived from subsurface discharge as opposed to immediate precipitation runoff." Hence, $\overline{X}_{12} = 0.40\,\overline{X}_1 = 3.3443836$

Combining knowledge items 16 and 17 yielded 15.7278965 $DR = 3.3443836$, or $DR = 0.2126402$.

18. Parameterized coefficients:

$$c_{10} = 3.0\,\overline{X}_{12} = 10.033151$$
$$c_{11} = c_{12} = 0.5\,DR = 0.106320$$

Substituting known values into Equation 4 provided $\overline{X}_{11} = 3.3487192$.

From knowledge items 13 and 18, 32.743156 $PR_1 = 3.3487192$, or $PR_1 = 0.1022723$ thus yielding

$$c_5 = c_6 = 0.5\,PR_1 = 0.051136$$
$$c_7 = c_8 = 0.3\,PR_1 = 0.030682$$
$$c_9 = PR_1 = 0.102272.$$

Average percolation delay for surface storage: $D_1 = 1/PR_2 = 16.30$ years.

Equation 7 $X_{13} = X_6 - \underbrace{(c_{13}X_1 + c_{14}X_2)}_{T_1} - \underbrace{(c_{15}X_3 + c_{16}X_4)}_{T_2}$ for unexplained surface loss.

Let LR_j be the loss rate for T_j, $j = 1, 2$. Then, $T_1 = G \cdot (X_1 + X_2)/2$ is a scaled average of two *stochastic forms*, where $G = LR_1$ gives the magnitude of physical correlation. $T_2 = 1/D_2[\alpha\, X_3 + (1-\alpha)\, X_4)]$ is a *continuous delay form* with $\alpha = 0.5$ and $1/D_2 = LR_2$.

19. "Each unit of average surface flow is subjected to a 30% atmospheric loss as it passes through the basin, and each unit of average surface storage experiences a 15% loss over the year." Thus, $LR_1 = 0.30$, $LR_2 = 0.15$, and explained surface loss was expressed as

$$0.3\left(\frac{\overline{X}_1 + \overline{X}_2}{2}\right) + 0.15\left(\frac{\overline{X}_3 + \overline{X}_4}{2}\right) = T_1 + T_2.$$

Associating like coefficients yielded

$$c_{13} = c_{14} = 0.15$$
$$c_{15} = c_{16} = 0.075.$$

Average evaporation delay for surface storage: $D_2 = 1/LR_2 = 6.67$ years.

11.4.3 The Tentative Model

Phase One began by structuring and calibrating a model for the Yellowstone at the $L_{2,2}$ level for which all 13 systems variables were included. The model consisted of balance equations and regression equations. Calibration involved reverse regression: Heuristic knowledge was interacted with five *primary* data and the remaining eight variables to produce regression coefficients. Although domain experts assisted with the calibration, heuristics at best are intelligent assumptions, and the model is tentative. The equations are:

1. $X_6 = -X_1 + X_2 + X_3 - X_4 + X_5 - X_{11} + X_{12}$
2. $X_7 = 0.005\, X_1 + 0.102718$
3. $X_8 = 0.005\, X_2 + 0.110814$
4. $X_{10} = X_7 - X_8 + X_9 - X_{11} + X_{12}$

$$X_9 = \begin{cases} X_9^0 & \text{initially,} \\ X_{10} & \text{of previous } \Delta t \text{ thereafter} \end{cases}$$

5. $X_{11} = 0.051136\, X_1 + 0.051136\, X_2 + 0.030682\, X_3 + 0.030682\, X_4 + 0.102272\, X_5$

6. $X_{12} = 10.033151 - 0.106320\, X_9 - 0.106320\, X_{10}$

7. $X_{13} = X_6 - 0.15\, X_1 - 0.15\, X_2 - 0.075\, X_3 - 0.075\, X_4$

A systems model can be validated for several possible applications. The status of the model remains tentative except for those applications in which validity has been established. The first application of the model is to complete the data base, that is, to supplement primary data with secondary data that are consistent with system knowledge. These data must pass a Turing Type (TT) test: Can the domain experts detect that these data could not have come from the system domain? If secondary data do not pass the TT test, then the regression structure and/or calibration heuristics are revised, and the calibration procedure is repeated. Thus, validation tests comprise the major component of the calibration process.

By calibrating to the average year, the resulting model has theoretically been calibrated to each of the 39 years. However, if the time series data generated by the model evoke doubt from the experts, the model fails the TT test. The model is restructured until all apparent aberrations are removed from the secondary data. Introducing perturbations in the primary time series is also effective in discovering structural deficiencies.

Additional tests are performed by inverting the roles of primary (exogenous) and secondary (endogenous) variables. Sensitivity of the model to inversion can be traced to the regression coefficient of the solution variable of the pair being inverted. The closer this coefficient is to zero, the greater will be the calculation error under matrix inversion.

Accepting and processing heuristic knowledge is far superior to its alternative: assigning zeros to secondary data. Heuristics do introduce approximation errors, but these are distributed over the secondary data, thus protecting the integrity of primary data. If primary data exist for all variables of a subsystem, two options become available: (1) Ignore primary values for the one least accurate variable and redesignate it as a secondary variable, or (2) define a new secondary variable to account for imbalance error.

Data based regression models are demonstrably valid or invalid via tests that examine residual (approximation) error. Systems-based models having a complete primary data base must also pass residual error tests. However, many systems models, just as for the Yellowstone, are synthesized without having a complete primary data base, and residual error tests are not applicable for two reasons.

- Regression error cannot be computed for secondary data simulated by a systems model; there are no primary data to serve as a standard.

- Regression error does not exist for primary data simulated by a systems model via role inversion, only calculation error that can be diminished by *distortion proofing* the model.

At each level of detail, a water-planning model must be capable of simulating historical time series without approximation error (and negligible distortion error) and of simulating "highly probable" hypothetical series that pass the inspection of domain experts. The model must be structured to provide validity for each intended application.

Consequently, models such as the Yellowstone can only be validated via a series of simulation tests: (1) *Simulate secondary data* to thereby complete Phase One, and (2) Perform Phase Two tests, *history matching* and *perturbation*, for each anticipated application of the model.

11.5 MODEL VALIDATION

Applications of the model require a complete data base. Thus, data base generation was the first of the tests. Role inversion is required by many of the other simulation tests: interchange of role of an exogenous primary variable with an endogenous secondary variable. One or more of the exogenous variables in the data base are altered to determine the response of the model to "What if?" questions. Potentially, an endogenous vector can be specified for any 7 variables of the 13-variable Yellowstone model. Some vectors of the possible 1,716 combinations (13 variables taken 7 equations at a time) are not mathematically feasible (nonsingular coefficient matrix). Of the feasible combinations, only those that aid in developing a management plan are of interest.

Professional experience and domain familiarity of the experts provided the basis by which all model generated data were judged for validity. When validity was questionable, model revision and recalibration became necessary. For any given application, testing terminated when generated data passed the experts' inspection.

11.5.1 Data Base Generation

Primary data were input to the model year by year for each of the 39 years, and output values were computed for each secondary variable to thereby generate a complete data base. The exogenous member of the IT solution pair, X_9, X_{10} was given a starting value: $X_9^0 = 31.455793$, the average value used to calibrate the model. X_9^0 was reset to X_{10}^{39} following each iteration until $X_9^0 = X_{10}^{39}$, thus satisfying the *zero trend* heuristic that required $\overline{X}_9 = \overline{X}_{10}$.

Upon completion of Phase One, the starting value for X_9 that resulted in zero trend was recorded for use in Phase Two: $X_9^0 = 30.623191$. Figure 11.5 presents the time series for X_6 and X_7. The 13-variable data base is displayed as Table 11.4

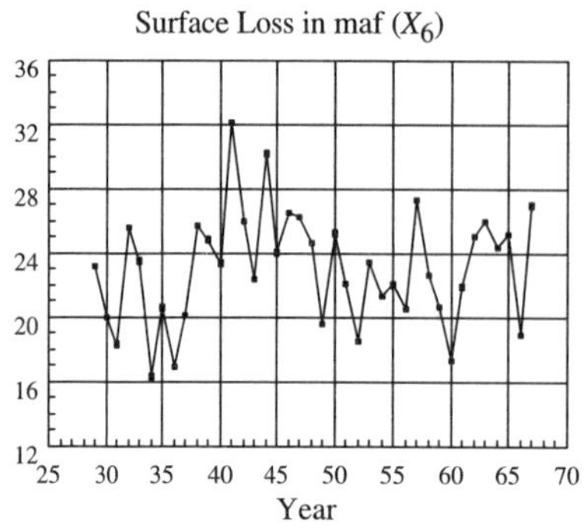

Figure 11.5 Secondary Time Series for Yellowstone River Basin

in Appendix H for selected years. More secondary time series plots follow Table 11.4 in Figure 11.13 of Appendix H.

11.5.2 History Matching for X_1

History matching for X_1 (surface outflow) required that some secondary variable exchange roles with X_1. X_{13} (unexplained surface loss) was a logical choice since both variables are related via their unity coefficients in Equations 1 and 7, indicative that inversion will induce the least-possible distortion.

$X_9^0 = 30.623191$ from Phase One and values for X_2, X_3, X_4, X_5, and X_{13} from the data base were input to the model to obtain values for endogenous variables X_1, X_6, X_7, X_8, X_{10}, X_{11}, and X_{12}. History matching for X_1 is presented in Table 11.5 of Appendix H for selected years. The "change" column records any deviation of the model's generated values for X_1 from "standard" historical values for X_1 of data base. Previously generated data base values also provided a standard of comparison for the other six endogenous variables. Zero or negligible deviation is observed for all endogenous variables for this inversion. For example, the cumulative algebraic value of change for X_1 is 2 acre-feet over the 39 years. Thus, the model reproduced the X_1 time series without approximation (residual) error and with but slight distortion.

11.5.3 Perturbation Tests

Although a systems model used for planning must be able to reproduce historic time series, the most rigorous tests involve applications of the model that require data base perturbations. TT tests are applied, and deficiencies originating with the model are noted and traced to their source. Model structure is revised, the model is recalibrated, and the application is retested. The model was further validated by two "What if?" tests and three design problems.

11.5.3.1 No Surface Storage Development

A pristine basin is one in which stream channels and flows have not been altered by human activity, prompting the question, "What if there had been no surface storage development within the basin?" To obtain an answer from the model, a copy of the Phase One data base, with $X_9^0 = 30.623191$, was perturbed by setting $X_3 = X_4 = 0$ for the entire 39 years. The endogenous vector for this simulation was (1, 6, 7, 8, 10, 11, 12). Values for the other exogenous variables (2, 3, 4, 5, 13) were supplied from the data base, which also provided a standard of comparison for the solution variables. Although validation of response magnitudes required the judgment of domain experts, response directions were as expected: (1) surface outflow averaged 0.4% higher, (2) maximum outflow increased 9.0%, occurring in 1966 when Yellowtail Reservoir would have started filling, (3) maximum surface loss decreased 0.051023 maf, coinciding with the completion of Yellowtail Reservoir filling in 1967, (4) groundwater available capacity showed a slight average increase, and (5) percolation and groundwater discharges showed slight average increases. Table 11.6 of Appendix H presents the results for three selected years.

Prior to this test, Equation 7 and X_{13} were not part of the model but were added to correct a structural deficiency that was uncovered by the test. Initial response to the surface storage perturbation was obviously incorrect: No change occurred in surface loss. Critical inspection verified that the model contained no mechanism to account for specific components of evaporation loss due to surface flow or surface storage. Consequently, Equation 7 and X_{13} were added to the model.

11.5.3.2 Total Drought from 1953 On

Many river basin models lack the structure to provide transient response to change in steady state. Cutting off input, cuts off output. Thus, the question was posed, "What if total drought conditions had prevailed from 1953 to 1967?" In order not to confound the effect of drought with surface water development on surface outflow, the X_3 and X_4 series were set to zero in a copy of the Phase One data base, with $X_9^0 = 30.623191$. This data base was further perturbed by removing all inflow and precipitation from 1953 on, that is, by setting X_2 and X_5 to zero. Furthermore, X_{13} (unexplained surface loss) was also set to zero because the source of unexplained loss (X_5) was removed. Again, the solution vector (1, 6, 7, 8, 10, 11, 12) was used.

Although there was a step loss in 1953, the model predicted a gradual decline in X_1 (and other flow activities). See Figure 11.6 for decline in X_1. By the end of seven years, surface outflow had dwindled to 5% of its previous average, viewed as a reasonable response by the domain expert. Correctly structured, groundwater discharge generated by the model maintains surface flow during periods of nonprecipitation, yet declines rapidly as the water table drops, indicated by the Malthusian growth in available capacity, X_{10} (see Figure 11.7).

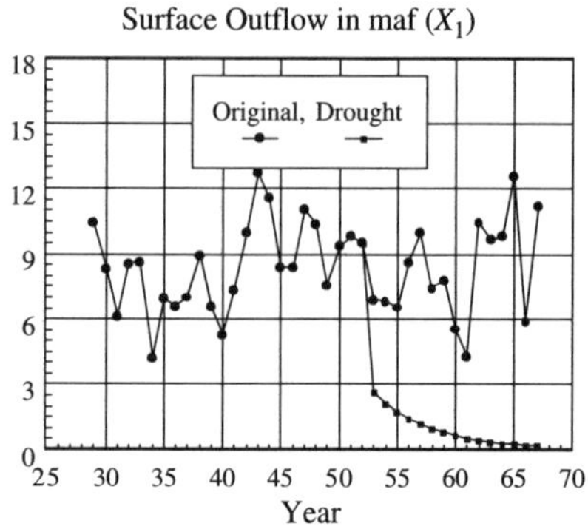

Figure 11.6 Exponential Decline in Flow

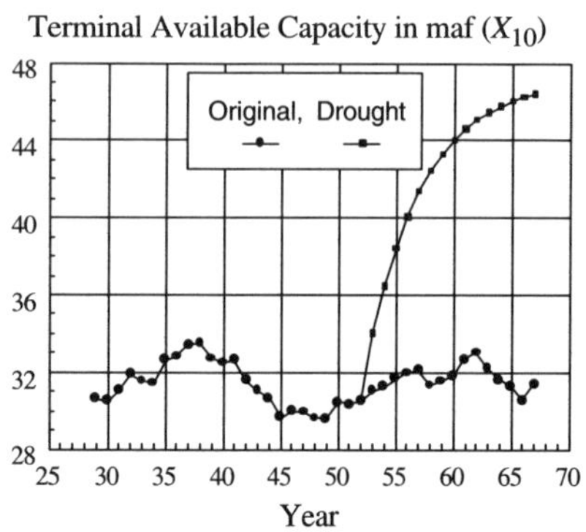

Figure 11.7 Malthusian Growth in Available Capacity

The phenomenon that groundwater discharge maintains surface flows during low or no precipitation periods was captured by the dynamic form for continuous delay. To see how, reconsider Equation 6:

$$\begin{aligned} X_{12} &= 10.033151 - (0.106320\,X_9 + 0.106320\,X_{10}) \\ &= K - \frac{1}{D}\,(0.5\,X_9 + 0.5\,X_{10}) \end{aligned}$$

Equating coefficients yields $D = 0.5/0.10632 = 4.7$ years. Thus, prior to 1953, a unit of surface water that became groundwater was delayed by the subsurface subsystem an average of 4.7 years. Then with recharge cut off during the drought, surface flow had exponentially declined to 5% average by the end of seven years.

11.5.3.3 Maximum Surface Storage

Design problem Given a maximum storage capacity of 10 million acre-feet (maf) subject to dual targets: (1) maintain *maximum capacity* as a target for surface storage, (2) except for those years in which storage must be reduced to maintain a target, *minimum surface outflow* at Sidney of 5 maf. Role inversion was used between X_1 and X_4 in the following procedure to maintain one target or the other:

- $X_1 = \left\{\begin{array}{ll}\text{Endo while} > 5, & \text{otherwise} \\ \text{Exo at } 5 & \end{array}\right\}$, that is, X_1 becomes *Exo* in response to minimum surface flow.
- $X_4 = \left\{\begin{array}{ll}\text{Endo while} < 10, & \text{otherwise} \\ \text{Exo at } 10 & \end{array}\right\}$, that is, X_4 becomes *Exo* in response to reservoir at full capacity.
- Start with solution vector (1, 6, 7, 8, 10, 11, 12), $X_3^0 = 10$, and $X_9^0 = 30.623191$.
- Iteratively, solve for X_1. If $X_1 < 5$, invert roles and repeat with $X_1 = 5$ and solution vector (4, 6, 7, 8, 10, 11, 12).
- Iteratively, solve for X_4. If $X_4 > 10$, invert roles and repeat with $X_4 = 10$ and solution vector (1, 6, 7, 8, 10, 11, 12).

Solution More than 5 maf of surface flow was maintained in all but 8 years; 1931, 1934–1937, 1940, 1960, and 1961; as depicted in Figure 11.8. Reservoir draw down to below 10 maf occurred in the same 8 years. See Figure 11.9. The cumulative algebraic effects of the changes are presented in Table 11.7 of Appendix H for three selected years.

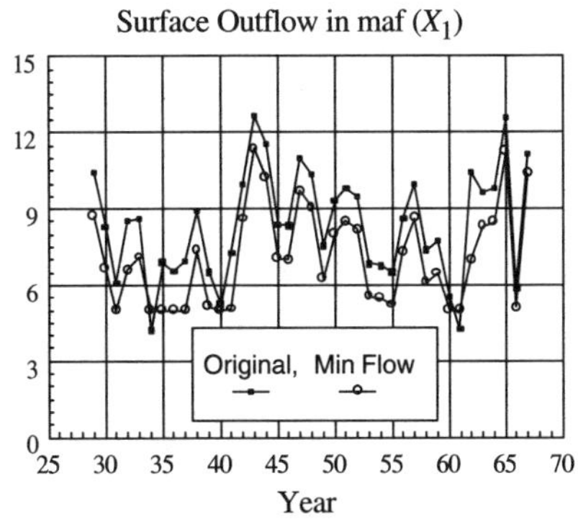

Figure 11.8 Maximum Surface Flow

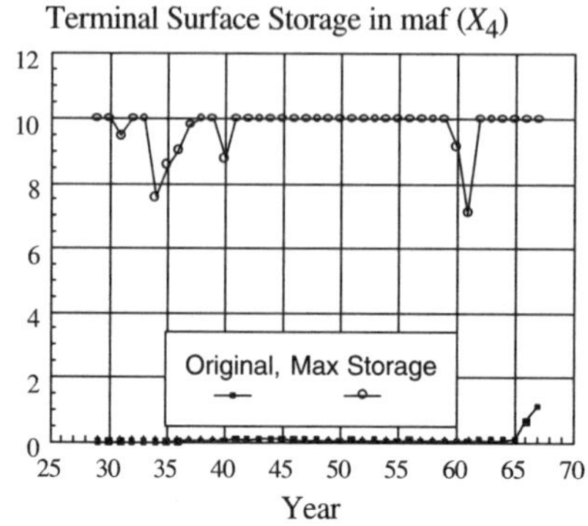

Figure 11.9 Reservoir Draw Down

11.5.4 Flood Control

Design problem Given a maximum storage capacity of 10 maf. What *minimum storage capacity* must be held in reserve to prevent outflow from exceeding flood stage threshold at 10 maf? The following procedure requires role inversion between X_1 and X_4 and automatic reinversion:

- $X_1 = \left\{\begin{array}{ll}\text{Endo while } < 10, & \text{otherwise} \\ \text{Exo at } 10 & \end{array}\right\}$, that is, X_1 becomes *Exo* in response to flood control.
- $X_4 = \left\{\begin{array}{l}\text{Endo while} > \text{lower bound, otherwise} \\ \text{Exo at lower bound } - \Delta \end{array}\right\}$, that is, X_4 becomes *Exo* in response to minimum storage capacity, or
- $X_4 = \left\{\begin{array}{ll}\text{Endo while } < 10, & \text{otherwise} \\ \text{Exo at } 10 & \end{array}\right\}$, that is, X_4 becomes *Exo* in response to reservoir at full capacity.
- Start with solution vector (1, 6, 7, 8, 10, 11, 12), initial flood reservation of 2 maf and $X_3 = X_4 = 8$ maf, and $X_9 = 30.623191$.
- Iteratively, solve for X_1. If $X_1 > 10$, invert roles and repeat with $X_1 = 10$ and solution vector (4, 6, 7, 8, 10, 11, 12).
- Iteratively, solve for X_4. If $X_4 <$ *minimum storage capacity*, reinvert to solution vector (1, 6, 7, 8, 10, 11, 12), or reinvert if $X_4 > 10$, reduce X_3 by $\Delta = 0.5$ maf, *automatically reinvert* and repeat from beginning.

Solution Establishing reserve storage capacity at 2.5 maf provides adequate flood control. Flooding would have occurred in 5 years (1943, 1944, 1947, 1965, and 1967) of the 39 years except for reserve capacity. Refer to Figure 11.10. Utilization of reserve capacity is shown in Figure 11.11. The cumulative algebraic effects are presented in Table 11.8 of Appendix H for three selected years.

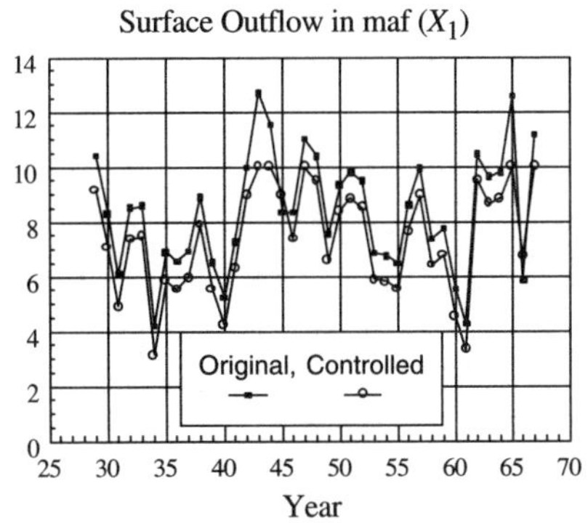

Figure 11.10 Flood Prevention

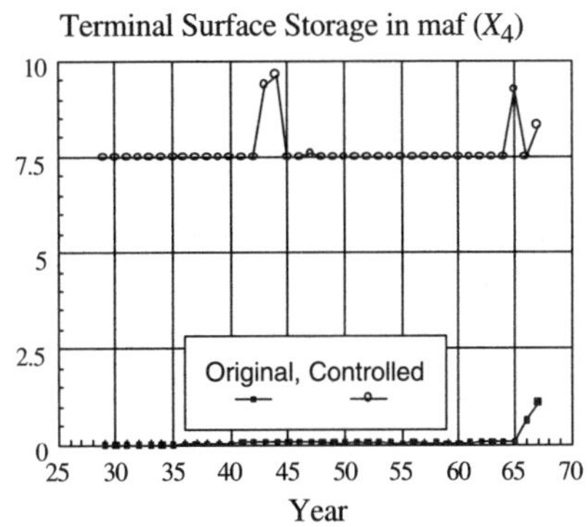

Figure 11.11 Reservoir Capacity Utilization

11.5.4.1 Firm Yield

Design problem Find the *maximum surface storage* required to produce a *maximum firm yield*, that is, constant outflow X_1 from the Yellowstone River Basin at Sidney, Montana.

Procedure

- Set solution vector to (4, 6, 7, 8, 10, 11, 12).
- Vary X_1 as a controlled *Exo* variable, starting with 8.36 maf, average surface outflow.
- Iteratively, solve for X_4, adjusting the IT pair X_3, X_4 to obtain zero trend starting from $X_3^0 = 15.0$ maf..
- If $X_4 < 0$, then step X_1 down by $\Delta_1 = -0.25$ maf and repeat from beginning.

Result $X_1 = 8.36 - 0.25 - 0.25 - 0.25 - 0.25 - 0.25 = 7.11$ maf.

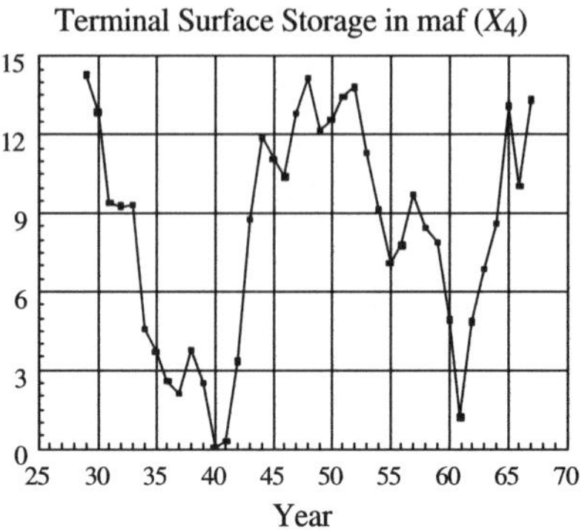

Figure 11.12 Surface Storage Requirements

For greater precision, a second run was made using a closer estimate, $X_1 = 7.11 + 0.25 = 7.36$, and smaller steps, $\Delta_2 = -0.03$.

Solution Surface storage requirements are shown in Figure 11.12. Maximum surface storage: $X_4 = 14.283148$ maf, occurring in 1929 just prior to the drought of the 1930's. Maximum firm yield: $X_1 = 7.27$ maf. The cumulative algebraic effects of the changes are presented in Table 11.9 of Appendix H for three selected years.

REMARKS The foregoing tests readily demonstrate the utility of even the $L_{2,2}$ level of the Montana State water planning model. $L_{6,j}$ level models were calibrated for many of the middle and lower sub-basins of the Yellowstone by the Montana DNRC and used to answer "what if" questions at public hearings during the diversion moratorium of 1976.

At whatever level of development, the model is composed of a system of linear equations, thereby permitting a fast and efficient method of solution. Furthermore, the matrices of adjacent systems are easily linked thereby providing an effective means for modeling systems interactions [9].

EXERCISES

1. Discuss the pros and cons of developing a water planning model as a systems model.

2. In Chapter 2, two principles were presented for calibration by reverse regression. Which principle was used to calibrate the water planning model and why was it selected?

3. How was knowledge for item 19 of the knowledge base obtained?

4. Outline the procedure for calibrating the $(L_{2,2})$ planning model.

5. Using library resources, write a short report to explain how activity times are obtained from "domain experts" for the PERT/CPM network analysis method when activity times are unknown, random variables.

6. Design a knowledge acquisition format that might very well have produced the calibration heuristics for the Yellowstone River Basin model.

7. In the heuristic of knowledge item 17, let 0.40 vary $\pm 10\%$. How sensitive is *average percolation delay for surface storage* in knowledge item 18 to variation in the heuristic?

8. No measure of *residual error* can be used in validating secondary data generated by a systems model. Explain why.

9. Compare the time series for X_{11} and X_{12} in Figure 11.13 of the Appendix in terms of appearance. Suggest an explanation.

10. In history matching for X_1, why did endo-exo role inversion with X_{13} induce only slight distortion?

11. How did a perturbation test result in adding Equation 7 to the Yellowstone River Basin model?

12. In the perturbation test "Total Drought from 1953 on," an expected phenomenon was observed. In seven years, X_1 followed an exponential decline to 5% of its previous average.

    a. Relate the exponential decline in X_1 to the Malthusian growth in available capacity by solving the model for X_1 in terms of the IT pair, X_9, X_{10}.

    b. Check your result in (a) against the value for X_1 in 1953 from Figure 6 by estimating the value for X_1 in 1953 using values for X_9 and X_{10} from Figure 7.

13. Explain how the perturbation test "Maximum Surface Storage" relates to a trade-off between conflicting uses for water.

14. Discuss the conflicting interests that must be balanced in operating a reservoir for flood control.

15. What water uses might be enhanced by operating a reservoir system to produce *maximum firm yield* from a river basin?

APPENDIX H

Table 11.3 Primary Data for Yellowstone River Basin

Year	X_1	X_2	X_3	X_4	X_5
1929	10.438997	6.533848	.000000	.000000	27.159348
1930	8.306996	5.858187	.000000	.000000	21.935013
1931	6.082995	4.326836	.000000	.000000	19.196838
1932	8.521447	6.650352	.000000	.000000	27.720383
1933	8.601413	5.584848	.000000	.000000	26.634201
1934	4.208997	3.245835	.000000	.000000	16.077698
1935	6.907496	5.162136	.000000	.000000	22.133698
1936	6.571898	5.726196	.000000	.014510	17.117645
1937	6.958398	5.171018	.014510	.038310	21.872055
1938	8.910496	5.903912	.038310	.039010	29.454987
1939	6.525395	5.017608	.039010	.041480	26.552109
1940	5.232897	3.866636	.041480	.031480	24.597427
1941	7.274997	5.245839	.031480	.080400	35.229401
1942	9.969498	6.323128	.080400	.061910	30.181931
1943	12.686196	8.541931	.061910	.076570	26.894196
1944	11.546497	6.752533	.076570	.077650	35.994186
1945	8.362996	6.490890	.077650	.078100	25.618942
1946	8.338296	5.948360	.078100	.065140	28.914124
1947	11.019997	7.572433	.065140	.061570	30.062897
1948	10.381398	6.727863	.061570	.061470	28.372025
1949	7.558398	5.785263	.061470	.057090	20.503021
1950	9.342497	6.670556	.057090	.054740	28.021835
1951	9.791495	7.399075	.054740	.065700	24.280090
1952	9.487098	6.011074	.065700	.055770	21.469208
1953	6.842694	4.904791	.055770	.065050	25.119553
1954	6.759698	5.098290	.065050	.056390	22.526855
1955	6.505996	4.695572	.056390	.038450	23.565720
1956	8.604396	6.104386	.038450	.060640	22.904755
1957	9.942197	6.823954	.060640	.044610	31.182785
1958	7.380197	5.479470	.044610	.054310	24.360825
1959	7.767197	5.184215	.054310	.043900	22.949249
1960	5.511196	4.137877	.043900	.021620	17.836655
1961	4.260325	3.549088	.021620	.049640	22.285599
1962	10.452196	6.733562	.049640	.070210	29.634003
1963	9.639397	6.426247	.070210	.060200	29.849716
1964	9.797896	6.678123	.060200	.063110	27.830688
1965	12.560699	8.159577	.063110	.076090	30.310486
1966	5.869895	4.569275	.076090	.646530	19.975677
1967	11.156598	8.044383	.646530	1.113440	31.344849

Table 11.4 Data Base for Yellowstone River Basin

Year	X_1	X_2	X_3	X_4	X_5
	X_6	X_7	X_8	X_9	X_{10}
	X_{11}	X_{12}	X_{13}		
1929	10.438997	6.533848	0.000000	0.000000	27.159348
	23.140901	0.154913	0.143483	30.623191	30.521323
	3.645564	3.532266	20.594974		
1930	8.306996	5.858187	0.000000	0.000000	21.935013
	20.005915	0.144253	0.140105	30.521323	31.045182
	2.967688	3.487400	17.881138		
1931	6.082995	4.326836	0.000000	0.000000	19.196838
	18.286740	0.133133	0.132448	31.045182	31.891928
	2.495616	3.341677	16.725265		
1932	8.521447	6.650352	0.000000	0.000000	27.720383
	25.524493	0.145325	0.144066	31.891928	31.568393
	3.610844	3.286049	23.248723		
1933	8.601413	5.584848	0.000000	0.000000	26.634201
	23.500439	0.145725	0.138738	31.568393	31.458183
	3.449362	3.332165	21.372500		
1934	4.208997	3.245835	0.000000	0.000000	16.077698
	16.306526	0.123763	0.127043	31.458183	32.646893
	2.025509	3.217499	15.188301		
1935	6.907496	5.162136	0.000000	0.000000	22.133698
	20.578335	0.137255	0.136625	32.646893	32.837521
	2.880850	3.070848	18.767891		
1936	6.571898	5.726196	0.000000	0.014510	17.117645
	16.863962	0.135577	0.139445	32.837521	33.440183
	2.379976	2.986505	15.018159		
1937	6.958398	5.171018	0.014510	0.038310	21.872055
	20.118337	0.137510	0.136669	33.440183	33.498486
	2.858769	2.916231	18.294963		
⋮					
1967	11.156598	8.044383	0.646530	1.113440	31.344849
	26.960814	0.158501	0.151036	31.420637	30.623192
	4.241560	3.436651	23.948669		

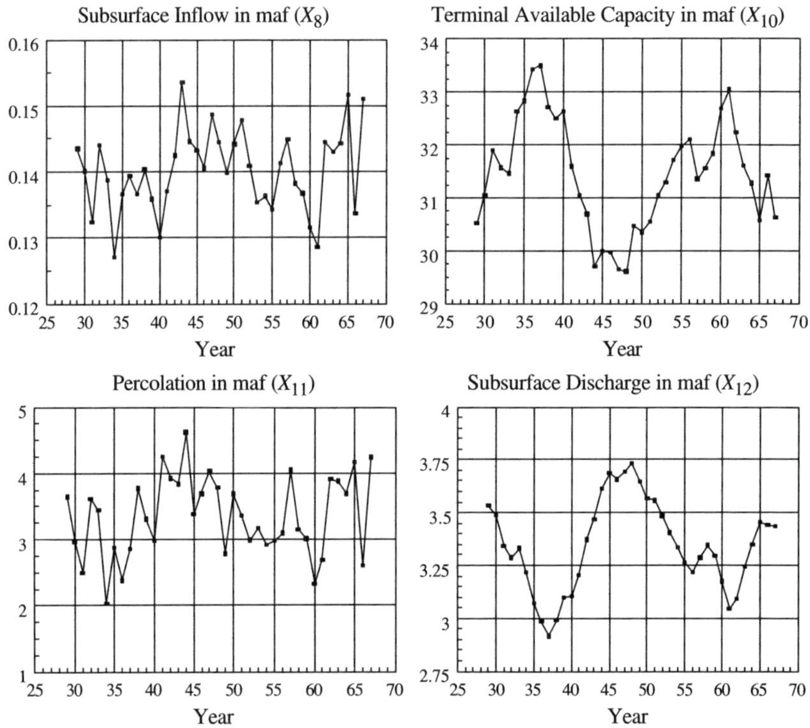

Figure 11.13 More Secondary Time Series for Yellowstone River Basin

Table 11.5 History Matching for X_1

Key: X_1 Surface Outflow
X_6 Surface Loss
X_7 Subsurface Outflow
X_8 Subsurface Inflow
X_{10} Terminal Available Capacity
X_{11} Percolation
X_{12} Subsurface Discharge

SOLUTION FOR THE PERIOD 1929

VARIABLE	VALUE	CHANGE	SUM*
X_1	10.438997	0.000000	0.000000
X_6	23.140901	0.000000	0.000000
X_7	0.154913	0.000000	0.000000
X_8	0.143483	0.000000	0.000000
X_{10}	30.521322	−0.000001	−0.000001
X_{11}	3.645564	0.000000	0.000000
X_{12}	3.532266	0.000000	0.000000

SOLUTION FOR THE PERIOD 1930

VARIABLE	VALUE	CHANGE	SUM*
X_1	8.306996	0.000000	0.000000
X_6	20.005916	0.000001	0.000000
X_7	0.144253	0.000000	0.000000
X_8	0.140105	0.000000	0.000000
X_{10}	31.045182	0.000000	−0.000001
X_{11}	2.967688	0.000000	0.000001
X_{12}	3.487400	0.000000	0.000000

SOLUTION FOR THE PERIOD 1967

VARIABLE	VALUE	CHANGE	SUM*
X_1	11.156598	0.000000	0.000002
X_6	26.960814	0.000000	0.000000
X_7	0.158501	0.000000	0.000007
X_8	0.151036	0.000000	0.000002
X_{10}	30.623192	0.000000	0.000000
X_{11}	4.241560	0.000000	0.000002
X_{12}	3.436651	0.000000	0.000000

* SUM = Cumulative algebraic value of change

Table 11.6 No Surface Storage Development

Key: X_1 Surface Outflow
X_6 Surface Loss
X_7 Subsurface Outflow
X_8 Subsurface Inflow
X_{10} Terminal Available Capacity
X_{11} Percolation
X_{12} Subsurface Discharge

SOLUTION FOR THE PERIOD 1929

VARIABLE	VALUE	CHANGE	SUM*
X_1	10.439002	0.000002	0.000002
X_6	23.140898	−0.000002	−0.000002
X_7	0.154913	0.000000	0.000000
X_8	0.143483	0.000000	0.000000
X_{10}	30.521322	0.000002	0.000002
X_{11}	3.645565	0.000001	0.000001
X_{12}	3.532266	0.000000	0.000000

SOLUTION FOR THE PERIOD 1930

VARIABLE	VALUE	CHANGE	SUM*
X_1	8.306992	−0.000004	0.000004
X_6	20.005917	0.000007	0.000004
X_7	0.144253	0.000000	0.000000
X_8	0.140105	0.000000	0.000000
X_{10}	31.045182	0.000002	0.000004
X_{11}	2.967688	0.000000	0.000001
X_{12}	3.487400	0.000000	0.000000

SOLUTION FOR THE PERIOD 1967

VARIABLE	VALUE	CHANGE	SUM*
X_1	11.696398	0.539798	1.365635
X_6	26.909787	−0.051023	−0.225398
X_7	0.161200	0.002699	0.006835
X_8	0.151036	0.000000	0.000002
X_{10}	30.656793	0.033603	0.390239
X_{11}	4.215165	−0.026395	−0.106174
X_{12}	3.432117	−0.004534	−0.079403

* SUM = Cumulative algebraic value of change

Table 11.7 Maximum Surface Storage

SOLUTION VARIABLES

X_1 Surface Outflow	X_8 Subsurface Inflow
X_4 Terminal Surface Storage	X_{10} Terminal Available Capacity
X_6 Surface Water Loss	X_{11} Percolation
X_7 Subsurface Outflow	X_{12} Subsurface Discharge

SOLUTION FOR THE PERIOD 1929

VARIABLE	VALUE	CHANGE	SUM*
X_1	8.722061	−1.716936	−1.716936
X_4	10.000000	10.000000	10.000000
X_6	24.383361	1.242460	1.242460
X_7	0.146328	−0.008585	−0.008585
X_8	0.143483	0.000000	0.000000
X_{10}	30.038262	−0.483061	−0.483061
X_{11}	4.171399	0.525835	0.525835
X_{12}	3.583625	0.051359	0.051359

SOLUTION FOR THE PERIOD 1934

VARIABLE	VALUE	CHANGE	SUM*
X_1	5.000000	0.791003	−7.060409
X_4	7.565707	7.565707	57.030520
X_6	17.742604	1.436078	7.678089
X_7	0.127718	0.003955	−0.035302
X_8	0.127043	0.000000	0.000001
X_{10}	30.794978	−1.851915	−7.493910
X_{11}	2.604901	0.579392	3.213223
X_{12}	3.589680	0.372181	1.396610

SOLUTION FOR THE PERIOD 1967

VARIABLE	VALUE	CHANGE	SUM*
X_1	10.384570	−0.772028	−50.103450
X_4	10.000000	8.886560	375.970600
X_6	28.213012	1.252198	48.963590
X_7	0.154641	−0.003860	−0.250510
X_8	0.151036	0.000000	0.000002
X_{10}	28.119359	−2.503833	−87.263570
X_{11}	4.761715	0.520155	20.542830
X_{12}	3.970065	0.533414	18.289520

* SUM = Cumulative algebraic value of change

Table 11.8 Flood Control

SOLUTION VARIABLES

X_1 Surface Outflow	X_8 Subsurface Inflow
X_3 Initial Surface Storage	X_{10} Terminal Available Capacity
X_4 Terminal Surface Storage	X_{11} Percolation
X_6 Surface Water Loss	X_{12} Subsurface Discharge
X_7 Subsurface Outflow	

SOLUTION FOR THE PERIOD 1929

VARIABLE	VALUE	CHANGE	SUM*
X_1	9.151295	−1.287702	−1.287702
X_3	7.500000	7.500000	7.500000
X_4	7.500000	7.500000	7.500000
X_6	24.072746	0.931845	0.931845
X_7	0.148474	−0.006439	−0.006439
X_8	0.143483	0.000000	0.000000
X_{10}	30.159027	−0.362296	−0.362296
X_{11}	4.039940	0.394376	0.394376
X_{12}	3.570785	0.038519	0.038519

SOLUTION FOR THE PERIOD 1944

VARIABLE	VALUE	CHANGE	SUM*
X_1	10.000000	−1.546497	−19.367760
X_3	9.363585	9.287015	121.479900
X_4	9.632141	9.554491	123.534400
X_6	31.370682	1.181139	15.470910
X_7	0.152718	−0.007732	−0.096836
X_8	0.144577	0.000000	−0.000001
X_{10}	27.760893	−1.939202	−23.000630
X_{11}	5.120674	0.499006	6.527044
X_{12}	4.013193	0.401123	4.684678

SOLUTION FOR THE PERIOD 1967

VARIABLE	VALUE	CHANGE	SUM*
X_1	10.000000	−1.156598	−39.867210
X_3	7.500000	6.853470	296.002300
X_4	8.299528	7.186088	295.688400
X_6	27.840291	0.879477	38.396720
X_7	0.152718	−0.005783	−0.199329
X_8	0.151036	0.000000	0.000002
X_{10}	28.639447	−1.983745	−68.387270
X_{11}	4.613173	0.371613	16.115370
X_{12}	3.863760	0.427109	14.330960

* SUM = Cumulative algebraic value of change

Table 11.9 Firm Yield

SOLUTION VARIABLES

X_1 Surface Outflow	X_8 Subsurface Inflow
X_4 Terminal Surface Storage	X_{10} Terminal Available Capacity
X_6 Surface Water Loss	X_{11} Percolation
X_7 Subsurface Outflow	X_{12} Subsurface Discharge

SOLUTION FOR THE PERIOD 1929

VARIABLE	VALUE	CHANGE	SUM*
X_1	7.270000	−3.168998	−3.168998
X_4	14.283150	14.283150	14.283150
X_6	24.736335	1.595434	1.595434
X_7	0.139068	−0.015845	−0.015845
X_8	0.143483	0.000000	0.000000
X_{10}	29.887755	−0.633567	−0.633567
X_{11}	4.330647	0.685083	0.685083
X_{12}	3.599627	0.067361	0.067361

SOLUTION FOR THE PERIOD 1940

VARIABLE	VALUE	CHANGE	SUM*
X_1	7.270000	2.037102	−0.027431
X_4	0.052207	0.020727	74.291600
X_6	23.856560	0.492063	12.137620
X_7	0.139068	0.010186	−0.000135
X_8	0.130147	0.000000	0.000000
X_{10}	31.438988	−1.196532	−18.324180
X_{11}	3.163644	0.180464	4.965637
X_{12}	3.370954	0.264442	3.769239

SOLUTION FOR THE PERIOD 1967

VARIABLE	VALUE	CHANGE	SUM*
X_1	7.270000	−3.886599	−42.547390
X_4	13.327308	12.213870	320.237900
X_6	27.997091	1.036277	41.737090
X_7	0.139068	−0.019433	−0.212730
X_8	0.151036	0.000000	0.000002
X_{10}	28.486719	−2.136473	−74.363700
X_{11}	4.705238	0.463678	17.509290
X_{12}	3.887523	0.450872	15.585550

* SUM = Cumulative algebraic value of change

CHAPTER 12

Prototyping A Forest Systems Model

12.1 INTRODUCTION

The combination of dwindling natural resources, revolutionary management ideas, and modern computer facilities has produced a proliferation in planning procedures and models. Both federal and state agencies have managed forests under a wide range of statutes for many years, while attempting to please special interest groups and their conflicting demands. For example, short-sighted economic interests compel some groups to seek maximum timber harvest levels without regard for the detrimental long-term impact publicly opposed by environmental and other groups. Systems analysis and modeling provide invaluable assistance.

Traditionally, modelers have started with the infinitesimals dx, dy, dz, and dt to describe a complex model of a minuscule geographic area in terms of point phenomenon. Model equations are derived by application of natural laws governing growth, flow, motion, and so on. The model is extrapolated from a homogeneous, micro region to encompass a not-so-homogeneous macro region and is attended by a multitude of variables and simplifying assumptions. The old adage, "One cannot see the forest for the trees" very literally describes starting at the micro level.

Rather than starting from a micro level, a macro level of entry presents distinct advantages. A relatively simple model of a large geographical area is considered first, making it possible to encounter and overcome problems at a simple level of articulation before proceeding on to the more micro and complex levels. Significant economies in time and effort are possible through this strategy.

Descending in the micro direction, quantifiers of space, geographical areas containing forests, include *planet earth, hemisphere, continent, country, national forest, ranger district, compartment, subcompartment, specimen stand, specimen substand, acre, ..., increment* $(\Delta x, \Delta y, \Delta z)$, *infinitesimal* (dx, dy, dz).

12.1.1 Prototyping

Occasionally, agency administrators (or resource managers) may not understand that system domain analysis and modeling can transform raw knowledge into useful information for management decision making. Consequently, features and possible applications of a systems model must be adequately demonstrated to them before they become enthusiastic about committing funds to the systems approach. To address their reluctance, an operational model can be developed in a relatively short time for the domain of a *generic* system and demonstrated prior to calibration and validation for a *specific* domain. Building a model of a scaled-down (or macro) system or portion (macro segment) of a system in relatively short time is called *prototyping* [51]. A prototype model can be upgraded to domain specificity by acquiring and compiling primary data. When primary data become available, the prototype is calibrated and revised as necessary to meet validation requirements. The model is tested and improved over several iterations.

This chapter features *rapid* prototyping of a forest systems model in *generic* form at a level easily comprehended by both knowledge engineers and domain experts. The various steps involved in obtaining a first operational model (systems analysis, model specification, data acquisition and compilation, calibration, and validation) are more simply performed at a macro level. Thus, an operational model is conceived, defined, designed, synthesized, debugged, and demonstrated at this relatively low cost level. Consequently, the knowledge engineer is better prepared to extend the model to finer levels of detail.

To illustrate prototyping, let's assume that *seasonal growth* of a *specimen stand* is to be modeled over a time frame incremented by *month*. However, prototyping must start at a level that does not overwhelm one's mind with minute detail. Accordingly, space and time quantifiers, *ranger district* and *decade*, were selected as compatible elements. Because size of these elements is relatively large, a systems analysis conducted at the ranger district level is termed macro, and, the corresponding $L_{1,j}$ model is a macro model of the forest system, where to enable rapid prototyping, $j \leq 3$ three subsystems.

12.1.2 Systems Analysis

Badger Ranger District is a generic district within *Galloping National Forest*, a caricature forest. Primary data, therefore, are totally nonexistent. Although much smaller than average, Badger Ranger District provided an example of prototype model development based on knowledge of system components, their relationships, and heuristics. Uniquely, this example illustrates dependency of the data base upon model definition for identification and interpretation of data categories, rather than dependency of the model upon available data for determination of its scope and detail. Paying "homage" to the variables, primary data were simulated for those system components that would most likely possess primary data if analyzing and modeling a real forest. Thirty decades were considered in the simulation.

The forest system, generic or real, is modeled in terms of what transpires from the vantage point of an observer located at its periphery—its boundary in time and geographical space. Thus, the forest system is viewed as an inventory of trees, within which trees multiply and grow, and from which trees are depleted by natural causes or by thinning practices and harvesting.

12.2 KNOWLEDGE BASE

Items that make up the system's knowledge base were classified as *hard* and *soft* knowledge. Hard knowledge items define the system in terms of component subsystems, rate and level variables, relationships, and data (when available). On the other hand, soft knowledge items are based on domain expertise expressed in heuristic form as rules of good judgment. Silviculture and timber management are mature fields with an abundant literature, widely accepted terminology and analytical techniques, and data-gathering methods that have not changed in hundreds of years. Furthermore, continued use of well-defined analytic techniques over a long period of time has allowed accumulation of a wealth of empirical knowledge. Data requirements are widely known, enabling domain experts to reckon data as reasonable or outside an expected range. Nonetheless, knowledge-based modeling represents a break with the traditional approach because of its unique systems terminology, model structure, and calibration based on knowledge items that take explicit form in terms of *heuristics* and/or *historical data*. Explicitly stated items are readily critiqued, can be easily corrected if in error, and provide a basis for perturbation tests in answer to "What if?" questions. Model failure owing to faulty structure often occurs during validation testing and is traceable to explicit items.

Calibration of this prototype model required 17 knowledge items. Hard knowledge items 1 through 6 are readily available for any system, generic or specific; they define the system, its component subsystems, and rate and level variables. For a specific system, soft knowledge items are normally acquired through sessions with one or more domain experts. But for this generic system, the remaining 11 knowledge items are *place holders*, selected on a likelihood basis to simulate domain expertise.

12.2.1 Hard Items

1. $L_{1,3}$ circumscription of Badger Ranger District.

2. Fundamental units of analysis:

   - Space: Badger Ranger District, 64.0 thousand acres (100 square miles).
   - Mass: One tree of nondescript specie (but similar to lodgepole pine).
   - Time: $\Delta t = 1$ decade.

3. Three Subsystems

   SS_1: Immature trees—a conglomerate of trees: seedlings, saplings, and pole timber, having diameter at breast height (dbh) less than five inches, quantified by numerical count in millions of trees.

   SS_2: Mature trees—all other trees: saw timber having dbh five inches or greater, measured by volume in millions of cubic feet.

   SS_3: Land area—shared by forested and meadowland, measured in thousands of acres.

4. System variables:

   SS_1: Immature trees
   X_1 = growth count
   X_2 = mortality count
   X_3 = initial standing count
   X_4 = terminal standing count
   X_5 = harvest count
   X_6 = ingrowth (to SS_2) count

   SS_2: Mature trees
   X_7 = growth volume
   X_8 = mortality volume
   X_9 = initial standing volume
   X_{10} = terminal standing volume
   X_{11} = harvest volume

   SS_3: Land area
   X_{12} = initial forested area
   X_{13} = terminal forested area
   X_{14} = initial meadow area
   X_{15} = terminal meadow area

5. Schematic: See Figure 12.1.

6. Description of variables:

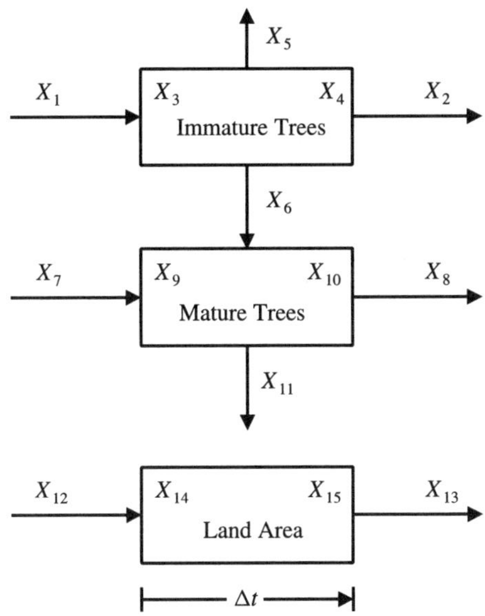

Figure 12.1 Block Diagram of Badger Ranger District

- Growth count (X_1) is an input rate for SS_1 that accounts for seedlings started by natural regeneration during each decade.
- Mortality count (X_2) is the rate of loss for a decade of SS_1 trees due to all causes except removal by direct harvest.
- IT pair (X_3, X_4) is the standing count or inventory level of SS_1 trees.
- Harvest count (X_5) is a lumped rate for SS_1 that may consist of commercial pole cutting and/or silvicultural thinning during a decade.
- Ingrowth count (X_6) is a rate that reclassifies each tree, whose dbh has reached five inches during the decade, from pole timber (SS_1) to mature tree (SS_2).
- Growth volume (X_7) is an input rate that accounts for the cumulative growth of mature trees during a decade.
- Mortality volume (X_8) is the rate of loss for a decade of SS_2 trees due to all causes except removal by direct harvest.
- IT pair (X_9, X_{10}) is the standing volume or inventory level of mature trees.
- Harvest volume (X_{11}) is a lumped rate that accounts for removal of mature trees due to commercial cutting and tree-thinning operations during a decade.
- Initial forested area (X_{12}), as an input "rate," is the composite of forest area covered by immature and mature trees at the beginning of a decade.

Table 12.1 Primary Data Availability

X_j	Code
X_1	A
X_2	A
X_3	B
X_4	B
X_5	B
X_6	A
X_7	B or C
X_7	A or B
X_8	B
X_9	A or B
X_{10}	A or B
X_{11}	A
X_{12}	A
X_{13}	A
X_{14}	A
X_{15}	A

- Terminal forested area (X_{13}), as an output "rate," is the composite of forest area covered by immature and mature trees at the end of a decade.
- IT pair (X_{14}, X_{15}) represents meadowland, nonstocked or clear areas whose cumulative level varies due to fluctuations in inventory levels of immature and mature trees during a decade.

12.2.2 Soft Items

Complete forest stand records are seldom available at the ranger district level. D. M. Cole, scientist, Montana State Forestry Lab, was asked the likelihood that each of the 15 variables would have primary data. His assessment is presented in Table 12.1. "A" indicates "very possible"; "B," "possible"; and "C," "some possibility." Based on the table, two "A" variables were selected to illustrate simulation of primary data for the generic Badger Ranger District.

12.2.2.1 Primary Data Simulation

Two stochastic "A" variables, X_1 and X_{11}, were assigned to exogenous dynamic forms. Data were synthesized for 30 decades via Monte Carlo simulated random sampling. As a result, soft knowledge items 7, 8, 9, and 10 were added.

7. Heuristic for simulation of X_1, growth count:

   In the long run, X_1 is impacted by random effects introduced by causal climatological factors such as moisture and temperature. Human harvesting activities and chance devastations such as forest fires, pests, and diseases also greatly impact growth count.

   To exploit this knowledge item, deviations from mean growth count were constructed from the normal distribution for Z with $\mu_z = 0$ and $\sigma_z = 1$, and using point estimators for μ_1 and σ_1:

$$X_1 = \mu_1 + \sigma_1 Z.$$

   Growth count for the average decade was assumed to be $\overline{X}_1 = 3.00$ million trees, with a coefficient of variation $CV = 1/5$, yielding a standard deviation of

$$S_1 = CV \cdot \overline{X}_1 = 0.60 \text{ million trees, so that}$$
$$\mu_1 \doteq 3.00$$
$$\sigma_1 \doteq 0.60.$$

8. Heuristic for simulation of X_{11}, harvest volume:

   Harvest volume is represented by a normal distribution, justifiable in that volume for each decade is composed of the sum of randomly occurring harvest activities. Within the span of a single year, cross-correlation may exist between X_1 and X_{11}, caused by weather and human intervention. However, cross-correlation over a 10-year Δt is nil and can be neglected.

   Similarly, normal deviations from mean harvest volume were constructed

$$X_{11} = \mu_{11} + \sigma_{11} Z.$$

   Harvest volume for the average decade was assumed to be $\overline{X}_{11} = 7.00$ million cubic feet, with $CV = 1/4$, yielding a standard deviation $S_{11} = 1.75$ million cubic feet, and thus point estimates:

$$\mu_{11} \doteq 7.00$$
$$\sigma_{11} \doteq 1.75$$

   Primary data for X_1 and X_{11} are presented in Table 12.2 of Appendix I.

9. Zero trend restrictions: Time series for X_1 and X_{11} are classified as *stationary* because time invariant means and standard deviations were used in their simulation. All three subsystems are driven by these series, and consequently exhibit equilibrium (zero trend) over the 30 decades. Therefore, zero trend was imposed on the following IT pairs:

- $\overline{X}_3 = \overline{X}_4$
- $\overline{X}_9 = \overline{X}_{10}$
- $\overline{X}_{14} = \overline{X}_{15}$

10. Vector table for model structure:

X_j	IT Pair	Exo	Endo	Equation Type	Equation No.
X_1		*		—	—
X_2			*	regression	1
X_3	I	*		—	—
X_4	T		*	balance	2
X_5			*	regression	3
X_6			*	regression	4
X_7			*	regression	5
X_8			*	regression	6
X_9	I	*		—	—
X_{10}	T		*	balance	7
X_{11}		*		—	—
X_{12}			*	balance	8
X_{13}			*	modification	9
X_{14}	I	*		—	—
X_{15}	T		*	modification	10

12.3 CALIBRATION

12.3.1 Model Equations

The vector table shows 5 *exo* variables: 2 have primary data, and 3 are *initial* members of IT pairs. Of the 10 *endo* variables, X_4, X_{10}, and X_{12} were assigned to balance equations, and X_{13} and X_{15} to modification equations. Exponential dynamic forms were identified for the remaining *endo* variables: *exponential decline* for X_2, X_5, X_6, and X_8, and *exponential growth* for X_7.
SS_1:

1. $X_2 = DR(w_1X_3 + w_2X_4) = c_1X_3 + c_2X_4$

2. $X_4 = X_1 - X_2 + X_3 - X_5 - X_6$

3. $X_5 = DR(w_1X_3 + w_2X_4) - c_3X_3 + c_4X_4$

4. $X_6 = DR(w_1X_3 + w_2X_4) = c_5X_3 + c_6X_4$

$$X_3 = \begin{cases} X_3^0 & \text{initially,} \\ X_4 & \text{of previous } \Delta t \text{ thereafter} \end{cases}$$

SS_2:

5. $X_7 = GR(w_1 X_9 + w_2 X_{10}) = c_7 X_9 + c_8 X_{10}$

6. $X_8 = DR(w_1 X_9 + w_2 X_{10}) = c_9 X_9 + c_{10} X_{10}$

7. $X_{10} = G_1 X_6 + X_7 - X_8 + X_9 - X_{11}$

$$X_9 = \begin{cases} X_9^0 & \text{initially,} \\ X_{10} & \text{of previous } \Delta t \text{ thereafter} \end{cases}$$

SS_3 :

8. $X_{12} = X_{13} - X_{14} + X_{15}$

9. $X_{13} = G_2 X_4 - G_3 X_7 + G_3 X_{10}$

10. $X_{15} = c_{11} - G_2 X_4 - G_3 X_{10}$

$$X_{14} = \begin{cases} X_{14}^0 & \text{initially,} \\ X_{15} & \text{of previous } \Delta t \text{ thereafter} \end{cases}$$

12.3.2 Calibration Heuristics

Heuristics, rules of good judgment, made it possible to add seven more knowledge items (11–17) and thereby to enable calibration via the principle of reverse regression: "Given system knowledge, regression coefficients can be found that generate data consistent with that knowledge." Furthermore, constant means and standard deviations used to generate data for X_1 and X_{11} validated the calibration principle for stationary time series: "If the systems model is calibrated to the average Δt period of the time series, then it is calibrated to any other Δt period." In keeping with this principle, the model was calibrated to the average decade.

11. Initial values for the average decade were selected on a likelihood basis for the 64-thousand acre district, as heuristically defensible land allocation and growth density figures:

    - Immature trees: 30% of land with relatively low density of 730 trees per acre.

    $$\overline{X}_3^0 = (0.3)(64.0)(730) = 14.016 \text{ million trees}$$

    - Mature trees: 50% of land with density of 2,875 cubic feet per acre.

    $$\overline{X}_9^0 = (0.5)(64.0)(2,875) = 92.0 \text{ million cubic feet}$$

    - Meadow area: 20% of land with zero density (no trees).

    $$\overline{X}_{14}^0 = (0.2)(64.0) = 12.8 \text{ thousand acres}$$

12. Scale factors were calculated based on *dbh* classification and densities. Because variables X_4, X_6, and X_{10} effect linkage between subsystems, scaling was required to maintain unit consistency.

$$G_1 = 1.75\,\frac{cubic\ feet}{tree} = \frac{(1.75)\,10^6\,cubic\ feet}{10^6\ trees}$$

to scale pole timber of SS_1, with a dbh of 5 inches, from *count* in millions of trees to *volume* in millions of cubic feet for SS_2.

$$G_2 = \frac{1\ acre}{730\ trees} = \frac{(1.3698630)\,10^3\ acres}{10^6\ trees}$$

to scale SS_1 trees from *count* in millions of trees to *area* in thousands of acres for SS_3.

$$G_3 = \frac{1\ acre}{2,875\ cubic\ feet} = \frac{(0.3478261)\,10^3\ acres}{10^6\ cubic\ feet}$$

to scale SS_2 trees from *volume* in millions of cubic feet to *area* in thousands of acres for SS_3.

Supported by the knowledge base plus its extension via calibration heuristics, the serial solution scheme was applied to calibration equations to obtain values for the "c" coefficients of Equations 1, 3, 4, 5, 6, and 10. Heuristics are enclosed in quotes.

Equation 1 $X_2 = DR(w_1X_3 + w_2X_4) = c_1X_3 + c_2X_4$ for mortality count.

13. Heuristic for SS_1 mortality: "For the average decade, mortality count is one sixth of the growth count." From knowledge item 7: $\overline{X}_1 = 3.00$ so that $\overline{X}_2 = 1/6\overline{X}_1 = 0.50$. Letting $w_1 = w_2 = 1/2$ and referring to knowledge items 9, 11, and 13 results in

$$c_1 = c_2 = \frac{\overline{X}_2}{\overline{X}_3 + \overline{X}_4} = \frac{0.50}{(2)(14.016)} = 0.0178368.$$

Equation 3 $X_5 = DR(w_1X_3 + w_2X_4) = c_3X_3 + c_4X_4$ for harvest count.

14. Heuristic for SS_1 harvest count: "For the average decade, harvest is restricted to removing diseased trees, estimated by the ratio of 4 in 10,000 applied to average standing count, $(\overline{X}_3 + \overline{X}_4)/2$." Setting $w_1 = w_2 = 1/2$ and $DR = 4/10,000$ yields

$$c_3 = c_4 = \frac{4}{10,000} \cdot \frac{1}{2} = 0.0002.$$

Then from knowledge items 9, 11, and 14,

$$\overline{X}_5 = 0.0002\,(\overline{X}_3 + \overline{X}_4) = 0.0002\,(2)(14.016) = 0.0056064.$$

Equation 4 $X_6 = DR(w_1X_3 + w_2X_4) = c_5X_3 + c_6X_4$ for ingrowth to SS_2.

15. Heuristic for ingrowth: "On average, it takes immature trees 56 years (5.6 decades) to reach maturity—that is, dbh = 5 inches." Taking $w_1 = w_2 = 1/2$ and $DR = 1/5.6$ produces

$$c_5 = c_6 = 0.0892857.$$

Then for the average decade, from knowledge items 9, 11, and 15,

$$\overline{X}_6 = (0.0892857)(2)(14.016) = 2.5028567.$$

Equation 5 $X_7 = GR(w_1X_9 + w_2X_{10}) = c_7X_9 + c_8X_{10}$ for growth volume.

16. Heuristic for growth of mature trees: "For the average decade, gain in growth volume of mature trees is two thirds that obtained by ingrowth of pole timber." Therefore from knowledge items 12, 15, and 16,

$$\overline{X}_7 = \frac{2}{3}G_1\overline{X}_6 = \frac{2}{3}(1.75)(2.5028567) = 2.92.$$

With $w_1 = w_2 = 1/2$ and knowledge items 9, 11, and 16

$$c_7 = c_8 = \frac{\overline{X}_7}{\overline{X}_9 + \overline{X}_{10}} = \frac{2.92}{(2)(92.0)} = 0.0158696.$$

Equation 6 $X_8 = DR(w_1X_9 + w_2X_{10}) = c_9X_9 + c_{10}X_{10}$ for mortality volume.

17. Heuristic for mortality volume of mature trees: "For the average decade, mortality volume is one-tenth as great as the growth volume." Applying the heuristic and knowledge item 16,

$$\overline{X}_8 = 0.1\,\overline{X}_7 = (0.1)(2.92) = 0.292.$$

Taking $w_1 = w_2 = 1/2$ and knowledge from items 9, 11, and 17 yields

$$c_9 = c_{10} = \frac{\overline{X}_8}{\overline{X}_9 + \overline{X}_{10}} = \frac{0.292}{(2)(92.0)} = 0.001587.$$

Equation 9 $X_{13} = G_2X_4 - G_3X_7 + G_3X_{10}$ for terminal forested area.

Although Equation 9 contains no "c" coefficients, its structure and scale factors require elaboration. By Equation 7, some of the increase in X_{10} (terminal standing volume of mature trees) is due to X_7 (growth volume of mature trees), which does not induce change in meadow area. Thus, net change in area induced by X_{10} is equal to $G_3(X_{10} - X_7)$. Values for G_2 and G_3 are provided by knowledge item 12. From Equations 8 and 9 and knowledge items 9, 11, 12, and 16,

$$\overline{X}_{12} = \overline{X}_{13} = 50.184349.$$

Equation 10 $X_{15} = c_{11} - G_2X_4 - G_3X_{10}$ for terminal meadow area.

Rewriting Equation 10, it is readily seen that

$$c_{11} = (\text{SS}_1 \text{ acres}) + (\text{SS}_2 \text{ acres}) + (\text{SS}_3 \text{ acres}) = 64.0.$$

12.3.3 The Prototype Model

Calibration produced a tentative prototype model—tentative because validation testing may reveal structural deficiencies that require revision of the model. A series of simulation tests were performed, and several modifications were required. Equations of the current prototype are as follows:

1. $X_2 = 0.0178368\, X_3 + 0.0178368\, X_4$
2. $X_4 = X_1 - X_2 + X_3 - X_5 - X_6$
3. $X_5 = 0.000200\, X_3 + 0.000200\, X_4$
4. $X_6 = 0.0892857\, X_3 + 0.0892857\, X_4$
5. $X_7 = 0.0158696\, X_9 + 0.0158696\, X_{10}$
6. $X_8 = 0.001587\, X_9 + 0.001587\, X_{10}$
7. $X_{10} = 1.75\, X_6 + X_7 - X_8 + X_9 - X_{11}$
8. $X_{12} = X_{13} - X_{14} + X_{15}$
9. $X_{13} = 1.3698630\, X_4 - 0.3478261\, X_7 + 0.3478261\, X_{10}$
10. $X_{15} = 64.0 - 1.3698630\, X_4 - 0.3478261\, X_{10}$

$$X_3 = \begin{cases} 13.393984 & \text{initially,} \\ X_4 & \text{of previous } \Delta t \text{ thereafter} \end{cases}$$

$$X_9 = \begin{cases} 91.757840 & \text{initially,} \\ X_{10} & \text{of previous } \Delta t \text{ thereafter} \end{cases}$$

$$X_{14} = \begin{cases} 13.736306 & \text{initially,} \\ X_{15} & \text{of previous } \Delta t \text{ thereafter} \end{cases}$$

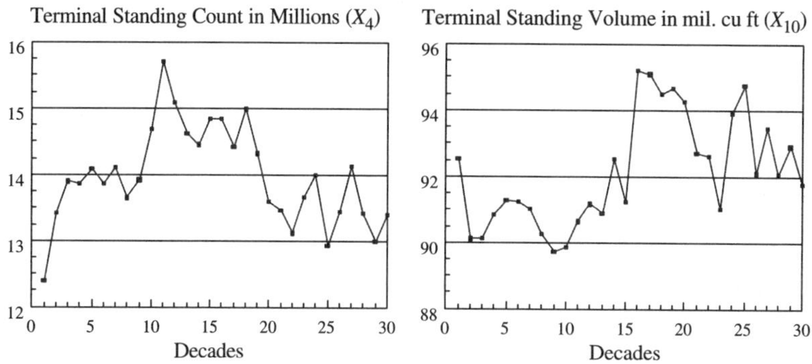

Figure 12.2 Secondary Time Series for Terminal Count and Volume

12.4 MODEL VALIDATION

Each simulation test provided grounds for revision of the model. The first test of the model, simulation of a complete data base, constituted Phase One. Phase Two involved use of the completed data base to perform three additional simulation tests, two of which required role inversion. As many other tests as there are other simulation scenarios can also be performed, each having the potential to serve as a catalyst for model revision.

12.4.1 Secondary Data Simulation

Primary data were processed through the prototype model to generate secondary time series data, thereby creating a complete data base for the 30-decade period. To obtain values for X_3^0, X_9^0, and X_{12}^0 as shown in the prototype model, initial values for the average decade, (as per knowledge item 11) were input, and the model was run for the 30 decades to obtain values for X_4^{30}, X_{10}^{30}, and X_{13}^{30}. For the next iteration, X_3^0, X_9^0, and X_{12}^0 were reset to these corresponding terminal values. Iteration continued until zero trends (as per knowledge item 9) were obtained.

These time series and their initial values were within realistic ranges of response. For example, the two graphs in Figure 12.2 present terminal count and volume for the immature trees and mature trees subsystems, respectively. Terminal count (X_4) ranged from 12.39 million trees to 15.72, averaging 14.0 million trees for the 30 decades. Terminal volume (X_{10}) ranged from 89.72 million cubic feet to 95.21, averaging 92.0 million cubic feet. Graphs for four more secondary variables, X_6, X_7, X_{13}, and X_{15} are displayed in Figure 12.13 of Appendix I. Initial values for zero trend and the completed data base were used to perform the validation tests that follow.

12.4.2 Commercial Harvesting of Poles

The heuristic of knowledge item 14 stated, "For the average decade, harvest is restricted to removing diseased trees, estimated by the ratio of 4 in 10,000 applied to average standing count, $(\overline{X}_3 + \overline{X}_4)/2$." What if SS_1 harvest were expanded to include commercial harvesting? To test this scenario, the heuristic was replaced by the following harvest policy:

> *Ten percent of SS_1 inventory, based on initial standing count, shall be removed by commercial harvesting of poles during each decade.*

Equation 3 was modified to accommodate harvesting of poles:

$$X_5 = 0.1X_3.$$

The purpose of this test was to perturb the immature trees subsystem and then to observe and evaluate model response for expected behavior. Although in practice, the harvest policy would be accompanied by a companion policy to finance planting of seedlings by sales receipts, the forest systems model allows both *isolation* and *superposition* of policy effects for study. The model was solved for the *endo* variables identified in the vector table.

Figure 12.3 displays the effect of commercial harvesting without modifying growth count (X_1). Note that the zero-trended, standing count (old X_4) has been replaced by a reduced standing count (new X_4). At the end of the first decade, the new count is down by a little less than 10%; and although countered by growth (X_1), trend analysis reveals a 1.41% decline.

Figure 12.4 shows the disastrous, long-term effect of the harvest policy on terminal standing volume of mature trees. Zero trend, exhibited by the *equilibrium series* (old X_{10}), has been replaced by a descending (new X_{10}) time series. Following a negative 2.68% trend, the new series ends with the volume 34.96 million

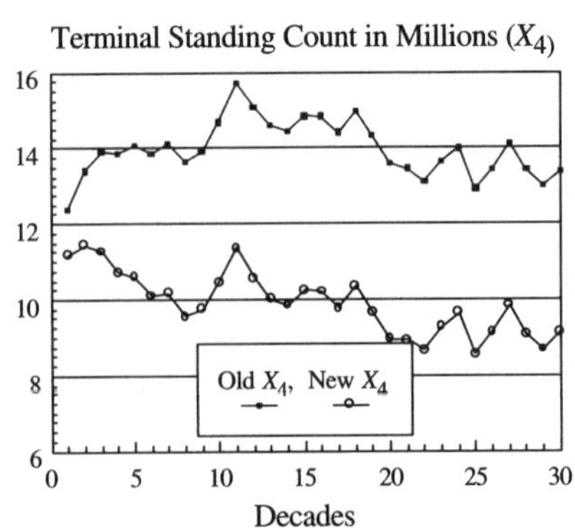

Figure 12.3 Effect of Harvest Policy on Terminal Standing Count

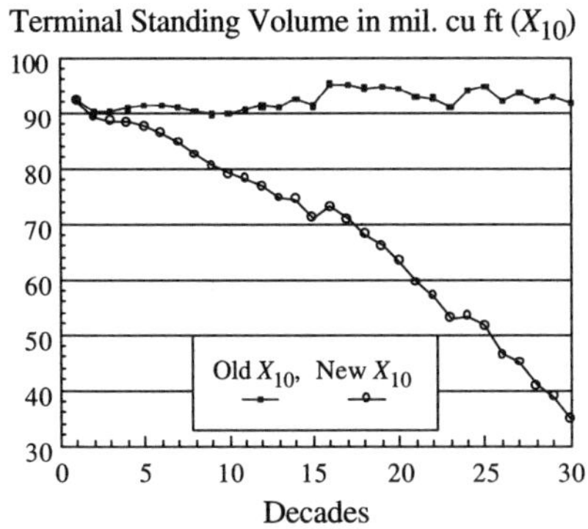

Figure 12.4 Effect of Harvest Policy on Terminal Standing Volume

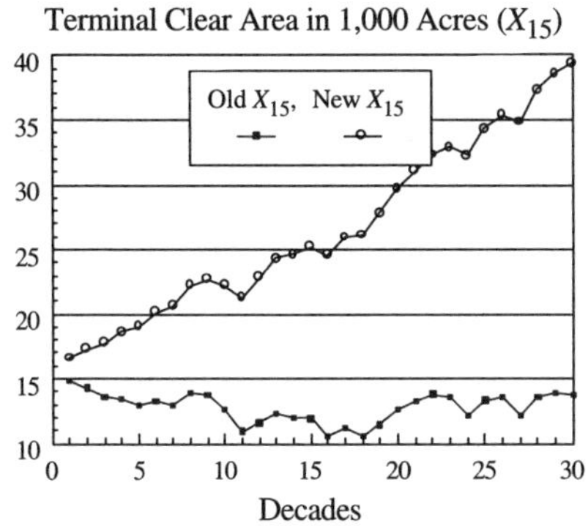

Figure 12.5 Meadow Response to SS1 Harvest Policy

cubic feet, down 62% from the closing volume, 91.58 mcf, of the equilibrium series.

Figure 12.5 presents the substantial loss of forested area to meadow (clear) area (new X_{15}). Meadow area displays a new 3.32% trend, ending with 39.33 thousand acres, an increase of 286% over the old area of 13.74 thousand acres.

REMARK Trends presented in Figures 12.3, 12.4, and 12.5 indicate the long-term effect of a management policy that favors harvesting of poles. Predicted response by the model provides evidence to challenge that policy. Consequently it may be concluded that if the existing acreage of forested land is to be subjected to severe commercial harvesting, management policy must be expanded to include (1) supplemental reseeding and (2) programs to counter various forms of deforestation.

12.4.3 Role Inversion

To accommodate two perturbation tests, the roles of *endo* variable X_{15} and *exo* variable X_1 were exchanged to include X_1 among the response variables. Consequently, role inversion was performed to test sensitivity of the model to distortion induced by inversion. Distortion (change) figures were accumulated by operating the model over the 30-decade period with (1, 2, 4, 5, 6, 7, 8, 10, 12, 13) as the solution vector. The solution for the last decade provides a summary:

X_j	Value	Change	Sum*
X_1	3.228713	.000000	−.000001
X_2	0.470743	.000000	−.000002
X_4	13.393984	.000000	−.000003
X_5	0.005278	.000000	−.000002
X_6	2.356399	.000000	−.000001
X_7	2.930651	.000000	.000000
X_8	0.293072	.000000	.000000
X_{10}	91.757840	.000000	.000009
X_{12}	49.103225	.000000	.000002
X_{13}	49.244337	.000000	.000000

*SUM = Cumulative algebraic value of change

Distortion in individual values appears in the CHANGE column and is accumulated in the SUM column. Average distortion (SUM $\div$ 30) was less than 10^{-6} in magnitude, and was, therefore, negligible.

12.4.4 Two-Stage Harvesting of Mature Trees

What if the stochastic series for harvest volume of mature trees (the old *primary* X_{11} series) were to be replaced by a perturbed series (a new deterministic X_{11} series)? What might be the compensatory change, such as a supplementary reseeding program (growth count X_1) that would preserve the same acreage of meadow area (X_{15})?

Figure 12.6 shows the old X_{11} (primary) series and the new X_{11} (deterministic) series. Mature trees are to be harvested in two stages. In stage 1, consisting of decades 1 through 15, commercial harvesting is to remove a volume of 6.0 million cubic feet during each decade, one-seventh *below* the 30-decade average of 7.0 mcf. During each decade of stage 2, decades 16 through 30, a harvest volume of 10.0 mcf is to be removed, three-sevenths *above* the average. Reduced harvesting during stage 1 permits volume to accumulate, followed by severe harvesting in stage 2, thereby removing old growth to make way for reforestation in ensuing decades.

Perturbation was implemented by replacing the stochastic series for X_{11} in the data base by the two-stage deterministic series. The model was rerun with (1, 2, 4, 5, 6, 7, 8, 10, 12, 13) as the solution vector. Figure 12.7 shows growth count (new X_1) dropping below the *equilibrium series* (old X_1) and surpassing the equilibrium

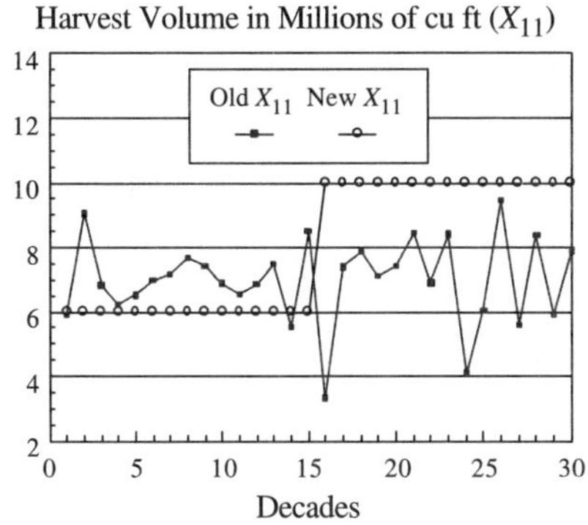

Figure 12.6 Perturbation in Harvest Volume

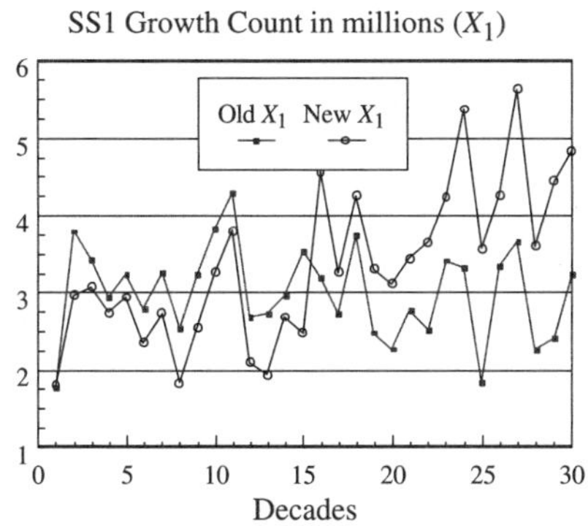

Figure 12.7 Reseeding Program to Compensate for Two-Stage Harvesting

series during decade 16, advancing to 4.83 thousand acres during stage 2, up 50% above the closing count, 3.23, of the equilibrium series.

Figure 12.8 indicates a substantial two-stage response in terminal standing volume (new X_{10}) of mature trees relative to the *equilibrium series* (old X_{10}). Under reduced harvesting, volume increased to a peak of 102.35 mcf by the end of decade 15, followed by heavy harvesting and a drop to 65.52 mcf, down 28.6% from the closing volume, 91.76 mcf, of the equilibrium series. The two-stage scenario produced an additional 30.0 mcf of total harvest for the 30 decades.

REMARK Figures 12.6, 12.7, and 12.8 show the need to establish a reseeding program to countermand a stepped-up level of harvesting that would otherwise deplete standing volume to a level below its equilibrium level, that is, if equilibrium between forested and meadow areas is to be maintained as a long-term goal.

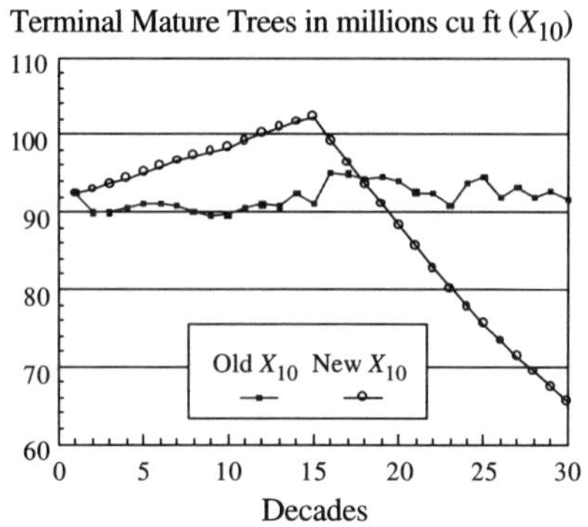

Figure 12.8 Effect of Two-Stage Harvesting on Terminal Standing Volume

12.4.5 Stepped Program to Maximize Forested Land

Badger Ranger District will potentially support forest growth on all but 5 percent of its area, due to outcroppings of rock. A 30-decade program was specified that expands the growth count (X_1) to extend the forest to all potential meadow area but that continues proportional harvesting of diseased, immature trees (original equation for X_5) and stochastic harvesting of mature trees (primary X_{11} series).

A simple solution is to decrement meadow area each decade by a fixed amount, from $X_{14}^{0} = 13.736306$ thousand acres to $X_{15}^{30} = (0.05)(64.0) = 3.2$ thousand acres:

$$\Delta X = (13.73606 - 3.2)/30 = 0.351210.$$

Consequently, the X_{14}, X_{15} IT pair was perturbed by replacing its two series by two 30-step series, offset by Δt, that introduce a simple linear decline in meadow area. Figure 12.9 exhibits the *equilibrium series* (old X_{15}) and a perturbed new X_{15} series that reduces meadow area to 3.2. The model was rerun with (1, 2, 4, 5, 6, 7, 8, 10, 12, 13) as the solution vector.

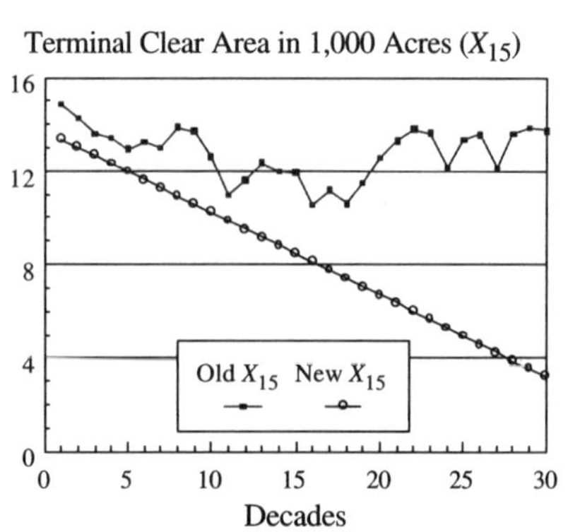

Figure 12.9 Programmed Reduction in Meadowland

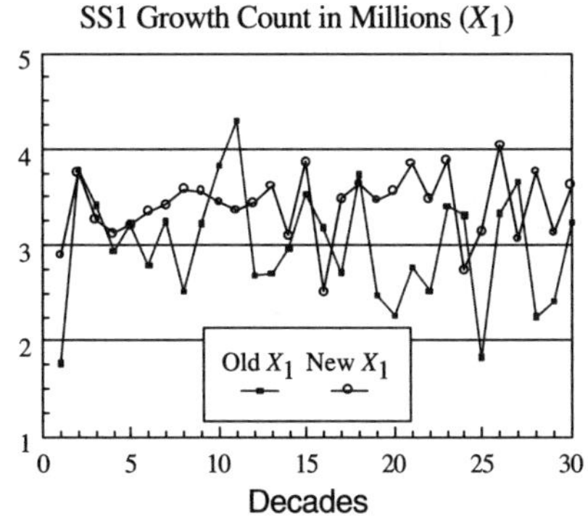

Figure 12.10 Immature Tree Growth Required by the Program

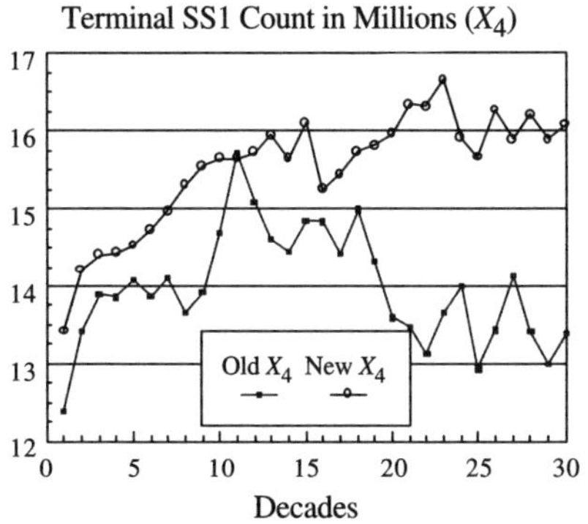

Figure 12.11 Growth in Terminal Standing Count

Figure 12.10 shows growth count (new X_1) striking an average of 3,412,268 trees, which is 412,268 above the average of the *equilibrium series* (old X_1). Thus, the program must monitor new seedling starts, and supplement as needed, to ensure this increase in immature tree growth.

Figure 12.11 presents the resulting terminal SS_1 count (new X_4 series) that surpasses the *equilibrium series* (old X_4) with a modest 0.581% trend, terminating at 16.1 million trees, 20% above the equilibrium series at 13.4 million trees. Growth of immature trees into mature trees also accounts for much of the reduction in meadow land.

Figure 12.12 similarly depicts growth in terminal volume of mature trees (new X_{10} series) that rises above the *equilibrium series* (new X_{10}) by a 0.667% trend, topping off at 111.4 million cubic feet, 21.4% above the equilibrium series at 91.8 mcf.

REMARK The preceding scenario indicates great potential for extending forested land to maximum coverage of the district. Moreover, Figure 12.10 shows that such

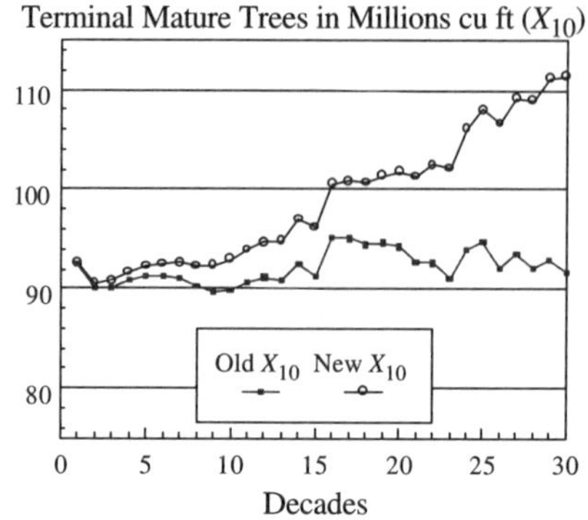

Figure 12.12 Growth in Terminal Volume of Mature Trees

a program is feasible if implementation is consistent. Required SS_1 growth count ranges from a low of 2.51 million trees to a high of 4.04 million and averages 3.41 million.

12.5 CONCLUDING REMARKS

The forest systems model of this chapter has a deterministic structure, although either deterministic or stochastic structure can be synthesized by the Mtm approach. However, stochastic model structure related to any given data base is more macro than its deterministic counterpart. For example, to internalize Monte Carlo simulation, data must be grouped into empirical frequency distributions or approximated by standard distributions. A summing process is involved in tallying the number in each class interval, thereby replacing individual values by the class mark of the interval.

Defining the generic Badger Ranger District made it possible to illustrate systems analysis and rapid prototyping of a forest systems model despite the absence of domain experts and primary data. Likelihood provided the basis for simulating primary data and for defining calibration heuristics. Iterative improvement of the model was prompted by three validation tests. The resulting 15-variable model was used to demonstrate prediction of forest systems dynamics over a 30-decade period. Procedural steps were established, and a demonstration model was presented with potential for application to any ranger district.

Admittedly, this prototype model does contain flaws, and Exercise 10 asks for a critique. However, rapid prototyping shortens the initial knowledge acquisition phase. Furthermore, prototyping effectively provides a starting point for *replacement programming*. For this procedure, heuristic knowledge items are initial *place holders* to be replaced by domain specific knowledge.

More detail than is provided by this $L_{1,j}$ prototype can be obtained by continued differentiation. Examples of additional details to incorporate at some level are as follows:

- Separate models for immature and mature trees.
- Immature tree model with four subsystems—seed, seedling, sapling, and pole timber—with Δt = one year.
- Mature tree model to accommodate commercial harvesting with subsystems comprised of harvest sites.
- Subsystems determined by age group: $SS_1 \sim$ age group 1, $SS_2 \sim$ age group 2, and so on.
- Replacing fixed rates by variable rates, either deterministic or stochastic.
- Monthly model to account for insect infestation, fire damage, and weather extremes.

EXERCISES

1. How can an operational systems model be obtained at relatively low cost?
2. What is a prototype model and what is its purpose?
3. Distinguish between *hard* and *soft* knowledge with respect to a system's knowledge base.
4. The forest system model may be viewed as a three-subsystem inventory model. Explain.
5. Superimpose (add) the output rate equations of SS_1 of the zero-trend model. From the resulting equation and the dynamic form for delay, calculate the expected longevity of trees in SS_1.
6. Viewing the mature trees subsystem as an inventory system, use steady-state values for the zero-trend model to calculate the expected longevity of mature trees.
7. What technique is used to effect linkage between subsystems that have different units?
8. For the Phase One forest systems model, superimpose a +1.0% linear trend on harvest volume (primary X_{11} series) and rerun the model. What effects does this perturbation have on model response?

9. Verify that $\overline{X}_{12} = \overline{X}_{13} = 50.184349$ thousand acres using model Equations 8 and 9 and knowledge items 9, 11, 12, and 16.

10. Provide a critique of the prototype forest systems model and suggestions for its upgrade.

APPENDIX I

Table 12.2 Primary Data for Forest Systems Model

Decade	X_1	X_{11}
1	1.766995	5.896019
2	3.791855	9.058089
3	3.414509	6.823586
4	2.937657	6.214650
5	3.221865	6.519177
6	2.785714	7.004723
7	3.248117	7.192943
8	2.521869	7.683091
9	3.225791	7.432079
10	3.831849	6.890451
11	4.296489	6.555775
12	2.680546	6.877498
13	2.712665	7.513362
14	2.956969	5.519033
15	3.532621	8.506738
16	3.180455	3.329886
17	2.716568	7.409877
18	3.730197	7.890239
19	2.478017	7.122586
20	2.267749	7.446247
21	2.772634	8.445102
22	2.509793	6.896072
23	3.410309	8.410675
24	3.309569	4.092164
25	1.824727	6.035004
26	3.330640	9.446935
27	3.653837	5.581307
28	2.250582	8.371820
29	2.410537	5.918587
30	3.228713	7.916320

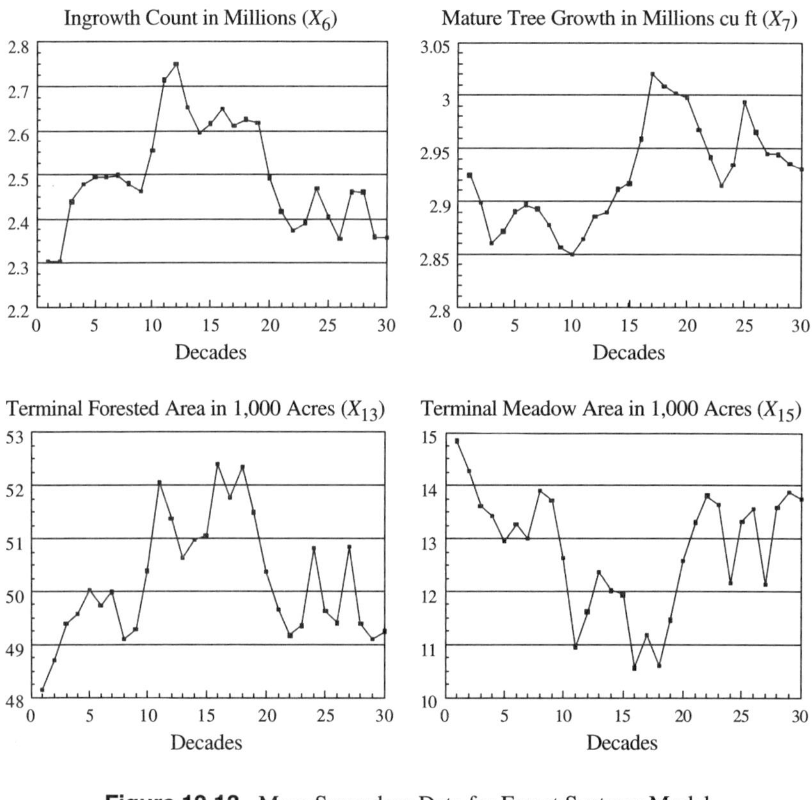

Figure 12.13 More Secondary Data for Forest Systems Model

CHAPTER 13

Applications Unlimited

13.1 INTRODUCTION

As demonstrated by a generous number of previous examples, variation in the application of the Mtm method of systems analysis and modeling is unlimited. The method is capable of yielding accurate models pursuant to the need for understanding the many interrelated features of a total system: interactions between its component subsystems and between component variables of each subsystem. This concluding chapter features four areas that support a myriad of additional applications that range from modeling a simple, machine-tool process to structuring a large-scale systems model.

13.2 VIRTUAL MANUFACTURING

In virtual manufacturing applications, modeling of *entire* processes requires a unifying medium to "standardize" process modeling. Standardization greatly improves cost effectiveness in that it simplifies software development. Systems analysis and modeling, as described in previous chapters, provide that standardized delivery medium.

Cutting tools used in manufacturing processes, such as machine turning, facing, chamfering, parting, and threading, are limited in life length because of wear. This first example revisits a well-known process model widely referenced in manufacturing books: the Taylor tool-life equation for predicting life length of a common lathe tool.

13.2.1 Tool-Life Test Example

Data were obtained from a tool-life curve presented in the *Machining Data Handbook* [33].

Machine test conditions

- Material: AISI (and SAE) 8640 steel, 190 BHN
- Tool: C-6 carbide

BRA: 0°	ECEA: 5°
SRA: 6°	Relief: 5°
SCEA: 0°	NR: .040 in.

- Depth of cut: .100 in.
- Feed rate: .010 in./rev.
- Cutting fluid: dry
- Wear land: .015 in.

Coordinate values for the six data points that comprise the tool-life curve were scaled from an enlarged ($\times$ 2) copy of the curve. Their values are listed in Table 13.1. These six points, shown connected by a smooth-line curve in Figure 13.1, provided a standard for comparing the tool-life curve obtained by Taylor's nonlinear equation and the tool-life curve obtained from the linear equations of a systems model.

Table 13.1 Speed versus Tool-Life Data

Point No.	Speed in ft./min.	Tool Life in mins.
1	800	4.69
2	700	7.19
3	600	11.25
4	500	17.5
5	400	27.5
6	300	44.0

13.2.2 Taylor's Tool-Life Equation

Almost a century ago, F. W. Taylor [50] analyzed the results of extensive lathe tool tests, similar to Figure 13.1, and found that when normal cutting speeds are plotted against tool life on log-log paper, the data approximate a straight line of the form $y = a + bx$.

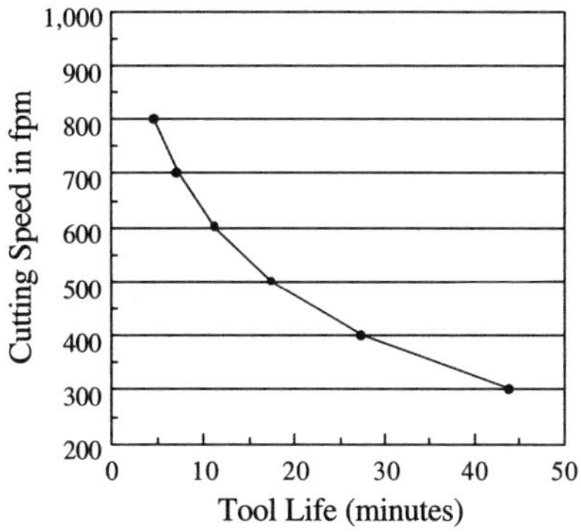

Figure 13.1 Tool-Life Data

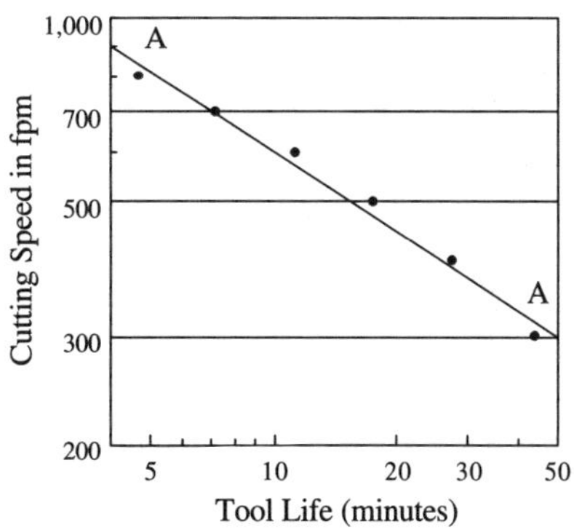

Figure 13.2 Log-Log Plot of Tool-Life Data

Taylor derived this standard tool-life equation from

$$\ln v = \ln C - n \ln T, \text{ from which}$$
$$v = CT^{-n}, \text{ or}$$
$$vT^n = C, \tag{13.1}$$

where v is the cutting speed in feet per minute (fpm), T is tool life (cutting time between changes) in minutes, C is a machining constant for a tool life of one minute, and n is the slope, $\Delta v / \Delta T$, calculated over the normal operating range of the tool.

A *log-log* plot of the six data points in Figure 13.2 was examined for linearity. The data do cluster about the line AA, although a concave downward tendency is apparent. Nonetheless, Equation 13.1 was calibrated to interpolate between point 1 (4.69,800) and point 6 (44.0, 300), which define the operating range.

$$n = \frac{\ln 800 - \ln 300}{\ln 44.0 - \ln 4.69} = 0.43811365, \text{ and}$$

$$C = (800)(4.69)^{0.43811365} = 1,574.49,$$

leading to

$$v = 1,574.49\,T^{-0.43811365}. \tag{13.2}$$

Predicted speeds were obtained from Equation 13.2 for $T = 1, 2, \ldots, 50$ minutes, and are plotted in Figure 13.3. Predictions differed from test data by 0.0 at $v = 800$, 36.6 at $v = 700$, 54.7 at $v = 600$, 50.7 at $v = 500$, 31.4 at $v = 400$, and 0.0 at $v = 300$. The total error measure, sum of the absolute deviations (SAD), was 173.4 *fpm*.

13.2.3 Systems Approach

Systems analysis and model synthesis provide an attractive, accurate alternative for modeling manufacturing processes. Again, the approach is to relate system variables, both levels and rates, via a system of linear equations capable of interacting with adjacent systems by linking their equation systems within a common matrix structure.

13.2.3.1 Knowledge Base

- Material removal via Lathe/Tool system.
- Fundamental units of analysis:

  Space: Lathe/tool domain

  Mass: Tool

  Time: Unit equals 1 minute, tool life in minutes and cutting speed represented implicitly in increments of $\Delta v = 10$ feet per minute

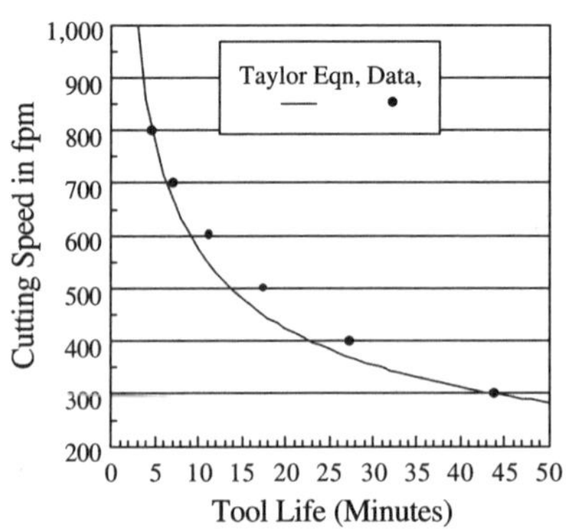

Figure 13.3 Tool Life Predicted by the Taylor Equation

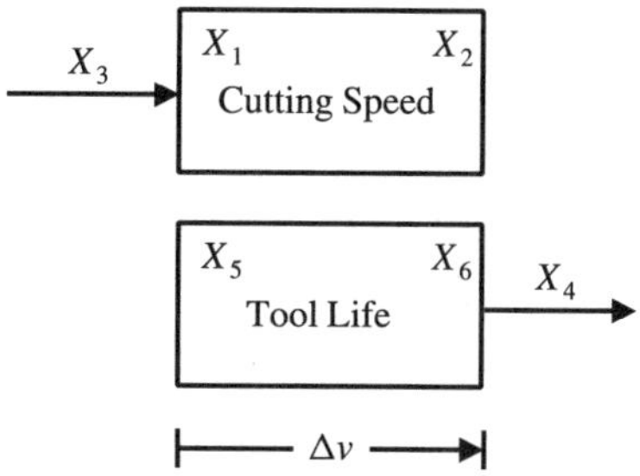

Figure 13.4 Block Diagram of Tool-Life Model

- Data: See data from tool-life test example.
- Number of subsystems: two
- System variables:

SS_1: Cutting speed

X_1 = initial level of cutting speed

X_2 = terminal level of cutting speed

X_3 = incremental decline in cutting speed (rate variable)

SS_2: Tool life

X_4 = incremental change in tool life (rate variable)

X_5 = initial level of tool life

X_6 = terminal level of tool life

- Schematic: See Figure 13.4.
- Vector table for model structure.

X_j	IT Pair	Exo	Endo	Equation Type	Equation No.
X_1	I	*		—	—
X_2	T		*	balance	1
X_3			*	constant	2
X_4			*	regression	3
X_5	I	*		—	—
X_6	T		*	balance	4

- Dynamic form for X_4.

13.2.3.2 Model Equations

SS_1:

1. $X_2 = X_1 + X_3$ (balance equation) where

$$X_1 = \begin{cases} X_1^0 = 0 & \text{initially,} \\ X_2 & \text{of previous } \Delta v \text{ thereafter} \end{cases}$$

2. $X_3 = \Delta v = 10\,fpm$ (rate equation)

SS_2:

3. $X_4 = c_1 X_5 + c_2 X_6$ (rate equation)

4. $X_6 = X_5 - X_4$ (balance equation) where

$$X_5 = \begin{cases} X_5^0 = T_0 & \text{initially,} \\ X_6 & \text{of previous } \Delta v \text{ thereafter} \end{cases}$$

Equation 3 is a linear regression (rate) equation that represents the dynamic properties of the tool-life curve of Figure 13.1. X_4 introduces curvilinear incremental decline in level into its balance equation (Equation 4). Each Δv decline in cutting speed from 800 feet per minute to 300 feet per minute results in an incremental increase in tool life. Consequently, the dynamic form for X_4 was defined as follows:

Tool-life data in Figure 13.1 were investigated for negative exponential decline relative to the horizontal or T axis. Conceptually, exponential curvature is validated to the extent that a *log-arithmetic* plot of the data relocates the points in a straight line, as in Figure 13.5.

Compared to the line BB, the six data points clearly exhibit linearity. Thus, the decline in tool life can be approximated by the *nonlinear* dynamic form for exponential decline

$$T = T_0\, e^{-DR\,v}, \tag{13.3}$$

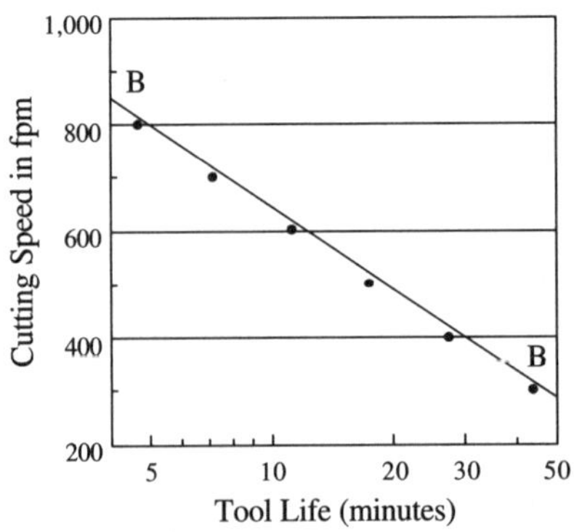

Figure 13.5 Log-Arithmetic Plot of Tool Life Data

where T_0 is the tool-life axis intercept, and DR is the exponential rate of decline in tool life. The method of linear goal programming was used to find the best two of the six points through which to pass the *log-arithmetic* form of the curve. See Appendix J for calibration details. $DR = 0.004469089$ and $T_0 = 164.3209627 = X_5^0$. Model Equation 3 for X_4 was calibrated via the *linear* dynamic form for *exponential decline*

$$X_4 = DR[w_1 X_5 + w_2 X_6] = c_1 X_5 + c_2 X_6, \tag{13.4}$$

where $w_1 = c_1/(DR)$ and $w_2 = c_2/(DR)$ are nonnegative weights that sum to 1.0. Regression coefficients c_1 and c_2 are related to DR via the calibration equations for exponential decline

$$c_1 + c_2 = DR \tag{13.5}$$

$$\frac{1 - c_1}{1 + c_2} = e^{-DR} \tag{13.6}$$

DR was replaced by $DR \cdot \Delta v = 0.04469089$ in Equations 13.5 and 13.6; their simultaneous solution yielded $c_1 = 0.022179008$ and $c_2 = 0.022511882$.

13.2.3.3 Results

The model was run to predict tool life for $N = 80\Delta v$'s, from $v = 10$ to 800 *fpm* in steps of 10. Predicted tool-life values differed from test data by -0.09 at $T = 4.69$, 0.01 at $T = 7.19$, 0.00 at $T = 11.25$, 0.09 at $T = 17.5$, 0.00 at $T = 27.5$, and -1.00 at $T = 44.0$. Figure 13.6 provides a visual inspection of these results. The SAD measure was 1.19 minutes. Converting to the speed axis, SAD was approximately 9.68 *fpm*, which is 94.4% lower than the SAD measure (173.4 *fpm*) for the Taylor equation.

In that only one set of data was used in the comparison, one may not infer accuracy superiority over the Taylor equation. What has been demonstrated, however, is the relative ease with which systems analysis and modeling can be applied to tool life. Although this linear model related only tool life and cutting speed, linkage with other interacting systems of the manufacturing process can be easily achieved.

13.3 TIME SERIES SYNTHESIS

Traditional time-series analyses may divide a series into as many as four components: (1) long-term trend, linear or curvilinear, (2) seasonal variation, (3) cyclical variation, and (4) uncorrelated "noise." Time series synthesis may combine components using either of two approaches: (1) multiplicative factors or (2) additive terms. The following example will illustrate synthesis of a stationary time series

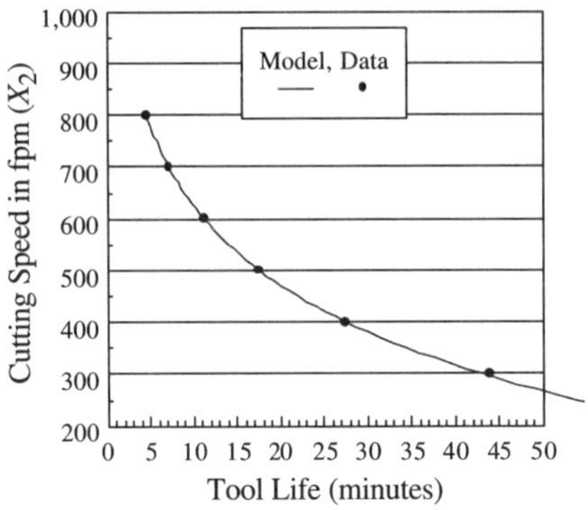

Figure 13.6 Tool Life from Systems Model

model using the Box-Jenkins additive method [40]. Using this method, a time series may be characterized as possessing (1) an *autoregressive (AR)* "memory" of previous levels reaching back p time periods, (2) a *moving average (MA)* "memory" of previous noise disturbances reaching back q time periods, or (3) a mixed memory, containing both AR and MA terms. A mixed model will be synthesized that contains both AR and MA characteristics.

13.3.1 Knowledge Base

- System: An *ARMA(2,2)* stochastic process that produces a stationary time series z_t, characterized by AR parameters ϕ_1 and ϕ_2, MA parameters θ_1 and θ_2, and AR constant δ

  $$z_t = \phi_1 z_{t-1} + \phi_2 z_{t-2} + \delta + a_t - \theta_1 a_{t-1} - \theta_2 a_{t-2},$$

  with parameter dependent mean, $E(z_t) = \mu = \delta/(1 - \phi_1 -, \phi_2)i$. More concisely,

  $$(1 - \phi_1 B - \phi_2 B^2)\tilde{z}_t = (1 - \theta_1 B - \theta_2 B^2)a_t, \qquad (13.7)$$

  where

  $\tilde{z}_t$, denotes deviations of the process from μ at equally spaced times, $t = 0, 1, 2, \ldots$, so that $\tilde{z}_t = z_t - \mu$,

  B^k is a *backshift operator* that simply shifts the subscript of z_t backward by k time periods, that is, $B^k z_t = z_{t-k}$, and

  a_t is a white noise process which imposes its variance on the series.

  Employing the backshift operator yields

  $$\tilde{z}_t = \phi_1 \tilde{z}_{t-1} + \phi_2 \tilde{z}_{t-2} + a_t - \theta_1 a_{t-1} - \theta_2 a_{t-2}.$$

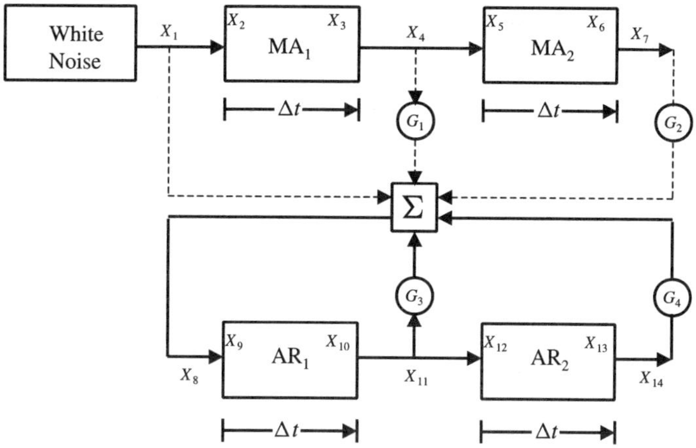

Figure 13.7 Block Diagram of an ARMA(2,2) Stochastic Process

An $L_{1,4}$ systems model will be synthesized for this *ARMA(2,2)* process.

- Fundamental units of analysis:

  Space: Process domain

  Mass: Unidentified for the example

  Time: $\Delta t = 1$ month, for visual emphasis.

- Four subsystems: MA_1, MA_2, AR_1, and AR_2
- System schematic: See Figure 13.7.
- System variables:

$X_1 = a_t$

$X_2 =$ initial MA_1 level, a_{t-1}

$X_3 =$ terminal MA_1 level, a_t

$X_4 = Ba_t$, that is, a_t delayed by one Δt

$X_5 =$ initial MA_2 level, a_{t-2}

$X_6 =$ terminal MA_2 level, a_{t-1}

$X_7 = B^2 a_t$, that is, a_t delayed by two Δt

$X_8 = \tilde{z}_t$

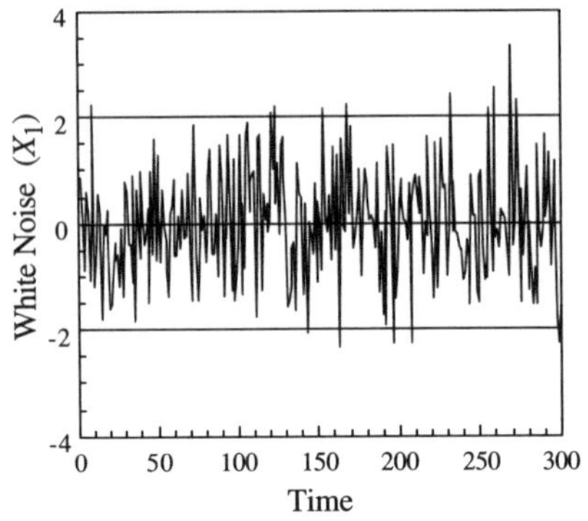

Figure 13.8 White Noise Time Series

X_9 = initial AR_1 level, $\tilde{z}_{t-1}$

X_{10} = terminal AR_1 level, $\tilde{z}_t$

$X_{11} = B\tilde{z}_t$, that is, $\tilde{z}_t$ delayed by one Δt

X_{12} = initial AR_2 level, $\tilde{z}_{t-2}$

X_{13} = terminal AR_2 level, $\tilde{z}_{t-1}$

$X_{14} = B^2\tilde{z}_t$, that is, $\tilde{z}_t$ delayed by two Δt

- Primary data series, an exogenous stochastic form for $X_1 = a_t$, $t = 1, 2, \ldots, 300$. White noise was generated via Monte Carlo random sampling from a normal distribution with $\mu = 0$ and $\sigma = 1.0$. See Figure 13.8.

- Vector table for model structure:

X_j	IT Pair	Exo	Endo	Equation Type	Equation No.
X_1		*		—	—
X_2	I	*		—	—
X_3	T		*	balance	1
X_4			*	unit delay	2
X_5	I	*		—	—
X_6	T		*	balance	3
X_7			*	unit delay	4
X_8			*	regression	5
X_9	I	*		—	—
X_{10}	T		*	balance	6

continued on next page

continued

X_{11}			*	unit delay	7
X_{12}	I	*		—	—
X_{13}	T		*	balance	8
X_{14}			*	unit delay	9

- Dynamic forms to introduce delay in a_t and $\widetilde{z}_t$

13.3.2 Generic Model

Based on the above knowledge items, a generic ARMA(2,2) model was defined by

1. $X_1 + X_2 = X_3 + X_4$
2. $X_4 = \frac{1}{D}[\alpha X_2 + (1 - \alpha)X_3]$
3. $X_4 + X_5 = X_6 + X_7$
4. $X_7 = \frac{1}{D}[\alpha X_5 + (1 - \alpha)X_6]$
5. $X_8 = X_1 - G_1 X_4 - G_2 X_7 + G_3 X_{11} + G_4 X_{14}$
6. $X_8 + X_9 = X_{10} + X_{11}$
7. $X_{11} = \frac{1}{D}[\alpha X_9 + (1 - \alpha)X_{10}]$
8. $X_{11} + X_{12} = X_{13} + X_{14}$
9. $X_{14} = \frac{1}{D}[\alpha X_{12} + (1 - \alpha)X_{13}]$

$$X_2 = \begin{cases} X_2^0 = a_{t-1} & \text{initially,} \\ X_3 & \text{of previous period thereafter} \end{cases}$$

$$X_5 = \begin{cases} X_5^0 = a_{t-2} & \text{initially,} \\ X_6 & \text{of previous period therafter} \end{cases}$$

$$X_9 = \begin{cases} X_9^0 = \widetilde{z}_{t-1} & \text{initially,} \\ X_{10} & \text{of previous period thereafter} \end{cases}$$

$$X_{12} = \begin{cases} X_{12}^0 = \widetilde{z}_{t-2} & \text{initially,} \\ X_{13} & \text{of previous period thereafter} \end{cases}$$

13.3.3 Calibration and Time Series Generation

Model Equations 2, 4, 7, and 9 are dynamic forms that delay X_4, X_7, X_{11}, X_{14}, and thus perform the function of the backshift operator B in Equation 13.7. B is equivalent to pure transportation delay and is realized with $D = 1$ period and $\alpha = 1$.

Parameter values were selected that ensured *stationarity* of the AR component and *invertibility* of the MA component: $\phi_1 = 0.5$, $\phi_2 = 0.3$, $\theta_1 = -0.7$, and $\theta_2 = 0.5$. These were inserted into the model as

$$G_1 = -\theta_1 = 0.7$$

$$G_2 = -\theta_2 = -0.5$$

$$G_3 = \phi_1 = 0.5$$

$$G_4 = \phi_2 = 0.3$$

Initial values were justifiably equated to 0.0, because transients resulting from initial conditions disappear after several periods.

The calibrated model is:

$$X_1 + X_2 = X_3 + X_4$$

$$X_4 = X_2$$

$$X_4 + X_5 = X_6 + X_7$$

$$X_7 = X_5$$

$$X_8 = X_1 + 0.7X_4 - 0.5X_7 + 0.5X_{11} + 0.3X_{14}$$

$$X_8 + X_9 = X_{10} + X_{11}$$

$$X_{11} = X_9$$

$$X_{11} + X_{12} = X_{13} + X_{14}$$

$$X_{14} = X_{12}$$

$$X_2 = \begin{cases} 0.0 & \text{initially,} \\ X_3 & \text{of previous period thereafter} \end{cases}$$

$$X_5 = \begin{cases} 0.0 & \text{initially,} \\ X_6 & \text{of previous period therafter} \end{cases}$$

$$X_9 = \begin{cases} 0.0 & \text{initially,} \\ X_{10} & \text{of previous period thereafter} \end{cases}$$

$$X_{12} = \begin{cases} 0.0 & \text{initially,} \\ X_{13} & \text{of previous period thereafter} \end{cases}$$

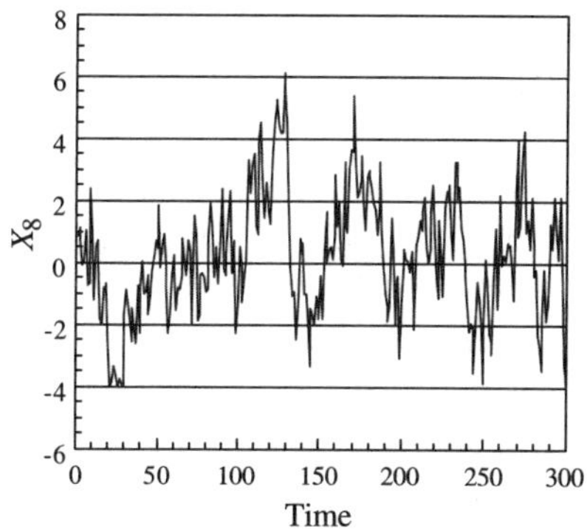

Figure 13.9 ARMA(2,2) Time Series

Given the primary X_1 white noise series and the calibrated model, secondary data were generated by the model over a duration of 300 months. The time series in Figure 13.9 displays the fluctuations of $X_8 = \widetilde{z}_t = z_t - \mu$ about its expected value: $E(\widetilde{z}_t) = E(z_t) - \mu = 0$. This 300-month, 25-year realization very adequately reflects the stationary nature of this *ARMA(2,2)* stochastic process in its breaking away and subsequent return to its mean of zero. If, for example, $\delta = 4$, then excursions of z_t would be centered around $\mu = \delta/(1 - \phi_1 - \phi_2) = 4/(1 - 0.5 - 0.3) = 20$.

REMARKS The X_1 series is the causal force that evokes the ARMA(2,2) effect from the system. Synthesizing this time series model from elemental knowledge items promotes a more basic understanding of the activities (or casual forces) that drive the stock market. The probability principle common to various expressions of the *central limit theorem* [27] indicates that random response, which is obtained by combining many activities via integration (or summation), tends in the limit to follow a normal distribution. Likewise, the combined activities of buyers and sellers, whose risk tolerances comprise a vast spectrum, shape price fluctuations (the effect) in the market. When viewed at the macro level, this causal force is best characterized by a normal distribution with a mean of zero and a derived value for standard deviation. A more micro approach will involve modeling individual classes of stock traders, characterized by levels of risk tolerance. Future modeling efforts that follow this lead will greatly enhance analysts' abilities to quantify and understand more of the dynamics of stock market response.

The foregoing example illustrated the synthesis of a systems model of a single time series for z_t. However, a time series will often be correlated with other time series. For example, if z_t and y_t are correlated, the linear-equation models of these series can be *linked* within a common matrix. Linking is featured in the next section.

13.4 OPTIMIZATION

In anticipation of the future, optimization may prove to be the ultimate justification for systems analysis and modeling. However, optimization might best be introduced during final stages of modeling. Advisedly, iterative stages of modeling should first be directed toward synthesis of valid, descriptive models. Management objectives can then be effectively incorporated into any systems model that has been synthesized via the Mtm approach. A linear or quadratic objective function can be superimposed over the linear equations of the model, to which have been appended linear constraints in the form of limits and bounds on decision variables and resources. Widely available linear and quadratic programming techniques, for example, as in reference [24], capably assist in identifying optimal policies in the model world. Implementation of policies derived in this manner will generally provide measurable improvement in the response (performance) of physical systems.

A matrix structure of an individual system (or subsystem) model augmented for optimization via linear programming is presented in Figure 13.10. The X_j variables are endogenous and exogenous elements of a column vector **X**, which have been transposed to provide headings for columns $1, 2, \ldots, n$. The $a_{i,j}$'s are elements of an $n \times m$ matrix **A**, containing coefficients of the model equations and coefficients of appended constraint relationships.

A linear objective function, $-Z + c_1X_1 + c_2X_2 + \cdots + c_jX_j + \cdots + c_nX_n = 0$, has been superimposed upon the $n \times m$ matrix by appending column "0" and row "0." The right-hand side is a column vector, **b**, containing constants of the model, limits on the decision variables, and bounds on resources. R is a column vector of relational operators: "=" for model equations, and "≤" or "≥" for constraint relationships.

The augmented matrix of Figure 13.10 accommodates an optimal transition

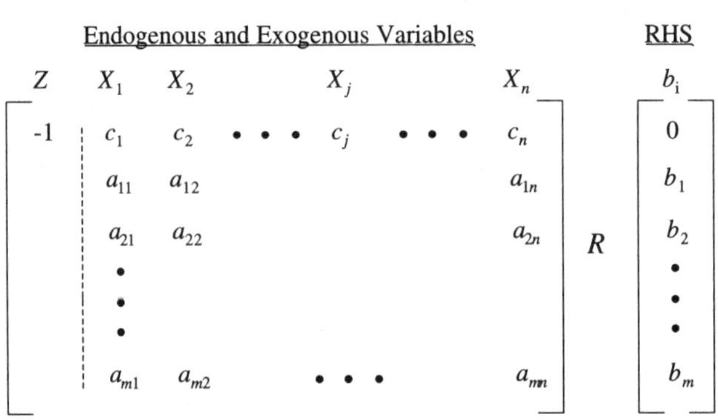

Figure 13.10 Matrix Structure for Single-Period Linear Programming

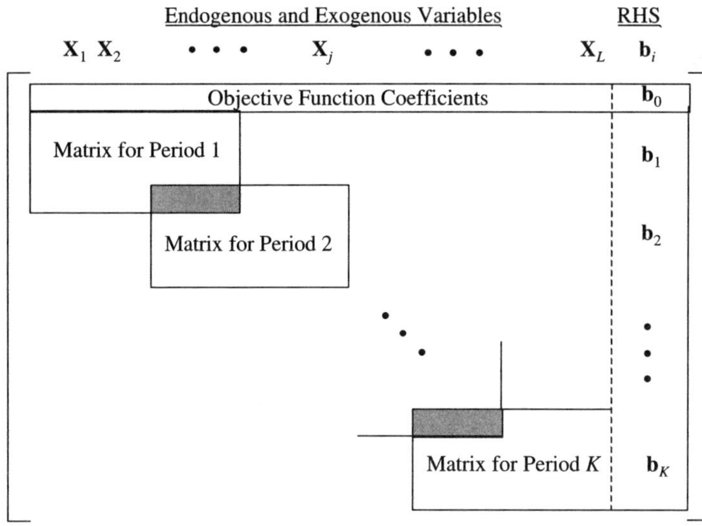

Figure 13.11 Matrix Structure for Multiple-Period Linear Programming

from an initial state over a single Δt period to a terminal state. This method can be validly applied to any linearly formulated systems model at any level on the Mtm scale. Furthermore, the method can be extended to enable optimization over multiple Δt periods, as indicated in Figure 13.11. The matrix for each period is identical to that of Figure 13.10 except that the objective function is extended to span K Δt periods instead of one Δt period. Coefficients of terminal state variables are located in the shaded regions to effect transfer of terminal state values to the corresponding initial state variables for the next period. Transfer in the model is effected by the inclusion of an X_I in a Period i balance equation and also in a Period i + 1 balance equation, thus linking all K matrices over the K periods.

13.4.1 Optimization Example

A dual-period model was synthesized to allocate water resources within the Gallatin Valley of Montana, which comprises the lower portion of the Gallatin River Basin [8]. During the time this research was performed, a suitable computer software/hardware system was not available for inverting large, sparse matrices. Consequently, each of 30 years was partitioned into two Δt periods: water surplus and water deficit. Linear programming was applied to a 35-variable systems model to allocate water resources among competing uses: agricultural, municipal, industrial, and recreational. A linear objective function, spanning one 2-period year, provided allocations that maximized value-in-use for each of the 30 years.

emergency evacuation of civilians from life-threatening circumstances. Forthcoming applications by those whose imaginations have been stimulated to action will provide the ultimate test of the macro-to-micro analysis and modeling techniques set forth in this book.

EXERCISES

1. Synthesize a systems model to match the Taylor tool-life equation, Equation 13.1, in the form $v_T = C\,T^{-n}$.

   a. Set up the knowledge base, defining $\Delta t = 1$ minute as the increment of time. *Hint*: Make X_3 exogenous. $X_3 = X_1 - X_2$ can then be presolved using values obtained from $X_2 = v_T = CT^{-n}$ for $T = 1, 2, \ldots, N \cdot \Delta t = 50$ minutes. You may assume $X_1^0 = 2,000\,fpm$.

   b. List the model equations.

   c. Calibrate the Taylor equation to the data of Table 13.1, using the best two points instead of points 1 and 6. Use a *log-log* transformation, and find values for n and C to provide the best-fitting linear equation.

   d. Generate data for X_3.

   e. Run the model for 50 periods, and display the results. Calculate the SAD measure for X_2, and compare it with 173.4, as obtained for Equation 13.2.

2. Synthesize a systems model for an *AR(1)* stochastic process with $\phi_1 = 0.5$.

   a. Set up the knowledge base, following the *ARMA(2,2)* example.

   b. List the model equations.

   c. Give the calibrated model.

   d. Run the model for 300 periods, and display the *AR(1)* time series generated by the model.

3. Using Chapter 6 for context, synthesize a systems model of an inventory system consisting minimally of *order delay*, *storage*, and *cost* subsystems. Append an objective function and constraints. Optimize with respect to cost (or profit = revenue - cost). Costs you may wish to include are (1) ordering, (2) purchasing, (3) storage and handling, (4) obsolescence, and (5) shortage.

4. Synthesize a large-scale systems model for one of the following:

   - World food
   - World population

- Arable-land loss
- Acid deposition

See D. H. Meadows's *The Limits to Growth* [34], for ideas concerning global modeling.

Appendix J

Linear goal programming [24] was used to calibrate a regression equation to Equation 13.3, $T = T_0 e^{-DR\,v}$, which contains two unknowns, T_0 and DR. Therefore, values from two points must be substituted. The *best two points* through which to pass T were identified by minimizing the SAD measure, defined as the sum of absolute deviations between T and the data. Values for v and T were substituted from all six points of Table 13.1 into the *log-arithmetic* form of Equation 13.3:

Table 13.1

$$\ln 4.69 = \ln T_0 - 800DR = 1.545433$$
$$\ln 7.19 = \ln T_0 - 700DR = 1.972691$$
$$\ln 11.25 = \ln T_0 - 600DR = 2.420369$$
$$\ln 17.5 = \ln T_0 - 500DR = 2.862201$$
$$\ln 27.5 = \ln T_0 - 400DR = 3.314186$$
$$\ln 44.0 = \ln T_0 - 300DR = 3.784190$$

Let $Y = \ln T_0$. Each of the six expressions was rewritten to include a negative and a positive imbalance (error). The expressions were solved for values for $Y,\ DR,\ n_i$, and p_i such that

$$Y - 800DR + n_1 - p_1 = 1.545433$$
$$Y - 700DR + n_2 - p_2 = 1.972691$$
$$Y - 600DR + n_3 - p_3 = 2.420369$$
$$Y - 500DR + n_4 - p_4 = 2.862201$$
$$Y - 400DR + n_5 - p_5 = 3.314186$$
$$Y - 300DR + n_6 - p_6 = 3.784190$$
$$Y, DR, n_i, p_i \geq 0, i = 1, 2, \ldots, 6$$

and such that $Z = n_1 + p_1 + n_2 + p_2 + \cdots + n_6 + p_6$ was minimized.
The goal programming solution indicated zero imbalance for points 3 and 5:

$DR = 0.004469$, or from the slope between points 3 and 5:
$= (\ln 27.5 - \ln 11.25)/(600 - 400) = 0.004469089$

$n_1 = 0.01888$

$p_2 = 0.0007693$

$n_3 = 0.000000$ (point 3)

$p_4 = 0.005077$

$n_5 = 0.00000$ (point 5)

$n_6 = 0.0231$

$Y = 5.102$

$Z = 0.04782$

$\text{SAD} = 4.69(1 - e^{-n_1}) + 7.19(e^{p_2} - 1) + 17.5(e^{p_4} - 1) + 44.0(1 - e^{-n_6})$
$= -0.0877 + 0.0055 + 0.0891 - 1.0047 = 1.1870$ minutes

T_0 was obtained by evaluating Equation 13.3 at point 5, (400, 27.5): $27.5 = T_0 e^{-(0.04469089)(400)}$, or $T_0 = 164.3209627 = X_5^0$, yielding the calibrated dynamic form: $T = 164.3209627e^{-0.004469089v}$.

References

[1] Aburdene, M.F., *Computer Simulation of Dynamic Systems*, Wm. C. Brown Publisher, Dubuque, IA, 1988.

[2] Armstrong, R.L., "Mass balance history of Blue Glacier,Washington, USA," *Glacier Fluctuations and Climatic Change*, Kluwer Academic Publishers, Boston, MA, 183–192, 1987.

[3] Astrand, P. and K. Rodahl, *Textbook of Work Physiology*, McGraw-Hill, New York, NY, 1970.

[4] Berenson, M.L., D.M. Levine, and M. Goldstein, *Intermediate Statistical Methods and Applications*, Prentice-Hall, Englewood Cliffs, NJ, 1983.

[5] Berry, R.S., *Understanding Energy: Energy, Entropy and Thermodynamics for Engineers*, World Scientific Publishing Co., Singapore, 1991.

[6] Bliss, G.L., *Mathematics for Exterior Ballistics*, John Wiley & Sons, New York, NY, 1944.

[7] Boyd, D.W. and T.T. Williams, "Development of a water planning model for Montana," Proceedings of ASCE Hydraulics Division 21*st* Annual Specialty Conference, Bozeman, MT, August 1973.

[8] Boyd, D.W., "Simulation via time-partitioned linear programming: a ground and surface water allocation model for the Gallatin Valley of Montana," Proceedings of the AIIE Joint Winter Simulation Conference, Washington, D.C., 1974.

[9] Boyd, D.W., "A macro-to-micro simulation methodology for producing linkable natural resources models," ORSA/TIMS Bulletin, No. 6, Los Angeles Joint National Meeting, November 1978.

[10] Boyd, D.W., "An integrative to differentiative methodology for producing mathematical models," Proceedings of the International Association for Mathematical Modeling, St. Louis, MO, July 1979.

[11] Boyd, D.W., "A macro-to-micro finite element method for producing water planning models," Proceedings Third International Conference on Finite Elements in Water Resources, Oxford, MS, May 1980.

[12] Boyd, D.W., "River basin modeling via systems analysis and artificial intelligence," *Hydrosoft*, Vol. 1, No. 2, 1988.

[13] Checkland, P., *Systems Thinking—Systems Practice*, Wiley, New York, NY, 1981.

[14] de Rosnay, J., *The Macroscope*, Harper & Row, New York, NY, 1979.

[15] Draper, N.R., *Applied Regression Analysis*, 2nd ed., Wiley, New York, NY, 1981.

[16] Einstein, A.,*Quotable Einstein*, Princeton University Press, Princeton, NJ, 1996, p 61.

[17] Eltis, W.A., *The Classical Theory of Economic Growth*, St. Martin's Press, New York, NY, 1984.

[18] Feather, N., *An Introduction to the Physics of Mass, Length and Time*, Edinburgh University Press, Edinburgh, Scotland, 1959.

[19] Forrester, J.W., *Principles of Systems*, Wright-Allen Press, Cambridge, MA, 1968.

[20] Garg, A., D. Chaffin, and G. Herrin, "Prediction of metabolic rates for manual materials handling jobs," *American Industrial Hygiene Association Journal*, Vol. 39, 1978, 661–674.

[21] Gupta, R.S., *Hydrology and Hydraulic Systems*, Prentice-Hall, Englewood Cliffs, NJ, 1989.

[22] Hagan, R., T. Strathman, R. Strathman, and L. Gettman, "Oxygen uptake and energy expenditure during horizontal treadmill running," *Journal of Applied Physiology*, Vol. 49, 1980, 571–575.

[23] Halliday, D. and R. Resnick, *Physics for Students of Science and Engineering*, Wiley, New York, NY, 1962.

[24] Hillier, F.S., and G.J. Lieberman, *Introduction to Operations Research*, McGraw-Hill, New York, NY, 1995.

[25] Hornbeck, R.W., *Numerical Methods*, Quantum, New York, NY, 1975.

[26] Huebner, K.H., *The Finite Element Method for Engineers*, John Wiley & Sons, Inc., New York, NY, 1975.

[27] Johnson, R.A., *Miller & Freund's Probability and Statistics for Engineers*, Prentice-Hall, Englewood Cliffs, NJ, 1994.

[28] LaChapelle, E., "Annual mass and energy exchange on the Blue Glacier," *J. of Geophysical Research*, 64(4), 443–449, 1959.

[29] Law, A.M., and W.D. Kelton, *Simulation Modeling & Analysis*, McGraw-Hill, New York, NY, 1991.

[30] Lindsay, R.B., and H. Margenou, *Foundations of Physics*, John Wiley & Sons, Inc., New York, NY, 1936.

[31] McGraw, K.L. and K. Harbison-Briggs, *Knowledge Acquisition: Principles and Guidelines*, Prentice-Hall, Englewood Cliffs, NJ, 1989.

[32] McShane, E.J., Kelly, and Reno, *Exterior Ballistics*, University of Denver Press, 1953.

[33] *Machining Data Handbook*, 2nd edition, Machinability Data Center, Metcut Research Associates Inc., Cincinnati, Ohio, 1972, p. 864.

[34] Meadows, D.H., et al., *The Limits to Growth*, Universe Books, New York, NY, 1972.

[35] Morton, R., "Two-dimensional short-term model of oxygen uptake kinetics," *Journal of Applied Physiology*, Volume 58 (5), 1985, 1736–1740.

[36] Nahmias, S., *Production and Operations Analysis*, Irwin, Homewood, IL, 1993.

[37] Neter, J.W. Wasserman, and M.H. Kutner, *Applied Linear Regression Models*, Irwin, Homewood, IL.1989.

[38] Novosod, J.P., *Systems, Modeling, and Decision Making,* Kendall/Hunt Publishers, Dubuque, IA, 1982.

[39] Oberg, E., F.D. Jones, and H.L. Horton, *Machinery's Handbook*, 23 edition, Industrial Press Inc., New York, NY, 1988.

[40] Pankratz, A., *Forecasting with Univariate Box-Jenkins Models*, Wiley & Sons, New York, NY, 1983.

[41] Pegden, C.D., R.E. Shannon, and R.P. Sadowski, *Introduction to Simulation Using SIMAN*, McGraw-Hill, New York, NY, 1990.

[42] Sandquist, G.M., *Introduction to System Science*, Prentice-Hall, Englewood Cliffs, NJ, 1985.

[43] Schmenner, R.W., *Production/Operation Management*, Macmillan, New York, NY, 1990.

[44] Sharp, J.J., and P.G. Sawden, *Basic Hydrology*, Butterworth & Co., London, 1984.

[45] Sharp, R. P., "Glaciers" (Condon lectures), Oregon State System of Higher Education, Eugene, Oregon, 1960.

[46] SoftTech., *ICAM Architecture Part II, Volume V, Information Modeling Manual (IDEF1)*, AFWAL-TR-81-4023, Materials Laboratory, Air Force Wright Aeronautical Laboratories, Air Force Systems Command, Wright-Patterson Air Force Base, Ohio 45433, June 1981.

[47] *Speer Reloading Manual*, No. 11, Blount, Inc.

[48] Starfield, A.M., K.A. Smith, and A.L. Bieloch, *How to Model It: Problem Solving for the Computer Age*, McGraw-Hill, New York, NY, 1990.

[49] Taha, H.A., *Operations Research: An Introduction*, Macmillan, New York, NY, 1992.

[50] Taylor, F. W., "On the art of metal cutting," *Trans. ASME*, 28, 1907, 31–350.

[51] Turban, E., *Decision Support and Expert Systems*, Macmillan, New York, NY, 1990.

[52] Turing, A.M., "Computing machinery and intelligence," *Mind*, Vol. 59, 1950, 433–460.

[53] Vardeman, S.B., *Statistics for Engineering Problem Solving*, PWS, Boston, MA, 1994.

[54] Wallack, L.R., *American Rifle Design and Performance*, Winchester Press, New York, NY, 1977.

[55] Wehrman, R.R., *Macro-to-Micro Modeling Applied to Work Physiology*, Unpublished Master's Thesis, Montana State University—Bozeman, 1994.

[56] *Western Ammunition Handbook*, Western Cartridge Company, East Alton, IL.

[57] Winston, W.L., *Operations Research: Applications and Algorithms*, Duxbury Press, Belmont, CA, 1994.

[58] Woolf, E., *Systems Analysis and Design*, Emile Woolf and Assoc., London, England, 1986.

Glossary

accretion The process of growth or enlargement of a *system* by gradual accumulation or buildup over time.

algorithm A repeatable procedure for solving data based mathematical models and often consisting of a series of iterative steps. May be applied in a forward- or backward-looking direction relative to the forward flow or indexing of the model.

articulate To define a *homogeneous* level or depth of *detail*.

autocorrelation A measure of the correlation within a single time series.

auxiliary equations Used only during the *reverse-regression calibration* process. See *calibration equations*.

back order An activity that allows for that portion of demand that exceeds supply to be carried over into the next period. Also called backlogging.

balance equation An equation, characterized by unity coefficients, that equates supply components to demand components of a *system*. (See *modification* for exception.) The components represent *rates* and *levels* of the system. Also called a *continuity equation*. Can be expressed in *difference equation* form to emphasize change in level.

birth A supply component in a population model.

calibration Process of assigning coefficients to *regression* and *modification equations* of the *model* via *forward* or *reverse regression*. See also *credibility*, *validation*, and *verification*.

calibration equations A system of equations that embody *knowledge items*, used to calibrate *dynamic forms* or *subsystems*. They are selected from *auxiliary equations* and model equations as may be required, one equation for each unknown *regression coefficient* and each *secondary variable*. They are solved for only one *calibration period*. The solution produces a tentatively *calibrated model* from which trial values are generated for all *secondary variables*.

calibration period A single, Δt period: the hypothetical, average period if the *system* is *homogeneous* and *stationary*; otherwise, a typical period if the system is *homogeneous* and *nonstationary*.

circumscription Selection of *system* boundaries for a given level of *articulation* and *scope*, thus enabling definition of *balance equations* and *regression equations.*

closed-loop system Current response of a *system* becomes an influence on future response of the system. See *feedback.*

complete data base A *data base* whose data categories are completely filled by some combination of *primary* and *secondary time series* data, thus covering all variables of a *systems model.*

confidence interval An interval estimate of a population parameter for which it can be asserted with $X\%$ of confidence that the unknown parameter is contained within the interval. Typical confidence levels are 95% and 99%.

conservative system A system in which mass is neither created or destroyed. See *continuity equation.*

continuity equation Accounts for transition of mass or its attributes through time. Also called a mass *balance equation.*

correlation A property that is present when the value of a variable is influenced by one or more variables.

credibility An intended user of a *systems model* must be convinced of its reliability. Credibility is established via close cooperation between the user of the model and the modeler during *systems analysis* and *model synthesis.* See *calibration*, *validation*, and *verification.*

cross-correlation A measure of the correlation between time series.

data base Comprised of one or more sequences of data. See also *time series.*

data base extension First application for a *systems model* if *primary data* are nonexistent. See also *Phase One.*

death A demand component in a population model.

decision maker One who synthesizes information into decisions. See *planning.*

decision variable A variable whose value may be set or sought, depending on the context, by a *decision maker.* Used in defining a design. See also *exogenous* and *primary variables.*

deep knowledge Exhibited by a *systems model* at relatively *micro* levels of *articulation.* Stands in contrast to *shallow knowledge.*

dependent variable In a *systems model*, a variable that represents "effect," as in "cause and effect"; a function of at least one *independent variable*, the "causal" agent(s).

depletion The process of decline or decrease in a system.

detail Expressed by *system circumscription* in compatible *units of analysis* and denominated in terms of space, mass, and time.

differentiative model A *model* of a *system* whose components are represented by relatively many, highly disaggregate, distributed variables, and that portrays relatively much detail.

domain The environment within which a *system* operates; a system's realm of influence.

domain expert An expert who is a source of *systems domain knowledge*. Typically helpful in judging *homogeneity*, defining *inference matrix correlations* and *regression coefficients*, and screening *secondary data* for aberrations.

dynamic forms Input or output *rates* in the form of *exogenous* data or *endogenous* variables that induce change in *level* of a *balance equation* of a *system* or *subsystem*. As endogenous rates, they are referred to as *regression equations*.

endogenous variables Variables that are solved for internally by the *model*. See *exogenous* for contrast.

entropy A term coined by the German physicist Rudolf Clausius in 1868, entropy provides a measure of the amount of energy no longer capable of conversion into work. According to the second law of thermodynamics, entropy, a *state* function of a system, always increases in any natural process; matter and energy can only be changed in one direction—from usable to nonusable, or from ordered to disordered. Change from order to disorder is inescapable. The law of increasing entropy is equivalent to the law of decreasing probability of change. The ever-changing levels of entropy provide a solid, natural basis for *systems analysis and modeling*.

envelope A curve used to profile the range of variation of a system variable as it evolves over time.

equation type Model equations are composed of *balance equations, equivalence equations, regression equations,* and *modification equations*.

exogenous variables Variables whose values are supplied externally to the *model*. Can be *input/output (I/O) rates* or *initial/terminal (I/T) levels*.

exponential change A nonlinear change that produces increasingly larger increments of change. Readily modeled by a set of linear equations consisting of a *balance equation* and a *dynamic form* for exponential change.

feasible A property of a solution or result that does not violate a constraint or set of constraints. For example, the coefficient matrix of a linear system of equations is feasible if its determinant does not equal zero.

feedback Information from *system* response is reintroduced as an input, providing the system with memory. Negative feedback enables *goal seeking*. Positive feedback accelerates growth or decline.

finite element method Also known as finite element analysis. Originating in the field of Structural Mechanics, the method mitigates mathematical problems by projecting upward from an infinitesimal (micro) element to a finite (macro) element, and relating finite elements by linear, algebraic equations that result in a matrix formulation of the problem.

forcing function A term from differential equation anlaysis that describes the manner in which force or energy is supplied as a function of time, for example, $a(dx/dt) + bx = f(t)$, where if no external force is acting, $f(t) = 0$.

forward regression A procedure that utilizes existing data and statistical methods to find coefficients of *regression equations*, that is, "given the data, find the coefficients."

goal seeking Property of a negative *feedback system.* "Goal" is its preferred, *steady-state* response; will seek to reestablish goal if "bumped" away from steady state.

heuristic A rule of thumb based on experience, or a rule of good judgment.

hierarchical organization The *systems domain* is divided into two or more levels of *articulation*, in a *macro-to-micro* succession of *levels.*

history matching A *validation* procedure carried out in *Phase Two* in which values for historical (*primary*) data are generated by inverting the roles of *exogenous* and *endogenous variables.*

homeostasis Maintenance of constancy of internal environment within an organism via goal-seeking, negative feedback mechanisms.

homogeneous Having the same structure, composed of similar parts, exhibiting consistently similar characteristics. Commensurable or measurable by the same standards. Application of the definition will vary for each *systems model.* Involves such notions as unit consistency; compatible units of space, time, and mass; exhibiting similar statistical properties.

independent variable In a *systems model*, a variable that represents "cause." as in "cause and effect." One or more independent variables are used as components in an equation expressing a *dependent variable*, or "effect."

inference matrix A *knowledge acquisition* tool, a mechanism for transferring *domain expertise* into a simple, weighted set of relationships known as *regression equations* that become *dynamic forms* in the *model. Physical* and/or *statistical correlations* are inferred between *dependent* and *independent variables.* Each matrix cell that pairs variables to be evaluated is coded by strength of the *correlation*, whether positive, negative, or lagged.

inertia The tendency of an object to resist change in state.

initial-terminal (IT) pair A pair of *model* variables, separated by one Δt period, that represents initial and terminal levels of a corresponding physical-system component. The initial level for any given period is identical to the terminal level for the previous period.

initial-terminal (IT) solution pair An IT pair such that the initial variable of the pair is *exogenous* and the terminal variable of the pair is *endogenous.*

integrative model A *model* of a *system* portraying relatively little detail whose components are represented by relatively few, highly aggregative, lumped variables.

knowledge base A list of knowledge items consisting of two types of knowledge: (*hard knowledge*) observable facts of the domain that include historical data and (*soft knowledge*) heuristics based on experience and rules of good judgment. See *domain expert*.

knowledge engineer A developer of knowledge-based decision support *systems*—for example, expert systems.

knowledge acquisition The process of extracting knowledge from various sources for the purpose of constructing a *knowledge base*.

large-scale system A *system* that impacts one or more human institutions and that is extensive relative to the *macro-to-micro* spectrum. Generally pertains to analysis and modeling of societal systems.

level (1) Growth in the level of *entropy*, caused by energy consumption. (2) An accumulation (or integration) achieved by a *systems model* in response to *rates* operating over time. The *state* or condition of a *systems model* as depicted by an *initial-terminal (IT) pair*. (3) In a broader sense as depicting the level of detail or *articulation* exhibited by *systems analysis* or a *model*.

longevity Life span or period of retention within the domain of a system.

macro Gross detail or *articulation*, fewer variables, employing a relatively high level of systems *integration*.

macro-to-micro (Mtm) approach A sequential *systems analysis* and *modeling* approach where the sequence is from a highly *integrative* model to progressively more *differentiative* models.

micro Highly definitive detail or *articulation*, many variables, employing a relatively high level of systems *differentiation*.

model A replica of reality that is comprised of symbols (imaginary constructs); for example, in systems engineering, models may consist of verbal descriptions, schematics, equations, computer code, and so on. Also see *simulate*.

model synthesis The process of constructing a *model* with one or more applications in mind.

modification A simple transformation that induces *gain* for a greater effect or *attenuation* for a lesser effect. Also used in equations that interact diverse *subsystems* to maintain unit consistency.

Monte Carlo sampling Randomly drawing a simulated observation or set of observations from an applicable cumulative probability distribution.

Mtm Acronym for macro-to-micro.

mtM Acronym for micro-to-macro.

nonstationary-homogeneous system A *system* whose *primary data* series exhibit time varying means and/or standard deviations yet exhibiting *homogeneous* behavior—for example, as in *exponential* growth or decline resulting from time invariant rates.

open-loop system *System* responds to input, but output is isolated from and has no influence on the input. Stands in contrast to a *closed-loop system*

parsimony Stinginess, extreme frugality.

parsimony principle Only those variables most directly *correlated* are included in a *regression equation.*

period One incremental (Δt) unit of time whose magnitude is fixed by circumscription.

perturbation A model *validation* technique in which different scenarios are obtained by altering values of designated *exogenous variables* of the *completed data base.* Carried out during *Phase Two* to find answers to "What if?" questions.

Phase One The initial phase in modeling within which *calibration* of the *model* and synthesis of *secondary data* comprise the first step toward *validation.* Also referred to as *data base extension.*

Phase Two Activities are devoted to validating the *model* for specific applications via *simulation tests.* As structural deficiencies appear, the model is recycled back through *Phase One.*

physical correlation A relationship between variable components of a physical *system* that embodies cause and effect. See also *statistical correlation.*

planning Selection of future courses of action based on a current body of information. Enhanced by *planning models* that transform raw data into information from which decisions are synthesized. See *planning model.*

planning model A computer *model* that transforms raw data into information. Used by *decision makers* to generate outcomes for alternative, future courses of action. See also *decision maker.*

primary data Data derived from historical records for *primary variables.* Stands in contrast to *secondary data.*

primary variable A variable of the *model* that possesses *primary data.*

rate A change in *level* of a *system* over time. If a system is at rest, rate is zero. A rate is an average and cannot be measured at a point in time. A rate stands in contrast to a level. See *regression coefficients.*

regression coefficients Coefficients of a *regression equation.* Each coefficient represents the average rate of change of the **dependent variable** with respect to its companion, *independent variable.* See *rate.*

regression analysis approach A modeling approach that requires collection of *primary data* for the purpose of fitting one or more *regression equations* to the data. A *shallow knowledge* or *hierarchical* approach.

regression equation An *endogenous dynamic form* that expresses the *dependence* of one variable on one or more *independent* variables. It need not have unity valued coefficients as do *balance equations*. It may occur in the form of an incomplete balance equation.

reverse regression A procedure that utilizes a *knowledge base* to find values of *regression equation* coefficients used to complete a *data base*—for example "given the coefficients, find the data." Stands in contrast to *forward regression*. See also *secondary data*.

role inversion Used in *Phase Two scenarios* in which roles of selected *exogenous* and *endogenous* variables are interchanged to establish various applications of the *model*.

SAD An acronym for "sum of absolute deviations," used as a measure of difference between two numeric sequences.

scenario A given set of conditions or assumptions under which a *simulation model* is operated.

scale Magnitude of a systems domain relative to the amount of space circumscribed by the system. See *scope*.

scope Breadth of coverage at a given level of *articulation*; expressed by the number of *subsystems* at that level.

secondary data Data generated by the *model* for *secondary variables* during *Phase One* to supplement and thereby to complete the *data base*. Stands in contrast to *primary data*.

secondary variable A variable of the *model* not having *primary data*. Values for secondary variables are obtained by *reverse regression*.

shallow knowledge Exhibited by a *regression equation model* that spans multiple levels of *articulation* or by a *systems model* at very *macro* levels of *articulation*.

simulate In systems engineering, the use of an imaginary construct, such as an equation or system of equations, to represent the physical world. Synonyms are *emulate, imitate,* and *model.*

simulation tests The *model* is operated over a given number of periods under preselected *scenarios* to answer "what if" questions. *Role inversion* is used with *history matching* and *perturbation* tests. *Secondary data* generated in this manner are subjected to *Turing Type tests.*

solution vector A subset of the *model* variables, one variable for each equation, corresponding to the *endogenous* partition of variables. Often designated by subscripts, as, for example, (2, 3, 4, 6, 7).

state of a system Characterized by the *entropy*-dependent levels exhibited by *system variables* as the system responds over time to *dynamic* factors that induce change.

stationary system A system whose *data series* exhibit time invariant means and standard deviations.

statistical correlation A relationship between variable components of a physical *system* that is purely statistical; that is, cause and effect is not involved.

steady-state response The response pattern of a *system* after *transient response* has died out. Characterized by a *time-persistent* pattern of variation, growth, decline, oscillation, and so on.

stochastic process Characterized by one or more *stochastic variables.*

stochastic variable Variation is governed by a probability distribution function (*pdf*).

subsystem Major, *homogeneous* component of a *system*, itself a system, and exhibiting a complete *balance equation.*

superposition Applies only to linear *systems.* Two or more causal agents can be applied simultaneously to obtain a combined effect that is equal to the sum of the individual effects produced by the causal agents acting independently. By superposition, the system *balance equation* can be obtained by adding its *subsystem* balance equations.

system A set of related components that function in unison within a common environment to achieve a common goal, end, or objective. Composed of the *subsystems* selected for *circumscription.*

systemic Of or affecting the whole.

systemic thinking Systematic (step-by-step) thinking.

systems analysis A technique by which the *systems domain* is divided into *homogeneous* groups (*subsystems*) of components. Components of the systems domain are set in one-to-one correspondence with variables of the *model.* Relationships among variables and among *subsystems* are identified. *Primary data* and applicable theories are collected.

systems analysis approach A *knowledge-based* approach to modeling that begins with the *systems domain.* Variables of the *model* are identified by one-to-one correspondence with components of the *systems domain.* Equations of the model are expressed in the form of *balance equations* and *regression equations* for each level of *articulation.* Each *model* is non*hierarchical*, the final *model* of the sequence exhibits ultimate detail, that is, *deep knowledge.* See *macro-to-micro approach.*

system component The most basic member of a *system* or *subsystem.* Each *system* component is identified in the *model* as a variable. See *parsimony principle.*

systems domain That portion of the physical world selected for *systems analysis and modeling.*

systems model A structured entity that *simulates levels* and *rates* of a *system* as its response evolves over time.

theoretical analysis approach An approach to modeling that commences with those laws that relate to action at a point. Equations of the *model* are expressed in the form of differential equations for an infinitesimal part of the *system*. An approximate solution is found by integrating over the whole of the *systems domain*. It often requires vastly simplifying assumptions.

time constant A system constant that establishes a reference point for marking the occurrence of a key system state.

time persistence Exhibited by *systems* whose *time series* are characterized by *homogeneous* variation during the period of observation.

time series Sequences of data quantified at uniformly spaced increments of time.

transfer function The process (i.e., function) performed by the model (or system) in shaping "velocities" (rates) and "displacements" (levels) relative to time.

transient response Introduced into a *system* by initial conditions that differ from the system's *steady-state response*. Characterized by a *non-time-persistent* patterns of growth, decline, oscillation, and so on.

trend The baseline about which a *time series* fluctuates. All series exhibit a trend, either zero or nonzero, linear or nonlinear. Trend can be reduced to a single measure by expressing the difference between the mean terminal value and the mean initial value of an *initial-terminal (solution) pair* as a percentage of the mean initial value.

trend analysis Computation of *trend* and adjustment of the starting value of the initial variable of an *initial-terminal solution pair* until a desired trend is exhibited.

triad A group of three items.

Turing Type (TT) test A *systems model* is said to have passed a TT test if the *domain expert(s)* cannot be sure if a set of data was generated by the model or by the physical system.

units of analysis Basis for *circumscribing* or extending *systems analysis* and *modeling*: space, mass, and time.

validation A *systems model* is validated via a series of *simulation tests*. Tests must be performed for each anticipated use of the model. See also *calibration*, *credibility*, and *verification*.

vector table A table that identifies all components of a *systems model*. All model variables are members of vectors: data, *initial-terminal solution pairs*, *exogenous*, and *endogenous*. Each *endogenous variable* is assigned an equation type.

verification A *systems model* is verified in terms of how well its design conforms to the *level* of *articulation* specified for its intended use.

Index

accumulator
  cubic, 117–118
  linear, 114–115
  quadratic, 115–117
  quartic, 119–120
accuracy, 9, 22, 24, 25, 27, 28, 67, 72, 83, 85, 87, 89, 95, 109, 110, 112, 125, 129, 139, 143, 230, 242, 244, 262 *see also* exact/precision
analysis *see also* principle, circumscription
  approach
    regression, 23–26 *see also* regression, analysis
    systems, 26–28
    theoretical, 20–23
  hierarchical, 13–16
  macro-to-micro, 17–19
  stochastic, 211–215
  systems *see* systems, analysis
  trend, 219–221
  units of extension, 5
applications, multidisciplinary. *see* model

balance equation. *see* linear equation
ballistics. *see* model, projectile
baseline, 36, 149

calibration
  equation(s), 48–49
    derivation, 157–158, 162
    exponential decline, 57
    exponential growth, 56
    Malthusian form, 60
  heuristic *see also* regression, reverse
    harmonic oscillation model, 110
    Badger Ranger District model, 311–314
  period, 48
  principles *see* principle, calibration
  procedure, 236–237
  schemes, 78–80
circumscription. *see* principle
comparison, analytic approaches
  regression, 29–30
  systems, 30–31
  theoretical, 28–29
continuity equation. *see* linear equation, balance
correspondence. *see* principle
cost
  containment, 149
  effectiveness, 230, 272
  inventory, 168, 178
  penalty, 150
  pricing policy, 187, 194–195
  raw materials, 190

data, 41
  secondary
    Badger Ranger District, 315
    Blue Glacier, 261
    work physiology, 238, 239, 241
  primary
    Badger Ranger District, 308–309
    Blue Glacier, 257–258
    work physiology, 232
data base generation
  asset growth, 188–189
  corporate assets, 198–199
  harmonic oscillation, 111–112
  inventory, 152, 154, 158, 162, 167, 170, 173, 175
  polynomial, 124–125
  population, 84
  projectile, 139
  Yellowstone River Basin, 285–286
design
  engineering, 3
  forest management plan, 316–322
  reservoir, 289–292
differential(s), 14, 15, 68, 271
differential equation, 9, 21, 24, 28–29, 31, 43, 50, 55, 59, 90, 104, 129, 176–177, 230
differentiation, 8–9, 13–17, 261
distortion reduction, 66–68
dynamic form, 50
  endogenous, 52
    constant, 52–53 *see also* accumulator, linear
    delay, 60–65
    exponential, 53–57
    logical, 53
    Malthusian, 57
    stochastic, 52
  exogenous, 51
    deterministic, 51
    stochastic, 50–51
  modeling, 103–125
dynamic response, shaping, 149

efficiency
  external work, 233–235
  inventory systems, 158–159
entropy
  agent of change, 10–12
  data and modeling, 13
equation. *see also* linear equation
  auxiliary, 48 *see also* calibration, equations
  differential *see* differential equation
  static relationships, 262–265
  Taylor's tool life, 328–330
error
  control, 88–90
  time–based, 8
exact. *see also* accuracy/precision
  error magnitude, 89
  match, 83
  model–generated data, 111
  primary value, 86
  reproduction, 31
  results, 88
  value, 105
expert, domain, 47, 198, 229, 255, 261, 265, 279–280, 283–285, 287, 304–305, 308, 322

feasibility. *see* role inversion
feedback, 107. *see* emphalso information
  negative, 103, 109, 150–152, 155, 157, 160, 163
  positive, 53, 109

forecast
  perturbation, 200–201
  sales, 187, 195

heuristic. *see* calibration
history matching. *see* test

information
  feedback, 66
  flow, 114, 151
initial condition(s). *see also* linear equation, initialization
  algorithm, 122
  Fortran program, 127–128
integration, 8–9, 13–17, 21–22, 29, 31, 40, 43–45, 50–52, 55, 61–62, 78, 129, 131, 261, 339
interact, 3. *see also* matrix
  interacting, 26, 28, 38
  interaction, 41, 89
IT pair. *see* linear equation, initialization

knowledge acquisition
  inference matrix, 222–223
  Yellowstone River Basin, 279–280
knowledge base
  ARMA(2,2) time series, 334–337
  asset growth, 186–188
  Badger Ranger District, 305–314
  Blue Glacier, 256–258
  corporate assets planning, 192–196
  inventory
    accumulation, 150–151
    ideal, 155–156
    order delay, 160–161
    order policy, 169
    random demand, 163–165
    reduction, 152–153
    three policy, 172
  pendulum, 106–108
  polynomial, 123–124
  population, 76–77
  projectile, 130–134
  tool life, 330–331
  work physiology, 231–233
  Yellowstone River Basin, 275–277, 280–283

large-scale system, 6. *see also* systems, large-scale
level. *see* variable
linear equation
  balance, 43
  equivalence, 44
  initialization, 42
  modification, 42
  regression, 44–45, 67–68
linked, 22, 122. *see also* matrix
linking, 3, 22. *See also* matrix

macro-to-micro approach. *see* Mtm approach
manufacturing, virtual, 327
materials requirement planning, offset for lead time, 201–203
matrix
  algebra, 27–28, 84, 94–95
  inference, 222–223
  interaction, 223
  linked, 28, 30, 292
  linking, 330, 339, 341, 342
  solution, 94–95
model
  ARMA(2,2) time series, 338
  asset growth, 188
  Badger Ranger District, 314
  Blue Glacier, 260
  corporate assets planning, 196–197

harmonic oscillation, 110
inventory
accumulation, 151
ideal, 156
order delay, 161
order policy, 169–170
random demand, 165–166
reduction, 154
three policy, 173
knowledge-based, 35
polynomial, 115, 116, 118, 120, 124
population, 19–20, 77
projectile, 139
tool life, 332
work physiology, 233–234
Yellowstone River Basin, 283–284
modification. *see* linear equation
Monte Carlo simulation. *see* simulation, Monte Carlo
Mtm approach, 16–19, 32, 36, 255, 263, 271–274, 322, 340

optimization, 340–341
goal programming, 346–347
linear programming
single-period, 340
multiple-period, 341

period. *see* calibration
perturbation. *see* test
phase, 75–76
precision, 69, 109, 127, 218, 242, 292. *see also* accuracy/exact
principle
building-block, 49–50
calibration, 46–49
circumscription, 36–37
conservation, 38
correspondence, 39
distortion reduction, 69
prototyping, 304

rate. *see* variable
regression
analysis, 224–225
forward, 46–47
linear, 24 *see also* linear equation, regression
nonlinear, 23–24
reverse, 47–48 *see also* calibration, heuristic
Yellowstone River Basin model, 280–283
role inversion, 82, 86
feasibility, 83, 91, 285

simulation. *see also* test
Monte Carlo, 215
random demand,165
Badger Ranger District, 308–309
physiological functions, 236
stochastic. *see also* analysis
modeling, simple inventory system, 215–216
conventional approach, 216–217
systems approach, 217–218
representation, 212–215
subsystem, 17–18, 43–44
building block, 49
number, by horizontal circumscription, 37
systems
analysis
Badger Ranger District, 304–305
Blue Glacier model, 256
pendulum, 104–106
work physiology, 231
Yellowstone River Basin, 274–275

approach, 32, 217–218, 304
see also Mtm approach
tool life, 330
component, 40
interactions, 292
large-scale, 342–343
matrix structure, 344
river basin model, 343

test
performance, 173–175
simulation
asset growth model, 188–191
corporate assets planning, 198–203
population model, 85–88
validation, 81–84
history matching, 83
Blue Glacier model, 262
population model, 85–87
work physiology model, 243–244
Yellowstone River Basin model, 286
perturbation, 84
Badger Ranger District, 316–322
Blue Glacier, 262
population model, 87–88
work physiology model, 244–245
Yellowstone River Basin model, 286–292
role inversion, 82–83
Badger Ranger District, 317–318
simulation, 82
Turing type, 81–82, 90
asset growth model, 187
population model, 84–85
Yellowstone River Basin model, 284
time series
classification, 45–46
synthesis, 333–334
triad, 4–10, 13, 36, 38–39, 41, 46, 49, 54, 56–57

validation, 81–84. *see also* test
asset growth model, 187–188
Badger Ranger District model, 315–321
Blue Glacier model, 262
population model, 84–88
projectile model, 139–143
work physiology model, 243–244
Yellowstone River Basin model, 285–292
variable(s)
endogenous and exogenous, 42
level and rate, 40–41
number, by conservation principle, 38
primary and secondary, 41